B1NA

ANALYTICAL PYROLYSIS

CHROMATOGRAPHIC SCIENCE

A Series of Monographs

Editor: JACK CAZES
Fairfield, Connecticut

Other Volumes in Preparation

ANALYTICAL PYROLYSIS

A COMPREHENSIVE GUIDE

William J. Irwin

Department of Pharmacy
University of Aston in Birmingham
Birmingham, England

MARCEL DEKKER, INC.   New York and Basel

Library of Congress Cataloging in Publication Data

Irwin, William J. [date]
    Analytical pyrolysis.

    (Chromatographic science; v. 22)
    Includes bibliographical references and indexes.
    1. Pyrolysis.   I.   Title.   II.   Series.
QD281.P9I78   1982   547.3'086   82-13060
ISBN 0-8247-1869-0

MARCEL DEKKER, INC.
270 Madison Avenue, New York, New York   10016

Current printing (last digit):
10 9 8 7 6 5 4 3 2 1

PRINTED IN THE UNITED STATES OF AMERICA

Accuse not Nature, she hath done her part;
Do thou but thine.

John Milton
(1608-1674)

Die Lust der Zerstörung ist zugleich eine schaffende Lust.

Michael Bakunin
(1814-1876)

Preface

The complex macromolecular state of many naturally occurring organic
constituents of the biosphere has meant that, in common with synthetic polymers, degradative analytical techniques are commonly employed for their characterization. Pyrolysis is one of the oldest
of these approaches and involves the thermal fragmentation of the
macromolecular component. A range of smaller molecules is produced,
and analysis of these products enables a profile of the original
sample to be constructed. Such analytical pyrolyses are usually
undertaken using an integrated pyrolysis-analysis system, with a
pyrolyzer interfaced directly to a gas chromatograph or to a mass
spectrometer. This arrangement enables pyrolysis products to be
analyzed immediately, so that transfer losses and secondary degradations are minimized. The resulting chromatogram or mass spectrum
may provide qualitative information concerning the composition or
identity of the sample, quantitative data on its constitution, or
it may enable mechanistic and kinetic studies of thermal fragmentation processes to be undertaken.

Advances in instrumental design, an appreciation of pyrolysis
mechanisms, and the current widespread availability of computing
facilities have combined to develop analytical pyrolysis into a
powerful modern analysis technique. Indeed, the combination of
pyrolysis-gas chromatography-mass spectrometry-data system, or
pyrolysis-mass spectrometry-data system, probably represents the

most effective approach to the initial study of complex, nonvolatile samples. These procedures require very small amounts of sample, and manipulations during sample preparation are minimal. Analysis times are short, and a direct mass spectrometric analysis may take one minute of instrument time. Chromatographic analyses require longer, depending upon the products to be separated and the conditions chosen, but nevertheless they compare favorably with alternative wet methods. Such systems may be totally automated and are suitable for unattended operation.

The compelling attributes of analytical pyrolysis are illustrated by the range of disciplines in which this approach has been of proven value. These include polymer chemistry, biochemistry, organic geo-chemistry, soil chemistry, forensic science, food science, toxicol-ogy, environmental studies and conservation, microbiology, pathology, extraterrestrial studies, and numerical taxonomy. Applications range from simple organic compounds to formulated synthetic polymers, from drugs to pathological conditions, from biological macromole-cules to the overwhelming complexity of whole organisms, and from ancient sediments and modern soils on earth to the totally automated analysis of martian soils on the surface of that planet.

This volume describes these various applications and discusses their significance and potential. The development and methodology of analytical pyrolysis are the subjects of Part A. Here, the vari-ous instrumental configurations are described and the design and control necessary to provide reliable results are discussed. Appen-dix 2 provides checklists to remind users of the important para-meters. The comparison of pyrolysis data is also a rapidly develop-ing field. Chapter 5 discusses the merits of available procedures, and Appendix 3 contains computer programs to illustrate some aspects of the computations.

Part B is a review of the applications and each chapter is fully referenced. The text has been extensively illustrated to give a visual representation of the important data. In many cases, for reasons of clarity or of space, these have been adapted from original sources. Tables of pyrolysis products have also been included wherever possible.

For permission to reproduce published figures I offer my grateful thanks to Elsevier Scientific Publishing Company, Preston Publications Inc., Heyden and Son Ltd., the Royal Society of Chemistry, the American Chemical Society, the American Geophysical Union, the American Institute of Physics, and the National Bureau of Standards. My thanks are also extended to R. A. Heller (Tandy Corp.) for allowing me access to line-printer facilities, and to K. Biemann, J. M. Bracewell, I. Ericsson, H. L. C. Meuzelaar, N. Sellier, W. Simon, W. C. Thompson, and VG Micromass and Extranuclear Laboratories for permission to reproduce figures.

Finally, I would like to thank my wife, Margaret, for typing this manuscript and for her forbearance during its production.

William J. Irwin

Contents

ANALYTICAL
PYROLYSIS

PART A

TECHNIQUES

Chapter 1

Historical Perspectives

1.1. INTRODUCTION

A large proportion of the organic substances found in nature are
unsuitable for direct analysis by powerful modern techniques such
as column chromatography and mass spectrometry. The molecules involved are usually polymeric with complex, variable structures, and
are frequently highly polar and nonvolatile. Developments involving
derivatization in gas chromatography, the introduction of affinity
and gel permeation liquid chromatography, and improvements in ionization and volatilization techniques in mass spectrometry, due to
mild desorption methods, have increased the range of these techniques
substantially. Nevertheless, the direct analysis of many natural
polymers is elusive. In common, therefore, with synthetic polymers,
the analysis of such materials, which may range from whole bacterial
or fungal organisms, through the complexity of organic geopolymers,
including humic substances and kerogen, to pure biopolymers such as
cellulose, is aided substantially by degradative techniques.

Many chemical and physical degradative methods, including hydrolysis, oxidation, reduction; photolysis, pyrolysis and radiolysis,
have been used for this purpose. The ease which pyrolysis (to yield
the degradation products) may be interfaced with analytical techniques such as gas chromatography or mass spectrometry (to separate
and identify these products), to produce an integrated system, has
ensured that pyrolysis methods have had a major influence in the
field of synthetic and biological polymer analysis.

3

Pyrolysis is the conversion of a sample into another substance or substances through the agency of heat alone. This process generally leads to molecules of lower mass, due to thermal fission (Greek: *pyr*,fire; *lysis*,dissolution), but may result in products with an increased molecular weight through various intermolecular events. The pyrolysis profile is characteristic of a particular sample, either in the appearance of unique components or in the relative distribution of the products. Petroleum fractions and polymeric materials of low volatility such as organic geopolymers, synthetic polymers and biopolymers, are the most frequent substrates for pyrolytic study, and, broadly speaking, two related and frequently complementary modes may be identified. These may be described as *applied pyrolysis*, in which the production of the pyrolysis products per se is the objective, and *analytical pyrolysis*, in which the characterization of the original sample, through analysis of the pyrolysis products, is achieved [1].

Applied pyrolysis is thus frequently a large-scale preparative operation with many important industrial applications. Of paramount importance is the production of raw materials, particularly petrochemicals [2], and the search for alternative supplies had led to pyrolysis studies on organic geopolymers such as coal, oil shales, and kerogen; or biopolymers such as wood and seaweed; and also on polymer wastes. Major emphasis has been placed upon maximization of yield by adjustment of the pyrolysis conditions, the elucidation of mechanistic and kinetic parameters and in the identification of products. In addition to these preparative applications, pyrolysis processes have a significant impact upon public health and environmental considerations. The pyrolysis products of synthetic polymers, for example, are important due to the widespread use of plastics as insulators or structural elements. Thermal degradation may release toxic or flammable compounds, and such possibilities are carefully monitored, particularly in aerospace applications. Pyrolysis is also of fundamental importance in studies on cellulosic materials. The use of these materials as textiles and as components of wood has initiated much work into thermal degradation pathways, flammability,

and flame retardation. Thermal processes may also lead to the production of carcinogenic components, as, for example, during smoking,
and here, too, pyrolysis studies provide useful information.

Analytical pyrolysis, in contrast, is a small-scale analytical
method. The technique is used to aid the analysis of relatively
high molecular weight compounds which are tedious or are difficult
to study by more conventional means. In essence, the sample is pyrolyzed by the rapid application of heat. The pyrolysis fragments
are then separated and quantified, and the product distribution is
used for the assessment of the original sample. This assessment may
be used to provide qualitative data on the identity or the composition of the sample and quantitative analysis of the components or
structure, and to investigate kinetic and mechanistic aspects of
thermal and pyrolytic fragmentation. In many instances, analytical
pyrolysis is complementary to the applied pyrolysis approach, and
thus may be a useful means to enable preliminary development and assessment to be accomplished. It must be emphasized, however, that
the scale and thermal profiles differ significantly between the two
approaches, so that an exact correspondence cannot be expected.

Of clear industrial importance is the characterization of feedstock for ethylene production, particularly as the heavier petroleum-
based fractions, which are increasingly being used for this purpose,
also yield larger molecules such as propylene, dienes, and aromatics.
Traditional assessment has been based upon cracking behavior in pilot
plants, and upon predictions based upon component ratios [e.g., paraf-
fin:olefin:naphthene:aromatic (PONA) or iso:normal:cyclo:aromatic
(INCA)]. Although useful, such methods may be relatively expensive
and time consuming. Analytical pyrolysis offers an alternative
method for assessing the feedstock [3]. This requires only a small
amount of sample (10 to 20 µl), the analysis is complete within 2
hr, the pyrolysis temperature may be readily varied, and information
on the total product distribution is obtained. A gas chromatogram
showing the products from a microscale naphtha pyrolysis is displayed
in Fig. 1.1. Variation in the pyrolysis temperature enables the
product-distribution:temperature profile to be measured, so that
optimization may be effected. These data are displayed in Fig. 1.2,

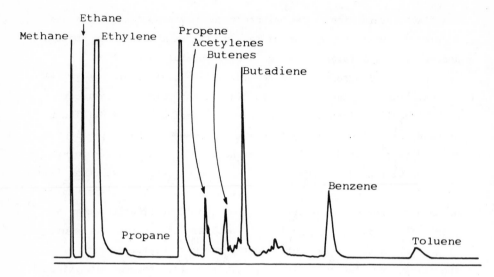

FIGURE 1.1 Analytical pyrolysis of KNG–CES LR-21081 naphtha at 860°:
product distribution. (From Ref. 3. Reproduced from the *Journal of
Chromatographic Science* by permission of Preston Publications, Inc.)

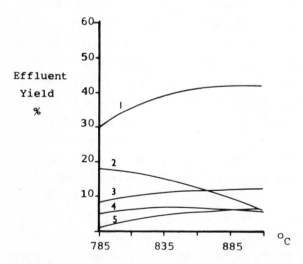

FIGURE 1.2 Analytical pyrolysis of KNG–CES LR-21081 naphtha: product
distribution as a function of temperature (1, ethylene; 2, propylene;
3, methane; 4, 1,3-butadiene; 5, benzene). (From Ref. 3. Reproduced
from the *Journal of Chromatographic Science* by permission of Preston
Publications, Inc.)

which shows the improvement in ethylene yield as the temperature is increased. The practical value in this approach is that the yields are directly correlatable with those from a commercial cracking furnace. The yield expected (y% ethylene) in the commercial cracker is calculated from that obtained using analytical pyrolysis (x) from

$$y = 0.83x - 4.03$$

for the conditions under test.

Table 1.1 illustrates some areas of study in which crossfertilization between the two approaches could be useful.

TABLE 1.1 Some Applications Common to Both Analytical and Applied Pyrolysis

Application	Applied pyrolysis Ref.	Analytical pyrolysis Ref.
Production based		
Coal: formation of volatile components	4	5
Kerogen: assessment of the potential of sedimentary rock as a petroleum source	6	7
Oil shales: assessment of oil-bearing potential	8	9
Organic synthesis	10	11
Petroleum cracking: theoretical and practical aspects	12	3
Polymer wastes: reclamation of raw materials	13	14
Degradation based		
Cooking: production and identification of thermally produced products	15	16
Flammability: structural, mechanistic and kinetic aspects of volatile, flammable compounds from wood and textiles	17	18
Smoking: modeling thermal processes and detection of carcinogenic products	19	20
Mechanistic aspects	21	22
Kinetic aspects	23	24

1.2. DEVELOPMENT OF ANALYTICAL PYROLYSIS

It has been claimed that the earliest report of analytical pyrolysis (Ca. 4000 B.C.) may be found in the Bible [25]. There, the process of "...Burnt sacrifice, an offering made by fire, of a sweet savour..." is described. This is clearly pyrolysis combined with an analytical technique (olfactory detection). An application is also described: "...Cain brought of the fruit of the ground an offering.. and Abel, he also brought of the firstlings of his flock.. And the LORD had respect unto Abel and his offering: But unto Cain and his offering he had not respect." Thus the differentiation of animal and vegetable material was effected.

The first scientific record of an analytical pyrolysis approach is probably that made by Greville Williams in 1860 [26]. Here, destructive distillation of rubber yielded isoprene and caoutchin (now known as *dipentene*). Williams was "...anxious to call attention to the fact that the atomic constitution of caoutchouc (rubber) appears to bear some simple relation to the hydrocarbons resulting from its decomposition by heat." He further showed that polymerization of isoprene (oxidation and heat) gave an elastic mass which, when burnt, yielded an odor characteristic of rubber. He thus had made significant advances in the structure determination of complex polymers. Moreover, two major present-day applications of analytical pyrolysis were foreshadowed; namely a detailed structural analysis of a polymer and the qualitative investigation into the identity of a polymer through the comparison of pyrolysis profiles. Williams displayed one more characteristic common in later work; he was highly critical of earlier studies.

Thereafter, the application of pyrolytic analysis techniques made little progress for almost 90 years. Other studies (1929) showed that rubber (7 kg pyrolyzed at 700°C at atmospheric pressure) yielded isoprene and dipentene [27], and that polystyrene (1935) (200 g pyrolyzed at 210 to 500°C) produced a complex mixture of hydrocarbons including monomeric, dimeric, and trimeric subunits of the polymer [28]. An early warning of some problems encountered in

later analytical pyrolysis work was also available, for it was found
that the product distribution varied with pressure, the pyrolysis
temperature and the rate of heating. In particular, under low-
pressure conditions a tetramer was also observed. The theoretical
aspects of polymer pyrolysis were also under development using concepts of free-radical chain reactions to model thermal reactivity
[29]. Thermal depolymerization reactions were widely reported, and
thermodynamic, kinetic, and synthetic studies were initiated. It
was also shown that monomer yield was dependent upon molecular weight
[30]. For analytical purposes, pyrolysates were comprised of many
closely related components, and this complexity limited the application of the technique because such mixtures could not be dealt with
satisfactorily.

The signficiant step in the evolution of analytical pyrolysis
was the combination of pyrolysis with a sophisticated physicochemical
technique for the efficient separation and/or identification of the
fragments. This development was aided by the recognition that polymer pyrolysis yielded hydrocarbon fragments and that mass spectrometry was a useful technique for the analysis of hydrocarbon mixtures
[31]. The way was now clear (1943) for the effective investigation
of complex molecular structures through the analysis of characteristic pyrolysis fragments. Thermal decomposition of polystyrene by
Madorsky and Straus [32], and a series of vinyl polymers by Wall [33]
were the first examples (1948) of this new approach. Initial studies
used milligram amounts of polymer and high-vacuum conditions, and
supposedly enabled higher molecular weight fractions to evaporate
from the matrix without secondary pyrolysis. The volatile pyrolysis
products were then fractionated and analyzed separately by mass
spectrometry. In 1949 this system was capable of generating a mass
spectrum which was characteristic of a particular polymer [34] and
which was largely due to the monomer and other small fragments (Fig.
1.3). Theoretical proposals to account for the observed modes of
fragmentation were also made. The scope of the method was extended
by Zemany (in 1952), who discussed the possibilities of using pyrolysis

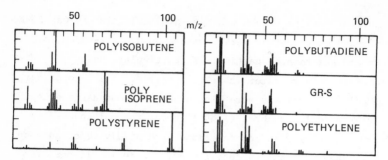

FIGURE 1.3 Mass spectra from polymer pyrolysis products.   (From Ref. 34.)

as an identification technique when he obtained characteristic spectra from homopolymers, copolymers and alkyds [35].   The first examples of the pyrolysis of biopolymers were also given when it was
shown that albumin and pepsin produced distinguishable pyrolysis
patterns.   Zemany also used punched cards for filing and retrieval
purposes--a foretaste of the importance of data-handling in the
modern applications of this technique.

     Integration of the pyrolysis-analysis system was achieved by
Bradt and co-workers (1953) [36].   They argued that only stable
products with significant volatility at room temperature would be
detected by the two-step process.   Thus valuable information concerning the polymer structure and its thermal degradation pathways
could be lost.   To overcome this, pyrolysis was effected within the
mass spectrometer (Py-MS) so that the products could be evaporated
directly into the ion-source region.   This method, in conjunction
with a more powerful mass spectrometer, allowed the detection of
pyrolysis products, up to a mass of about 520, from polystyrene.
At this time mass spectrometry was a rather specialized analytical
tool; the resolution of the components of a mixture required the
availability of the mass spectrum of the pure components, and considerable quantities of data could be generated rapidly.   For various reasons, therefore, an alternative to mass spectrometry as the
analytical technique was sought.   It was to be 20 years before pyrolysis mass spectrometry was a truly competitive technique again.

One such competitor was infrared spectroscopy. This technique was proving useful for the characterization of polymeric materials capable of being cast into thin films. The physical properties of many polymers, however, denied this approach, and pyrolysis followed by infrared spectroscopic characterization of the pyrolysis mixture was found useful in these cases. An early example was reported by Barnes and co-workers [37], but the scope of the method was illustrated in 1953 by Harms [38] who reproduced almost 30 spectra derived from the pyrolysis products of polymers. Rapid identification was effected in many cases, but the technique involved separate pyrolysis and analysis steps and relatively large scale (up to 2 g) manipulations. Thus, although pyrolysis infrared spectroscopy (Py-IR) has continued to find applications in polymer analysis, the major impact in analytical pyrolysis has been elsewhere.

The introduction of gas-liquid chromatography by Martin and James in 1952 [39] was the prelude to a rapid and continuing growth in pyrolysis applications. A technique was at last available which could separate and quantify the complex mixture of products obtainable from pyrolysis and which could produce a pyrolysis profile characteristic of a given polymer. This fact was first recognized in 1954 by Davison, Slaney, and Wragg of the Dunlop Company [40]. These workers suggested that comparison of the gas chromatographic patterns derived from the separation of the pyrolysis products of polymers would enable characterization to be effected. Using the newly described Katharometer detector [41], various profiles were presented to show that, for characterization purposes, the retention time was more important than peak intensities, which were affected by processes such as compounding or vulcanization. The possibility of identification through fraction collection was also mentioned. The introduction of integrated reaction gas chromatography (1955), enabling reaction products to be analyzed without prior extraction, rapidly followed [42], and the catalytic cracking of hydrocarbons in a hydrogen stream was thus demonstrated.

Progress was maintained by Haslam and Jeffs (1957) [43] who
applied gas chromatographic analysis to the products of the vacuum
depolymerization of synthetic polymers. Identification and estimation of the products enabled them to quantify styrene, methyl acrylate, and ethyl methacrylate polymers in polymethyl methacrylate
samples. A further advance in methodology (1958) was the recognition that the use of different columns for the analysis of a wide
range of polymers offered clear advantages [44]. The usefulness of
pyrolysis analysis was now firmly established. The next significant
refinement was the design of an integrated one-step system to overcome the need for prior isolation of the pyrolysate.

In 1959, combined pyrolysis-gas chromatographic (Py-GC) systems
and the application to polymer analysis were described almost simultaneously by three groups [45-47]. As in the case of Py-MS the integrated Py-GC system allowed a more effective sampling of the pyrolysis products and a greater reproducibility in operation. In one
application described by Radell and Strutz [45], the pyrolysis unit
was a heated tube which contained acrylic polymers. The pyrolysis
profile was found not to be reproducible above 550°C, whereas below
500°C insufficient degradation occurred. Between these temperatures
characteristic chromatograms of the pyrolysate were obtained. Components were identified by retention-time comparisons, and it was
shown that monomeric ester and alcohol components predominated
(Figs. 1.4 and 1.5). This paralleled the observations made in Py-MS
analysis [34] (cf. Fig. 1.3).

Concurrently, Lehrle and Robb [46] used a dielectric discharge
pyrolyser and a heated-filament method to illustrate two important
applications of the technique. The first of these involved a quantitative estimation of copolymer composition. Characteristic fragments (hydrogen chloride from polyvinyl chloride; acetic acid from
polyvinyl acetate) were used to construct calibration lines from
homopolymers. These data were used to estimate the relative proportions of each monomer in the copolymer; comparison with infrared
analysis was excellent. The second application illustrated the

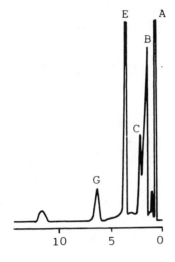

FIGURE 1.4 Pyrolysis products from polymethyl methacrylate. (A,air; B,methanol;
C,ethanol; E,methyl acrylate; G,methyl
methacrylate.) (From Ref. 45. Reprinted
with permission from E.A. Radell and H.C.
Strutz, Identification of Acrylate and
Methacrylate Polymers by GC, *Anal. Chem.*
31: 1890-1891. Copyright 1959 American
Chemical Society.)

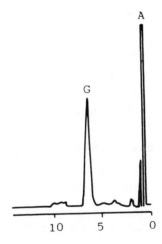

FIGURE 1.5 Pyrolysis products from polymethyl methacrylate and from polymethyl
acrylate. (A,air; G,methyl methacrylate.)
(From Ref. 45. Reprinted with permission
from E.A. Radell and H.C. Strutz, Identification of Acrylate and Methacrylate
Polymers by GC, *Anal. Chem. 31:* 1890-1891.
Copyright 1959 American Chemical Society.)

potential for mechanistic studies of the thermal degradation of
polymers. Variable temperature pyrolysis evidenced that, in contrast to the acrylate polymers, the degradation of polyvinyl acetate
showed little depolymerization breakdown, was characterized by acetic
acid loss at intermediate temperatures, and the resulting polyacety-
lene was degraded mainly at very high temperatures.

At this time Martin also described the application of his pyrolysis-gas chromatography system to the analysis of cellulosic polymers [47]. Pyrolysis was achieved within the carrier gas stream by means of high-intensity radiation, and the pyrolysis products were passed through a cold trap to remove condensable vapors prior to analysis.

This success heralded in 1960 a period of intense interest in pyrolysis gas chromatography. The ease with which pyrolyzers could be constructed and interfaced to a chromatograph led to the introduction of many different designs and analytical systems. This in turn meant that comparisons between these systems were difficult and that results were not of universal applicability. The development of analytical pyrolysis proceeded through two pathways. One sought to understand the processes involved in order to design efficient and reproducible systems. The other continued to expand the applications of the technique and to illustrate its widespread utility in many diverse fields.

Pioneering work in both areas was undertaken by Janak [48] when he examined the first pyrolytic analysis of drugs and biochemicals. The system was designed to eliminate as far as possible any secondary pyrolysis products, and to this end direct application of very small amounts of material (20 to 170 µg) to a heated filament was used. Flame-ionization detection (introduced in 1958 by McWilliam and Dewar [49]) enabled sufficient sensitivity to be achieved. Janak examined variables such as the heating current, the heating time of the wire, and the carrier gas flow rate. Quantitative reproducibility as a function of sample loading was also considered, and it was shown that quantitative analysis of amino acid mixtures was possible and that individual barbiturates could be recognized by their characteristic pyrolysis profiles. This was used in perhaps the first forensic application of analytical pyrolysis, when veronal was identified in a case of accidental drug consumption.

Pariss and Holland also designed a heated-wire pyrolyzer for direct on-column pyrolysis; in this case for the study of phenol-formaldehyde resins [50]. The term *pyrogram*, to describe the

chromatograms produced by this direct Py-GC analysis, was also introduced at this time. The quantitative reproducibility of pyrograms
was examined by Strassburger, Brauer, and co-workers [51] with excellent promise. Retention times were consistent to ±5 s and methyl
methacrylate copolymer composition could be estimated with a precision of better than ±0.5%. Pyrolyzer design also continued when
Szymanski, Salinas and Kwitowski introduced induction heating of
ferromagnetic metals [52]. This enabled a more controlled temperature to be achieved, and the method has developed into an important
modern pyrolysis technique.

In 1961 Jones and Moyles listed the basic requirements for an
effective pyrolysis unit when they described their heated filament
system [53]. The importance of rapid heating rates and small sample
sizes was clearly recognized. In the same year Robb's group used
mass spectrometry of the trapped pyrolysis products to identify the
components when they extended their mechanistic work on variable
temperature pyrolysis. They also compared quantitative evaluation
of polymers and polymer mixtures by pyrolysis and infrared spectroscopy [54]. Typical results were

Vinyl chloride-vinyl acetate: $y = 0.92x_1 + 5.66$ ($r^2 = 99.93\%$)
Binary polymer mixtures: $y = 1.02x_2 - 0.64$ ($r^2 = 99.85\%$)

where x_1 indicates the percentage composition by IR analysis, x_2 is
the composition of standard mixtures, and y is the corresponding
value from Py-GC determinations.

Improvement in the analysis conditions was also obtained when
the high-resolution capabilities of capillary columns, introduced
by Golay in 1958 [55], were demonstrated in an isothermal Py-GC
analysis of polymer mixtures by Stanley and Peterson in 1962 [56].
Sensitive flame-ionization detection was used because of the low
capacity of these columns, and one set of analytical conditions was
found to be appropriate for a range of polymers. Forensic applications were also extended when Kirk and co-workers reported the differentiation and identification of over 30 plastics [57].

In 1963 the range of applications of analytical pyrolysis was imaginatively extended by Oyama [58], who demonstrated the possibility of extraterrestrial life detection using this system. Pyrograms of three microorganisms isolated from soil samples were exhibited, and good resolution of the peaks, despite the complexity of the analyte, was obtained. Garner and Chi-Yuan [59] were also active at this time, but did not report their results in full. Geochemical applications were introduced by Nagar [60] in a Py-GC study of humic substances in soil.

The following year (1964) saw the recognition by Simon that inductive heating enabled the temperature of pyrolysis to be limited to the Curie-point of the ferromagnetic sample holder [61]. This enabled rapid and reproducible heating of samples to a well-defined end temperature to be achieved. Also reported was the comparison between temperature-programmed, packed columns and capillary columns by Cieplinski et al. [62], which showed the advantages of the coated capillary method. Various attempts to identify pyrolysis products were also described; mass spectrometry holds most promise [63].

The progress of analytical pyrolysis was such that an international meeting was held in Paris in 1965, and in the same year Brauer was able to write a lengthy review describing polymer analysis [64]. This year also heralded reports by Reiner on the advantages of using pyrograms, which were found to be characteristic of a particular microorganism, as a taxonomic tool [65]. A commercial pyrolyzer was also used in this work. Quantitative applications in drug analysis were described by Stanford [66], and a statistical approach to alkaloid identification was devised by Kingston and Kirk [67]. One other improvement was the development of pyrolysis-hydrogenation GC to simplify pyrograms, by converting olefins to the corresponding hydrocarbons [68].

The instrumentation of pyrolysis gas chromatography was further extended by Simon in 1966, when a rapid-scanning mass spectrometer was interfaced directly to his Py-GC system comprising a Curie-point pyrolyzer and capillary-column gas chromatograph [69]. This equipment allowed Simon's group to identify the fragments from amino acids, and to introduce a means to further a more thorough understanding of pyrolysis processes. A landmark in pyrolysis literature was achieved when a thorough review of pyrolysis gas chromatography was presented by R. L. Levy [70]. In view of the large number of different pyrolyzers in use at this time, this contribution, which described many of these systems and discussed the control of the variables, was particularly welcome.

Pyrolysis-gas chromatography was now in a rapid growth phase. In 1967, R. L. Levy further discussed the variables of pyrolyzer design and briefly described a significant advance in this field when he introduced a boosted heated-filament system [71]. The final temperature of this system was adjustable by means of the controlled discharge of a large capacitor, and very rapid temperature rise times could be achieved. Further Py-GC-MS work was reported by Merritt [72], quantitative measurements were used in dental research by Stack [73], and Oyama extended his work by examining reproducibility and the effect of growth medium [74]. Recognition of the problems in comparing pyrograms and of the necessity for coding and pattern-recognition techniques was also made at this time. Further extensions of the technique were made as Leplat showed that differences between nonvolatile petroleum fractions could be readily exposed by Py-GC methods [75], and E. J. Levy and D. G. Paul showed that identification of components separated by gas chromatography could be identified by online Py-GC [76].

By 1968, applications of pyrolysis gas chromatography as a fingerprint technique or as a quantitative tool were proving to be very successful. The wide diversity of pyrolyzer design and analytical conditions, however, was causing problems with reproducibility-- particularly when a comparison of results between different laboratories was attempted. To resolve this situation, a pyrolysis-gas chromatography subgroup of the Gas Chromatography Discussion Group of the Institute of Petroleum was formed. This group initiated trials aimed at the rationalization and standardization of methodology

in Py-GC, with the aim of controlling the technique sufficiently so
that a pyrogram library for identification or calibration purposes
could have interlaboratory significance. Frustration with variations in literature reports also led Shulman and Simmonds to re-
examine the pyrolysis of amino acids [77]. They designed a pyrolyzer
suitable for efficient pyrolysis onto a capillary column, and examined the effect of pyrolysis chamber material, pyrolysis temperature,
and analytical conditions on the pyrograms from various amino acids.
It was concluded that closely similar decomposition pathways were
followed, despite the apparent discrepancies in published work.

Basic work on pyrolysis rates by Farre-Rius and Guiochon revealed that in most systems pyrolysis was complete well before the
final pyrolysis temperature was achieved [78]. A typical half-life
for polytetrafluoroethylene at 600°C was 26 ms, indicating that the
temperature rise time of the pyrolyzer is a very important parameter
in ensuring reproducible pyrolysis. The thermomicrotransfer and
application of substances (TAS) procedure was introduced by Stahl to
include thin-layer chromatography in the analytical techniques available, and Biemann illustrated the application of Py-GC-MS to the
analysis of the organic components in the Murray and Holbrook meteorites [79]. Quantitative applications were extended when Szilagyi
began his work on the determination of quaternary ammonium compounds
such as acetylcholine in biological tissue [80]. Pyrolysis on the
mass spectrometer probe was under development at this time, and the
field-ionization mass spectrometry of the pyrolysis products of synthetic polymers was reported by Schuddemage and Hummel. This work
illustrated the simplification of spectra obtained with this technique, compared with the corresponding electron-impact data, and
probably catalyzed the resurgence of Py-MS as a powerful analytical
tool [81].

Laser-induced pyrolysis (1969) offered the hope that rapid temperature rise times could be obtained, together with surface degradation, so that primary pyrolysis products could escape without reaction. Such an apparatus was presented by Folmer and Azarraga, who

compared the performance of their instrument with conventional filament and furnace pyrolyzers [82]. This year also saw the first comprehensive identification of pyrolysis fragments from complex bio-
organic polymers, when Simmonds et al. reported Py-GC-MS studies on
the Murray meteorite, desert soil, and shale [83]. The origin of
the various components was assigned on the known pyrolysis behavior
of biological macromolecules, and it was concluded that analytical
pyrolysis was able to distinguish contemporary biological organic
matter from that of fossil or meteoric sources. Thus, a possible
extraterrestial life-detection system was available. The application
of analytical pyrolysis was extended to pathological systems when
Myers and Watson showed that pyrograms from the leaves of monocoty-
ledonous and dicotyledonous plants with viral or fungal infections
were distinguishable from healthy specimens [84]. Tsuge and Takeuchi
were also active at this time, and were showing how a knowledge of
the mechanism of pyrolytic fragmentation of synthetic polymers could
be used to establish the block versus random structure of copolymers
[85]. Soil components were under study by Wershaw and Bohner [86],
and Stack was able to write a critical review of the pyrolysis of
biological macromolecules [87].

In 1970, we saw continued interest in extraterrestrial samples
with the analysis of lunar material [88] and the Allende meteorite
[89] and the extension of Simmond's work to the identification of
bacterial pyrolysis products [90]. Vincent and Kulik extended the
chemotaxonomic potentialities to the differentiation of fungal
species and strains [91], while interest in technical aspects were
reflected in a report by Walker and Wolf on a comparison of the performance of different pyrolyzers [92]. At this time Vanderborgh
began to publish his work on laser pyrolysis, which was to involve
carbonaceous rocks and the quantitative assessment of fuel-bearing
potential [93]. Coupe, Jones, and Perry also made the first report
of correlation trials between various laboratories which illustrated
the range of equipment in use at that time, and the considerable
difficulties in ensuring pyrogram standardization [94].

In 1971, the correlation trials continued with attempts to produce fingerprint pyrograms from various synthetic polymers [95].
Reproducibility decreased as the complexity and intractability of the
sample increased; in particular, a gloss paint yielded widely diverse
results, while a simple copolymer was analyzed satisfactorily by most
participants. Successful new applications of analytical pyrolysis
included the structure determination of Actinomycin by Mauger [96],
and the measurement of the isotacticity of polypropylene by pyrolysis-
hydrogenation GC by Dimbat [97]. Bracewell began his work on the
classification of soils at this time, initially using Py-MS [98] and
Zeman described his technique of temperature-programmed probe pyrolysis directly into the ion source of the mass spectrometer [99]. This
approach allowed controlled degradation of a polymer to be undertaken
with the production of oligomeric fractions.

Progress in pyrolysis techniques was reflected by the convening
in 1972 of a second international conference in Paris. A wide diversity of reports were presented, including a comparison of the performance of various pyrolyzers [100], a review of forensic applications
[101], the detection of human pathological conditions [102], and the
use of a nitrogen-specific detector to enhance protein fragments in
pyrograms [103]. Up to 1972 the development or applications of appropriate technologies had not kept pace with the broad range of applications of analytical pyrolysis. Too many systems with too many
variables had created an environment of suspicion; results applicable
to one laboratory would not necessarily be directly applicable to
another. This year saw the beginning of the modern phase of analytical pyrolysis when pyrolysis, analysis and data-handling techniques
were developed and integrated to produce a powerful analytical system.

The necessity for data-handling to be placed on a rigid mathematical basis was recognized by various groups of workers. Kulik and
Vincent used a similarity index to identify *Penicillium* species [104].
Sekhon and Carmichael used a taxometric map procedure to differentiate
dermatophytic fungi [105], Merritt and Robertson used set theory to
aid fingerprint classification [106], and Reiner and Menger described

the results from a computer-based library-search system for the identification of unknown bacteria [107]. With pyrolyzer design, attention was drawn to the factors affecting interlaboratory reproducibility [100,107], and a discussion of the temperature rise time and true pyrolysis temperature led to the development of units with temperature rise times of 7 ms [108]. Efficient laser-pyrolysis systems [109] and photolysis-gas chromatographic analyses [110] were also reported. Complex biological samples were the interest of Meuzelaar and his group in Amsterdam. These workers and their collaborators were to revolutionize analytical pyrolysis and to provide an impetus for the rapid consolidation of advances in methodology--a factor sadly lacking up to this period. The first contribution was concerned with development of a Curie-point pyrolyzer, designed so that fast and reproducible pyrolyses onto a capillary column could be achieved [111]. The pyrolyzer also enabled the reaction chamber to be changed between each run to reduce background effects. A method for coating the ferromagnetic wires with microgram amounts of sample was also described. This enabled thin, uniform coatings of insoluble materials to be achieved, and significantly increased the reproducibility of the pyrogram.

The value of this approach was soon revealed when, in 1973, pyrograms from bacteria which differed in one antigenic component were shown to be distinct. Differences could be directly accounted for by pyrolysis of the purified antigen [112]. In parallel with this approach the Amsterdam group were also reconsidering the value of pyrolysis mass spectrometry in this field. A considerable volume of data was now available to enable efficient pyrolysis to be achieved, and mass spectrometry was a well-established analytical method in which significant instrumental advances had been made--not least of which were the fast-scanning capabilities of contemporary instruments. Also available at this time were signal averagers (now developed into online computer systems), which enabled a time-averaged rather than an instantaneous mass spectrum to be presented. This conjunction of pyrolysis, mass spectrometry, and computing now enabled pyrolysis

mass spectrometry to challenge the supremacy of Py-GC. In particular, when complex biological systems were studied, many products
were produced and the analytical profile of the mixture was highly
dependent upon the chromatographic conditions used. Furthermore,
Py-MS offered a much more rapid throughput of samples, a uniform
mass scale for fragment identification and a flat base line to facilitate automatic data collection.

Equipment to perform Py-MS analyses was described by Meuzelaar
and Kistemaker, and the application to bacterial fingerprinting was
illustrated [113]. Although the quadrupole mass spectrometer used
in this early work was only capable of unit resolution up to m/z 50,
characteristic and highly reproducible mass pyrograms were obtained.
Low-ionization-voltage mass pyrograms were quickly introduced in an
attempt to increase the proportion of molecular ions in the mass pyrogram [114]. Pyrolysis of bacteria in combination with field-
ionization mass spectrometry was reported by the same group in conjunction with Schulten and Beckey [115]. The high-resolution capabilities of this system and the spectral simplification afforded by
the mild ionization process enabled the identification of a large
number of products, many of which had been observed by Simmonds in
his Py-GC-MS work [90], although both technique and organisms differed. Field-desorption techniques were also applied by the same
groups to the analysis of nucleic acids [116]. Advances in other
areas were the use of ferromagnetic tubes as sample supports by
Simon's group [117], the design of a new pyrolyzer allowing variable
conditions of pyrolysis [118] and Luderwald's work on probe-Py-MS of
synthetic polymers [119]. Evaluation of the microstructure of copolymers was described by Alexeeva [120]. A study of the identification of cockroaches by Py-GC introduced internal standards to calibrate pyrograms [121], and the report on a third set of interlaboratory correlations using more carefully specified conditions was encouraging [122].

By 1974 the development of instrumentation for the proposed
Viking expeditions to Mars had continued for several years. As
part of the molecular analysis experiment to analyze atmospheric

and soil components GC-MS was proposed for the analysis of the organic components, with pyrolysis to volatilize higher molecular weight materials. The work of Oyama and Simmonds had shown that significant evidence concerning the origin of organic polymers in the biosphere was available through Py-GC-MS analysis, and it was expected that the proposed experiments could also answer questions concerning the existence of life forms on Mars. The final flight-configured Py-GC-MS system was described by Biemann, and it was shown to have excellent capabilities to handle synthetic mixtures and meteoric fragments, although some changes in column conditions were proposed [123]. The data handling was computer based to enable rapid recall and transmission of data, and specific ion current scans could be extracted.

In the same year Quinn described a capillary-column Py-GC system for the differentiation of various microorganisms [124]. Considerable discussion was given to optimization of the variables involved and over 200 components were detected. The main problem observed was the long-term degeneration of the column. Pyrolysis mass spectrometry was also undergoing development with a comparison of low-resolution electron-impact and high-resolution field-ionization techniques in the production of nucleic acid mass pyrograms [125].

An automated version of the Amsterdam Py-GC was described in 1975 [126]. This system was capable of unattended operation for 24 samples, and excellent performance was obtained. Also welcome was the report of the fourth interlaboratory correlation trial which suggested that, with a well-defined Py-GC system a high degree of reproducibility was obtainable [127]. Procedural improvements were the use of similarity coefficients [128] or linear learning machine methods [129] for pyrogram classification, the extended use of internal standards [130], and a description of laser Py-MS [131]. Reaction mechanisms were also studied by Py-GC-MS [22], and after a long history of reports on the thermal degradation of carbohydrates the Curie-point pyrolysis of cellulose at last appeared [132].

In 1976 the conjunction of the Viking landing on Mars and the third international symposium on analytical pyrolysis in Amsterdam highlighted the progress and broad range of applications of the

modern technique. As the ultimate in automated technology was re-
porting the condition of the martian atmosphere and surface (alas
with almost no organic component [133]) automation on earth was progressing with a report of a fully automated Py-MS system from Meuzelaar's group [134]. Collisional activation mass spectrometry was
used by Levsen and Schulten to examine the pyrolysis products of DNA
and to establish the structure of various components in the mass pyrogram [135]. Also appearing at this time was the pyrolysis-gas
chromatographic peak identification system of Uden [136], a method
for the production of pure microbial cells by culture on membrane
filters [137] and the temperature-programmed probe-pyrolysis-mass
spectrometry of whole organisms [138].

The published proceedings of the Amsterdam conference (1977)
gave a contemporary account of analytical pyrolysis reflected by 34
full papers and 16 summaries of contributions [139]. A monograph on
pyrolysis gas chromatography also appeared at this time, reflecting
the confidence now held by practitioners of the technique [140].
Wheals and co-workers described a system rapidly convertible from
Py-GC-MS to Py-MS [141], pyrolysis chemical-ionization mass spectrometry was found by Manura and Saferstein to be useful in simplifying
mass pyrograms for forensic purposes [142], and pyrolysis with selected ion-monitoring following GC enabled quantification in drug metabolic studies [143]. Pyrolyzer design also continued with a description of a very high-vacuum Curie-point system by Simon, which substantially eliminated secondary reactions [144], and of a vertical
furnace instrument by Tsuge and Takeuchi which was capable of producing reproducible and characteristic pyrograms from various polymers
[145]. The results of three correlation trials organized by the
American Society for Testing and Materials were also discussed by
Walker at this time [146]. The concept of tuning pyrolysis and
analysis conditions to a standard polymer pyrogram was developed and
proved capable of enabling useful interlaboratory correlations to be
made.

In 1978, following the application of nonlinear mapping to bacterial taxonomy by Meuzelaar's group [147], canonical variates analysis for Py-GC data [148] and a FIT factor for Py-MS data [149] were proposed to aid pyrogram classification. Curie-point pyrolysis inside the ion source of a field-ionization mass spectrometer was described [150], laser-desorption mass spectrometry experiments appeared [151], and successive pyrolyses at increasing temperatures was shown to provide information on decomposition kinetics [24,152]. Applications involved the determination of multipolymer structure [153], quantitative determination of tryptophan in proteins [154] and the direct detection of drugs and metabolites in urine [155]. A technique was also proposed by Needleman [156] to overcome the problem exposed by earlier workers of long-term column degeneration in the analysis of microorganisms [124,126].

The final year of this brief survey saw three major events; each confirming analytical pyrolysis had come of age and was now a powerful technique applicable to many fields of study. The first of these was the Fourth International Symposium on Analytical and Applied Pyrolysis held in Budapest, with current trends being reflected in the presentation of over 40 contributions covering a wide variety of topics. The first Gordon Research Conference was held in New Hampshire a little later, and emphasized the emergence of pyrolysis mass spectrometry as a viable technique in many areas of study. The third event of this year was the launching of a journal devoted solely to analytical and applied pyrolysis [*Journal of Analytical and Applied Pyrolysis* (Elsevier), first issue June 1979]. At last a unifying force was available to focus the pyrolysis literature, which had hitherto been dispersed throughout a wide variety of sources.

It was encouraging to note that with the acceptance of Py-MS techniques work was underway to develop standardized procedures [157] and to measure reproducibility [158]. Hopefully, the proliferation of different methodologies, which so plagued the development of Py-GC,

could be averted by timely study and recommendations. As the decade
closed, reports on data handling [159], quantitative analysis [160],
mechanisms [161], kinetics [162], synthetic polymers [163], organic
geochemistry [164], soil science [165], microorganisms [166], pathological conditions [167], and biopolymers [168] underlined the
power of analytical pyrolysis.

1.3. PYROGRAM VARIABILITY: A CASE STUDY

Variations in pyrolysis data may be due to the pyrolysis conditions,
to the analysis conditions or, in the case of Py-GC, to the lack of
precise identification of the products. These factors have conspired
to make earlier literature reports a mine field for the unwary reader.
The pyrolysis of amino acids serves as an illustration of these effects, and in view of the simplicity of these compounds the elucidation of their pyrolytic behavior has been surprisingly complex. Not
only have different pyrograms been observed, but variations in the
products identified has caused considerable concern.

  Initial reports on the pyrolysis behavior of amino acids were
made by Janak [48] and Ulehla [169]. These indicated that decarboxylation was a significant reaction pathway and that not all naturally
occurring amino acids gave characteristic pyrograms. Pyrograms were
obtained from the pyrolysis of microgram amounts of the potassium
salts of aliphatic amino acids analyzed isothermally at 30°C on a
squalane column and were characterized by the appearance of the
primary amine fragment (Table 1.2). Winter and Albro [170] also
found predominant amine fragments from amino acids, but their procedure allowed an equilibration period before analysis of the pyrolysis products. This ensured the occurrence of complex secondary reactions and products from phenylalanine included ammonia, methylamine,
dimethylamine, ethylamine, tripropylamine, dipropylamine, and tribu-
tylamine. In view of later work, almost a unique set of products.

  With a highly refined system, Simon and co-workers were able to
characterize 28 amino acids unambiguously, and found toluene (78%)
as the main product from phenylalanine pyrolysis with styrene (15%),

TABLE 1.2 Pyrolysis of Leucine and Phenylalanine: Some Major Products

	CHARACTERISTIC PRODUCT	RETENTION TIME	PROGRAM	SCALE	PYROLYSIS TEMP	FORM	MEANS OF IDENTIFICATION	REF.
FROM LEUCINE	2-Methylpropylamine	18.5 m	Packed column isothermal (30°C)	µg	-	K salt	Retention time	48
	3-Methylbutyronitrile	13 m	Capillary column 50 - 160°C	µg	700°C	Na salt	Mass spectrometry	69
	3-Methylbutanal	-	Packed column -180 - 125°C	-	-	Zwitterion	Mass spectrometry	73
	3-Methylbutyronitrile	-	Packed and capillary columns	mg	500°C	Zwitterion	Mass spectrometry	16
	2-Methylpropylamine	-						
	Amines and aldehydes but no nitriles	-	Packed column	g	270°C off-line	Zwitterion	Mass spectrometry	173
PHENYLALANINE	Various amines	-	Packed column 100-200°C	mg	300°C off-line	Zwitterion	Retention time	170
	Toluene	9 m	Capillary column 50 - 160°C	µg	700°C	Na salt	Mass spectrometry	69
	Styrene	15 m						
	Benzene	-	Packed column -180 - 125°C	-	-	Zwitterion	Mass spectrometry	73
	Toluene	9 m	Capillary column 50 - 200°C	mg	500°C	Zwitterion	Mass spectrometry	77
	Styrene	18 m						
	Toluene	-	Packed column	g	850°C off-line	Zwitterion	Retention time UV and IR spectroscopy	174
	Phenanthrene and anthracene	-						

benzene (5%) and ethyl benzene (2%) also identified [171]. Nitriles
were a common feature of the pyrograms, and a higher column temperature enabled benzyl cyanide to be detected in the phenylalanine pyrolysate [69]. Although no amines were detected in this work it was
noted that the stainless steel column may have resulted in the loss
of very polar fragments. Further work using ^{14}C-labeled compounds
and pyrolysis-radio-gas chromatography confirmed that decarboxylation was the predominant decomposition reaction for phenylalanine
[172], but the results of Merritt and Robertson were difficult to
correlate with this data [72]. These authors found that acetonitrile and acrylonitrile were fragments common to many amino acids,
but found also that each amino acid yielded a characteristic fragment. The surprising observations were made that benzene was characteristic of phenylalanine, despite this being a minor product in
other systems [171], and that toluene was specific for tyrosine,
despite this being reported as the largest component of phenylalanine
pyrolysis. Other results were also at variance, although agreement
that pyrrole was a characteristic proline fragment was evident.
Pyrolysis undertaken on a large scale without immediate analysis
also results in significant secondary reactions and increased pyrogram complexity. Thus, leucine yielded a range of hydrocarbons,
aldehydes, amines and imines [173], while phenylalanine produced the
anticipated toluene, styrene, and benzene, but complex products, including phenanthrene, naphthalenes, biphenyl, stilbene, and quinoline
were also detected [174].

The variability in the observed results may be due to several
factors. Differences in sample form, sample size, and pyrolyzer
performance would cause differences in product distribution. In
particular, analytical pyrolysis should avoid the secondary reactions favored by the slow pyrolysis of large amounts of samples,
which give rise to complex products. Secondly, an inefficient transfer of products from the pyrolyzer to the GC column, and the failure
of the column to transmit all components may lead to significant
variations in the product distribution observed. The appearance of
pyrograms will also depend upon the analytical conditions, such as

the nature of the GC column and its operating temperature. Finally,
the identity of the components may be wrongly ascribed, unless a definitive technique such as mass spectrometry is applied. Retention-
time comparisons alone are particularly dubious.

To expose the true pyrolysis behavior of amino acids, Simmonds,
Shulman, and co-workers undertook a mechanistic study of the controlled degradation of these compounds using Py-GC-MS [16,77,175].
Variations in the pyrolysis and analytical conditions, combined with
mass spectral identification of the products, allowed the origin of
the major fragments to be established. This indicated that all of
the amino acids studied, with the possible exception of glycine,
undergo pyrolysis via analogous reaction pathways.

In these studies, the earlier work of Janak and Simon was confirmed, with simple amines and nitriles among the important fragments. Phenylalanine (Fig. 1.6) in this system [77] yielded a broad
phenylethylamine peak, but otherwise a pyrogram similar to that reported by Simon was obtained [69]. Toluene and styrene were the
other large components. Tyrosine was shown to give phenol, *p*-cresol,
and *p*-ethylphenol; products not transmitted by the system used by
Simon; and also toluene as reported by Merritt [72]. Other products
from phenylalanine detected here [77], but not reported elsewhere
[69], were of minor importance, with one exception. Simmonds and
Shulman also identified 1,2-diphenylethane as the last peak in their
pyrogram. This may have been missed earlier, due to an insufficiently
high column temperature, but also may be due to the pyrolysis conditions used. This product clearly results from two molecules of
phenylalanine and has therefore been produced by a series of intermolecular events after the initial pyrolytic fragmentation, a secondary reaction. Such effects are observed particularly when the pyrolyzer is loaded with large amounts of sample. The Simmonds and
Shulman system involved the pyrolysis of 1 mg of sample (cf. 3 µg
[69]) and is thus a possible source of such products. This view
is supported by the identification of aldimines [16] and the detection of diketopiperazines [175] as important fragments from aliphatic
amino acids.

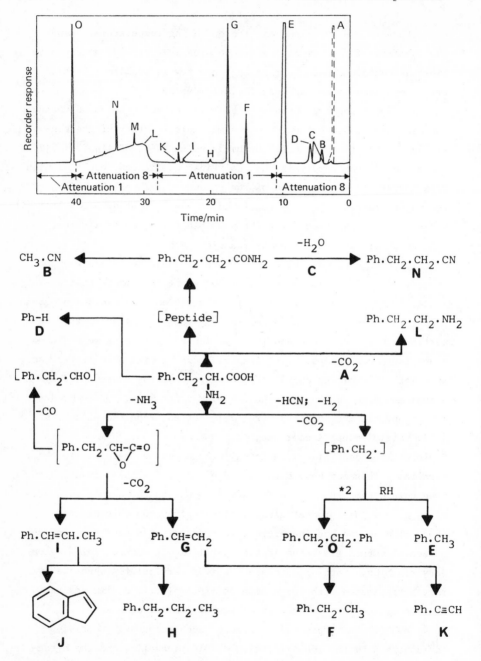

FIGURE 1.6 Pyrogram and possible fragmentation of phenylalanine: broken line, TCD; solid line, FID. (Pyrogram from Ref. 176.)

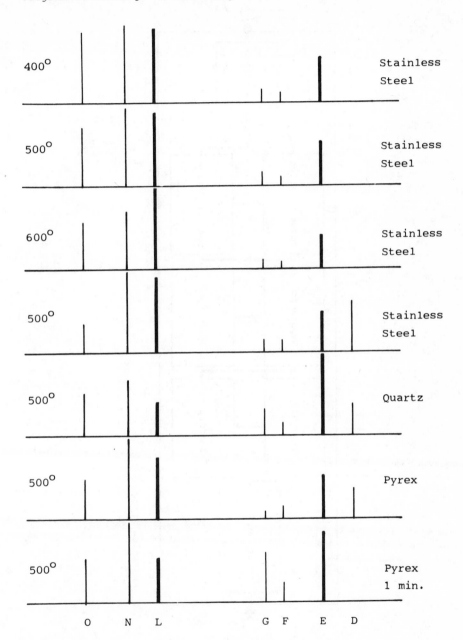

FIGURE 1.7 Reproducibility of pyrograms from phenylalanine. (For identification of fragments see Fig. 1.6; thickened peaks are attenuated sixfold.) (From Ref. 77.)

32

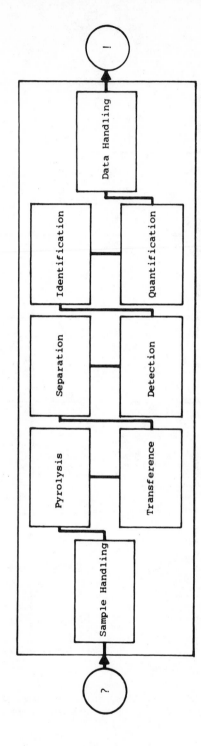

FIGURE 1.8 Analytical pyrolysis.

Differences may also be apparent when other variables such as the pyrolysis temperature or pyrolyzer equipment is changed. Figure 1.7 illustrates these effects for phenylalanine pyrolysis [77]. Although the same major products were obtained in each case, significant variations in their relative proportions are evident. Clearly, such discrepancies, if produced at random by poor pyrolysis control, are unacceptable in an analytical context.

To overcome these potential problems, the modern technique of analytical pyrolysis is an integrated system comprising pyrolysis, analysis, and data handling (Fig. 1.8). This combination is designed to provide rapid, controlled, and reproducible pyrolysis; efficient transfer, separation, and identification of the pyrolysis products and mathematical manipulation of the data, so that the reliability of the analysis may be assessed. The design and optimization of these systems is the subject of following chapters.

REFERENCES

1. Irwin, W.J., Analytical Pyrolysis--An Overview, *J. Anal. Appl. Pyrol.*, *1*(1979)3-25,89-122.

2. Szepesy, L., Welther, K. and Szalai, O., Production of Olefins by Pyrolysis of Liquid Hydrocarbons (Parts I-IV), *Hung. J. Ind. Chem.*, *5*(1977)161-281.

3. Greco, M., Ethylene Feedstock Characterisation by Microscale Py-GC, *J. Chromatogr. Sci.*, *16*(1978)36-45.

4. Campbell, J.H., Pyrolysis of Sub-bituminous Coal in Relation to *in-situ* Coal Gasification, *Fuel*, *57*(1978)217-224.

5. Hanson, R.L., Vanderborgh, N.E. and Brookins, D.G., Characterisation of Coal by Laser Py-GC, *Analyt. Chem.*, *49*(1977)390-395.

6. Harwood, R.J., Oil and Gas Generation by Laboratory Pyrolysis of Kerogen, *Amer. Assoc. Petrol. Geol. Bull.*, *61*(1977)2082-2102.

7. Gallegos, E.J., Terpane-Sterane Release from Kerogen by Py-GC-MS, *Analyt. Chem.*, *47*(1975)1524-1528.

8. Granoff, B. and Nuttall, Jr., H.E., Pyrolysis Kinetics for Oil-shale Particles, *Fuel*, *56*(1977)234-240.

9. Hanson, R.L., Brookins, D. and Vanderborgh, N.E., Stoichiometric Analysis of Oil Shales by Laser Py-GC, *Analyt. Chem.*, *48*(1976)2210-2214.

10. Blaslavsky, S. and Heicklen, J., The Gas Phase Thermal and Photochemical Decomposition of Heterocyclic Compounds Containing N, O, or S, *Chem. Revs., 77*(1977)437–472.

11. Wentrup, C., Flash Pyrolysis--Synthetic and Mechanistic Aspects, *Chimia, 31*(1977)258–262.

12. Albright, L.F. and Crynes, B.L. (Eds.), *Industrial and Laboratory Pyrolyses*, American Chemical Society Symposium Series No. 32, Washington, D.C., 1976.

13. Roy, M., Rollin, A.L. and Schreiber, H.P., Value Recovery from Polymer Wastes by Pyrolysis, *Polym. Eng. Sci., 18*(1978)721–727.

14. Brodowski, P.T., Wilson, N.B. and Scott, W.J., Chromatographic Analysis of Gaseous Products from Pyrolysis of Organic Wastes with a Single Column, *Analyt. Chem., 48*(1976)1812–1813.

15. Kosuge, T., Tsuji, K., Wakabayashi, K., Okamoto, T., Shudo, K., Iitaka, Y., Itai, A., Sugimura, T., Kawachi, T., Nagao, M., Yahagi, T. and Seino, Y., Isolation and Structure Studies of Mutagenic Principles in Amino Acid Pyrolysates, *Chem. Pharm. Bull., 26*(1978)611–619.

16. Simmonds, P.G., Medley, E.E., Ratcliff, Jr., M.A. and Shulman, G.P., Thermal Decomposition of Aliphatic Mono-amino Monocarboxylic Acids, *Analyt. Chem., 44*(1972)2060–2066.

17. Min, K., Vapour-phase Thermal Analysis of Pyrolysis Products from Cellulosic Materials, *Combust. Flame, 30*(1977)285–294.

18. Urbas, E. and Kullik, E., Py-GC Analyses of Untreated and Flame-proofed Wools, *Fire and Materials, 2*(1978)25–26.

19. Higman, E.B., Severson, R.F., Arrendale, R.F. and Chortyk, O.T., Simulation of Smoking Conditions by Pyrolysis, *J. Agric. Food Chem., 25*(1977)1201–1207.

20. Rosa, N., Py-GC Estimation of Tobacco Alkaloids and Neophytadiene, *J. Chromatogr., 171*(1979)419–423.

21. Benjamin, B.M., Raaen, V.F., Brown, L.L., Maupin, P.H. and Collins, C.J., Thermal Cleavage of Chemical Bonds in Selected Coal-related Structures, *Fuel, 57*(1978)269–272.

22. Merritt, Jr., C., DiPietro, C., Hand, C.W., Cornell, J.H. and Remy, D.E., Elucidation of Reaction Mechanisms for Thermal Elimination and Ring Closures by Py-GC-MS, *J. Chromatogr., 112* (1975)301–309.

23. Tran. D.Q. and Rai, C., A Kinetic Model for Pyrolysis of Douglas Fir Bark, *Fuel, 57*(1978)293–298.

24. Ericsson, I., Sequential Py-GC Study of the Decomposition Kinetics of cis-1,4-Polybutadiene, *J. Chromatogr. Sci., 16* (1978)340–344.

25. Jones, C.E.R. and Guichon, G., Pyrolysis-Gas Chromatography, *Chromatographia, 5*(1972)483.

26. Williams, C.G., On Isoprene and Caoutchin, *J. Chem. Soc. 15* (1862)110-125.

27. Midgley, Jr., T. and Henne, A.L., Natural and Synthetic Rubber. 1. Products of the Destructive Distillation of Natural Rubber, *J. Amer. Chem. Soc., 51*(1929)1215-1226.

28. Staudinger, H. and Steinhofer, A., Uber Hochpolymere Verbindungen. Beitrage zur Kenntnis der Polystyrole, *Liebig's Ann. Chem., 517*(1935)35-53.

29. Rice, F.O. and Rice, K.K., *The Aliphatic Free Radicals,* John Hopkins Press, Baltimore, 1935.

30. Washburn, H.W., Wiley, H.F. and Rock, S.M., The Mass Spectrometer as an Analytical Tool, *Ind. Eng. Chem. Analyt. Ed., 15* (1943)541-547.

31. Bachman, G.B., Hellman, H., Robinson, K.R., Finholt, R.W., Kahler, E.J., Filar, L.J., Heisey, L.V., Lewis, L.L. and Micuci, D.D., A New Method of Preparing Substituted Vinyl Compounds. Depolymerisation Studies on Vinyl Polymers, *J. Org. Chem., 12*(1947)108-121.

32. Madorsky, S.L. and Straus, S., High Vacuum Pyrolytic Fractionation of Polystyrene, *Ind. Eng. Chem., 5*(1948)848-852.

33. Wall, L.A., Mass Spectrometric Investigation of the Thermal Decomposition of Polymers, *J. Res. Nat. Bur. Stand., 41*(1948) 315-322.

34. Madorsky, S.L., Straus, S., Thompson, D. and Williamson, L., Pyrolysis of Polyisobutene (Vistanex), Polyisoprene, Polybutadiene, GR-S, and Polyethylene in a High Vacuum, *J. Res. Nat. Bur. Stand., 42*(1949)499-514.

35. Zemany, P.D., Identification of Complex Organic Materials by MS Analysis of their Pyrolysis Products, *Analyt. Chem., 24*(1952) 1709-1713.

36. Bradt, P., Dibeler, V.H. and Mohler, F.L., A New Technique for the MS Study of the Pyrolysis Products of Polystyrene, *J. Res. Nat. Bur. Stand., 50*(1953)201-202.

37. Barnes, R.B., Gore, R.C., Stafford, R.W. and Williams, V.Z., Qualitative Organic Analysis and IR Spectrometry, *Analyt. Chem., 20*(1948)402-410.

38. Harms, D.L., Identification of Complex Organic Materials by IR Spectra of their Pyrolysis Products, *Analyt. Chem., 25*(1953) 1140-1155.

39. James, A.T. and Martin, A.J.P., Gas Liquid Partition Chromatography. A Technique for the Analysis of Volatile Materials, *Analyst, 77*(1952)915-932.

40. Davison, W.H.T., Slaney, S. and Wragg, A.L., A Novel Method of Identification of Polymers, *Chem. Ind.* (1954)1356.

41. Ray, N.H., Gas Chromatography. The Separation and Estimation
     of Organic Compounds by Gas-Liquid Partition Chromatography,
     J. Appl. Chem., 4(1954)21-25, 82-85.

42. Kokes, R.J., Tobin, Jr., H. and Emmett, P.H., New Microcataly-
     tic-Chromatographic Technique for Studying Catalytic Reactions,
     J. Amer. Chem. Soc., 77(1955)5860-5862.

43. Haslam, J. and Jeffs, A.R., GLC in a Plastics Analytical Labora-
     tory, J. Appl. Chem., 7(1957)24-32.

44. De Angelis, G., Ippoliti, P. and Spina, N., Vapor-Phase Chroma-
     tography Applied to Polymer Analysis, Ricerca Sci., 28(1958)
     1444-1450.

45. Radell, E.A. and Strutz, H.C., Identification of Acrylate and
     Methacrylate Polymers by GC, Analyt. Chem., 31(1959)1890-1891.

46. Lehrle, R.L. and Robb, J.C., Direct Examination of the Degrada-
     tion of High Polymers by GC, Nature, 183(1959)1671.

47. Martin, S.B., GC Applications to the Study of Rapid Degradative
     Reactions in Solids, J. Chromatogr., 2(1959)272-283.

48. Janak, J., Identification of Organic Substances by the GC
     Analysis of their Pyrolysis Products, in R.P.W. Scott (Ed.),
     Gas Chromatography, Butterworths, London, 1960, pp. 387-400.

49. McWilliam, I.G. and Dewar, R.A., Flame Ionisation Detector for
     GC, Nature, 181(1958)760.

50. Pariss, W.H. and Holland, P.D., New Uses for GLC in Plastics,
     Brit. Plastics, 2(1960)372-375.

51. Strassburger, J., Brauer, G.M., Tryon, M. and Forziati, A.F.,
     Analysis of Methyl Methacrylate Co-polymers by GC, Analyt.
     Chem., 32(1960)454-457.

52. Szymanski, H., Salinas, C. and Kwitowski, P., A Technique for
     Pyrolysis or Vapourizing Samples for GC Analysis, Nature, 188
     (1960)403-404.

53. Jones, C.E.R. and Moyles, A.F., Rapid Identification of Polymers
     Using a Simple Pyrolysis Unit with a GLC, Nature, 189(1961)222-
     223.

54. Barlow, A., Lehrle, R.S. and Robb, J.C., Direct Examination of
     Polymer Degradation by GC. 1. Applications to Polymer Analysis
     and Characterisation, Polymer, 2(1961)27-40.

55. Golay, M.J.E., Theory and Practice of Gas-Liquid Partition
     Chromatography with Coated Capillaries, in V.J. Coates (Ed.),
     Gas Chromatography, Academic Press, New York, 1958, pp. 1-13.

56. Stanley, C.W. and Peterson, W.R., Polymer Analysis Using GC of
     Pyrolysates on a Capillary Column, Soc. Plastics Engrs. Trans.,
     2(1962)298-301.

57. Nelson, D.F., Yee, J.L. and Kirk, P.L., The Identification of
     Plastics by Pyrolysis and GC, Microchem. J., 6(1962)225-231.

58. Oyama, V.I., Use of GC for the Detection of Life on Mars, *Nature, 200*(1963)1058-1059.

59. Garner, W. and Chi-Yuan, F., quoted by P.A. Quinn in H.H. Johnston and S.W.B. Newsom (Eds.), *Rapid Methods and Automation in Microbiology*, Learned Information (Europe) Ltd., Oxford, 1976, p. 178.

60. Nagar, B.R., Examination of the Structure of Soil Humic Acids by Py-GC, *Nature, 199*(1963)1213-1214.

61. Giacobbo, H. and Simon, W., Methodik zur Pyrolyse und anschliessenden GC Analyse von Probemengen unter einem Mikrogramm, *Pharm. Acta. Helv., 39*(1964)162-167.

62. Cieplinski, E.W., Ettre, L.S., Kolb, B. and Kemmner, G., Py-GC with Linearly Programmed Temperature Packed and Open Tubular Columns. The Thermal Degradation of Poly-olefins, *Z. Anal. Chem., 205*(1964)357-365.

63. Gibson, C.V., The Identification of Polymers and other High Molecular Weight Compounds by a Combined Py-GC Technique, *Proc. Soc. Analyt. Chem., 1*(1964)79-81.

64. Brauer, G.M., Pyrolytic Techniques, *J. Polym. Sci.*, Part C, *8* (1965)3-26.

65. Reiner, E., Identification of Bacterial Strains by Py-GC, *Nature, 206*(1965)1272-1274.

66. Stanford, F.G., An Improved Method of Py-GC, *Analyst, 90*(1965) 266-269.

67. Kingston, C.R. and Kirk, P.L., Some Statistical Aspects of Py-GC in the Identification of Alkaloids, *Bull. Narcotics, 17* (1965)19-25.

68. Kolb, B., Kemmner, G., Kaiser, K.H., Cieplinski, E.W. and Ettre, L.S., Py-GC with Linearly Programmed Temperature Packed and Open Tubular Columns. The Thermal Degradation of Poly-olefins (II), *Z. Anal. Chem., 209*(1965)302-312.

69. Vollmin, J., Kriemler, P., Omura, I., Seibl, J. and Simon, W., Structural Elucidation with a Thermal Fragmentation GC-MS Combination, *Microchem. J., 11*(1966)73-86.

70. Levy, R.L., Py-GC. A Review of the Technique, *Chromatogr. Revs., 8*(1966)48-89.

71. Levy, R.L., Trends and Advances in Design of Pyrolysis Units for GC, *J. Gas Chromatogr., 5*(1967)107-113.

72. Merritt, Jr., C. and Robertson, D.H., The Analysis of Proteins, Peptides and Amino-Acids by Py-GC-MS, *J. Gas Chromatogr., 5* (1967)96-98.

73. Stack, M.V., Quantitative Resolution of Protein Pyrolysates by GC, *J. Gas Chromatogr., 5*(1967)22-24.

74. Oyama, V.I. and Carle, G.C., Py-GC. Application to Life Detection and Chemotaxonomy, J. Gas Chromatogr., 5(1967)151-154.

75. Leplat, P., Application of Py-GC to the Study of the Nonvolatile Petroleum Fractions, J. Gas Chromatogr., 5(1967)128-135.

76. Levy, E.J. and Paul, D.G., The Application of Controlled Partial Gas Phase Thermolytic Dissociation to the Identification of GC Effluents, J. Gas Chromatogr., 5(1967)136-145.

77. Shulman, G.P. and Simmonds, P.G., Thermal Decomposition of Aromatic and Heteroaromatic Amino-acids, Chem. Communs., (1968)1040-1042.

78. Farre-Rius, F. and Guiochon, G., On the Conditions of Flash Pyrolysis of Polymers as used in Py-GC, Analyt. Chem., 40(1968) 998-1000.

79. Hayes, J.M. and Biemann, K., High Resolution MS Investigations of the Organic Constituents of the Murray and Holbrook Chondrites, Geochim. Cosmochim. Acta, 32(1968)239-267.

80. Szilagyi, P.I.A., Schmidt, D.E. and Green, J.P., Microanalytical Determination of Acetylcholine, other Choline Esters and Choline by Py-GC, Analyt. Chem., 40(1968)2009-2013.

81. Schuddemage, D.R. and Hummel, D.O., Characterisation of High Polymers by Pyrolysis within the Field-Ionization Mass Spectrometer, Adv. Mass Spectrom., 4(1968)857-866.

82. Folmer, Jr., O.F. and Azarraga, C.V., A Laser Pyrolysis Apparatus for GC, J. Chromatogr. Sci., 7(1969)665-670.

83. Simmonds, P.G., Shulman, G.P. and Stembridge, C.H., Organic Analysis by Py-GC-MS. A Candidate Experiment for the Biological Exploration of Mars, J. Chromatogr. Sci., 7(1969)36-41.

84. Myers, A. and Watson, L., Rapid Diagnosis of Viral and Fungal Diseases in Plants by Py-GC, Nature, 223(1969)964-965.

85. Tsuge, S., Okumoto, T. and Takeuchi, T., Py-GC Studies on Sequence Distribution of Vinylidene Chloride-Vinyl Chloride Copolymers, Makromol. Chemie, 123(1969)123-129.

86. Wershaw, R.L. and Bohner, Jr., G.E., Pyrolysis of Humic and Fulvic Acids, Geochim. Cosmochim. Acta, 33(1969)757-762.

87. Stack, M.V., A Review of Py-GC of Biological Macromolecules, in C.L.A. Harbourn (Ed.), Gas Chromatography, 1968, Institute of Petroleum, London, 1969, pp. 109-118.

88. Oro, J., Gilbert, J., Updegrove, W., McReynolds, J., Ibanez, J., Gil-Av, E., Flory, D. and Zlatkis, A., GC and MS Methods Applied to the Analysis of Lunar Samples from the Sea of Tranquility, J. Chromatogr. Sci., 8(1970)297-308.

89. Levy, R.L., Wolf, C.J. and Oro, J., A GC Method for Characterisation of the Organic Content Present in an Inorganic Matrix, J. Chromatogr. Sci., 8(1970)524-526.

90. Simmonds, P.G., Whole Micro-organisms Studied by Py-GC-MS: Significance for Extra-terrestial Life Detection Experiments, *Appl. Microbiol., 20*(1970)567-572.

91. Vincent, P.G. and Kulik, M.M., Py-GC of Fungi: Differentiation of Species and Strains of Several Members of the Aspergillus Flavus Group, *Appl. Microbiol., 20*(1970)957-963.

92. Walker, J.Q. and Wolf, C.J., Py-GC: A Comparative Study of Different Pyrolysers, *J. Chromatogr. Sci., 8*(1970)513-518.

93. Ristau, W.T. and Vanderborgh, N.E., Simplified Laser Degradation Inlet System for Gas Chromatography, *Analyt. Chem., 42* (1970)1848-1849.

94. Coupe, N.B., Jones, C.E.R. and Perry, S.G., Precision of Py-GC of Polymers. A Progress Report, *J. Chromatogr., 47*(1970)291-296.

95. Jones, C.E.R., Perry, S.G. and Coupe, N.B., Precision of PY-GC of Polymers. Part II. The Standardisation of Fingerprinting. Preliminary Assessment of Second Correlation Trial, in R. Stock (Ed.), *Gas Chromatography, 1970,* Institute of Petroleum, London, 1971, pp. 399-406.

96. Mauger, A.B., Degradation of Peptides to Diketopiperazines: Application of Py-GC to Sequence Determination in Actinomycins, *Chem. Commun.* (1971)39-40.

97. Dimbat, M., Pyrolysis-Hydrogenation-Capillary GC. A Method for the Determination of the Isotacticity and Isotactic and Syndiotactic Block Length of Polypropylene, in R. Stock (Ed.), *Gas Chromatography, 1970,* Institute of Petroleum, London, 1971, pp. 237-246.

98. Bracewell, J.M., Characterisation of Soils by Pyrolysis combined with MS, *Geoderma, 6*(1971)163-168.

99. Zeman, A., The Identification of High Polymers by Thermal Degradation in the MS, in *Thermal Analysis (Volume 3),* Proc. Third ICTA (Davos), Birkhauser Verlag, Basle, 1971, pp. 219-227.

100. Walker, J.Q., A Comparison of Pyrolysers for Polymer Characterisation, *Chromatographia, 5*(1972)547-552.

101. Wheals, B.B. and Noble, W., Forensic Applications of Py-GC, *Chromatographia, 5*(1972)553-557.

102. Reiner, E. and Hicks, J.J., Differentiation of Normal and Pathological Cells by Py-GC, *Chromatographia, 5*(1972)525-528.

103. Myers, A. and Smith, R.L.N., Application of Py-GC to Biological Materials, *Chromatographia, 5*(1972)521-524.

104. Kulik, M.M. and Vincent, P.G., Py-GC of Fungi: Observations on Variability among Nine Penicillium Species of the Section Asymmetrica, Subsection Fasciculata, *Mycopath. Mycologia Applicata, 51*(1973)1-18.

105. Carmichael, J.W. and Sekhon, A.S., Py-GLC of Some Dermatophy-
      tes, Can. J. Microbiol., 18(1972)1593-1601.

106. Merritt, Jr., C. and Robertson, D.H., Qualitative Analysis of
      GC Eluates by Means of Vapour-phase Pyrolysis. II. Classifi-
      cation by Set Theory, Analyt. Chem., 44(1972)60-63.

107. Menger, F.M., Epstein, G.A., Goldberg, D.A. and Reiner, E.,
      Computer Matching of Pyrolysis Chromatograms of Pathogenic
      Micro-organisms, Analyt. Chem., 44(1972)423-424.

108. Levy, R.L., Fanter, D.L. and Wolf, C.J., Temperature Rise-Time
      and True Pyrolysis Temperature in Pulse-mode Py-GC, Analyt.
      Chem., 44(1972)38-42.

109. Fanter, D.L., Levy, R.L. and Wolf, C.J., Laser Pyrolysis of
      Polymers, Analyt. Chem., 44(1972)43-48.

110. Juvet, Jr., R.S., Smith, J.L.S. and Li, K.-P., Polymer Iden-
      tification and Quantitative Determination of Additives by
      Photolysis-GC, Analyt. Chem., 44(1972)49-56.

111. Meuzelaar, H.L.C. and in't Veld, R.A., A Technique for Curie-
      point Py-GC of Complex Biological Samples, J. Chromatogr. Sci.,
      10(1972)213-216.

112. In't Veld, J.H.J.H., Meuzelaar, H.L.C. and Tom, A., Analysis
      of Streptococcal Cell Wall Fractions by Curie-Point Py-GC,
      Appl. Microbiol., 26(1973)92-97.

113. Meuzelaar, H.L.C. and Kistemaker, P.G., A Technique for Fast
      and Reproducible Fingerprinting of Bacteria by Py-MS, Analyt.
      Chem., 45(1973)587-590.

114. Meuzelaar, H.L.C., Posthumus, M.A., Kistemaker, P.G. and Kis-
      temaker, J., Curie-point Pyrolysis in Direct Combination with
      Low Voltage Electron Impact Ionisation MS. A New Method for
      the Analysis of Non-volatile Organic Materials, Analyt. Chem.,
      45(1973)1546-1549.

115. Schulten, H.-R., Beckey, H.D., Meuzelaar, H.L.C. and Boerboom,
      A.J.H., High-Resolution FIMS of Bacterial Pyrolysis Products,
      Analyt. Chem., 45(1973)191-195.

116. Schulten, H.-R., Beckey, H.D., Boerboom, A.J.H. and Meuzelaar,
      H.L.C., Py-FDMS of Deoxyribonucleic Acid, Analyt. Chem., 45
      (1973)2358-2362.

117. Oertli, Ch., Buhler, Ch. and Simon, W., Curie-point Py-GC Using
      Ferromagnetic Tubes as Sample Supports, Chromatographia, 6
      (1973)499-502.

118. Tyden-Ericsson, I., A New Pyrolyser with Improved Control of
      Pyrolysis Conditions, Chromatographia, 6(1973)353-358.

119. Luderwald, I. and Ringsdorf, H., Untersuchungen von Polymeren
      im MS Fragmentierungsreaktionen von oligomeren β-Alaninen,
      Angew Makromol. Chem., 29,30(1973)441-452.

120. Alexeeva, K.V., Khramova, L.P. and Solomatina, L.S., Determination of the Composition and Structure of Polymers by Py-GC, *J. Chromatogr., 77*(1973)61-67.

121. Hall, R.C. and Bennett, G.W., Py-GC of Several Cockroach Species, *J. Chromatogr. Sci., 11*(1973)439-443.

122. Coupe, N.B., Jones, C.E.R. and Stockwell, P.B., Precision of Py-GC of Polymers. Part III. The Standardisation of Fingerprinting--Assessment of the Third Correlation Trial, *Chromatographia, 6*(1973)483-487.

123. Biemann, K., Test Results on the Viking GC-MS Experiment, *Origins of Life, 5*(1974)417-430.

124. Quinn, P.A., Development of High Resolution Py-GC for the Identification of Micro-organisms, *J. Chromatogr. Sci., 12* (1974)796-806.

125. Posthumus, M.A., Nibbering, N.M.M., Boerboom, A.J.H. and Schulten, H.-R., Py-MS Studies on Nucleic Acids, *Biomed. Mass Spectrom., 1*(1974)352-357.

126. Meuzelaar, H.L.C., Ficke, H.G. and den Harink, H.C., Fully Automated Curie-point Py-GC, *J. Chromatogr. Sci., 13*(1975)12-17.

127. Gough, T.A. and Jones, C.E.R., Precision of the Py-GC of Polymers. Part IV. Assessment of the Results of the Fourth Correlation Trial Organised by the Pyrolysis Sub-group of the Chromatography Discussion Group(London), *Chromatographia, 8*(1975) 696-698.

128. Meuzelaar, H.L.C., Kistemaker, P.G. and Tom, A., Rapid and Automated Identification of Micro-organisms by Curie-point Pyrolysis Techniques. I. Differentiation of Bacterial Strains by Fully Automated Curie-point Py-GC, in C.-G. Heden and T. Illeni (Eds.), *New Approaches to the Identification of Microorganisms*, John Wiley, New York, 1975, pp. 165-178.

129. Kullik, E., Kaljurand, M. and Koel, M., Analysis of Pyrolysis-Gas Chromatograms Using the Linear Learning Machine Method, *J. Chromatogr., 112*(1975)297-300.

130. Voorhees, K.J., Hileman, F.D. and Einhorn, I.N., Generation of Retention Index Standards by Pyrolysis of Hydrocarbons, *Analyt. Chem., 47*(1975)2385-2389.

131. Kistemaker, P.G., Boerboom, A.J.H. and Meuzelaar, H.L.C., Laser Py-MS: Some Aspects and Applications to Technical Polymers, *Dynamic Mass Spectrom., 4*(1975)139-152.

132. Ohnishi, A., Kato, K. and Takagi, E., Curie-point Pyrolysis of Cellulose, *Polymer J., 7*(1975)431-437.

133. Biemann, K., Oro, J., Toulmin III, P., Orgel, L.E., Nier, A.O., Anderson, D.M., Simmonds, P.G., Flory, D., Diaz, A.V., Rushneck, D.R. and Biller, J.A., Search for Organic and Volatile

Inorganic Compounds in Two Surface Samples from the Chryse
Planitia Region of Mars, *Science, 194*(1976)72-76.

134. Meuzelaar, H.L.C., Kistemaker, P.G., Eshuis, W. and Boerboom,
A.J.H., Automated Py-MS; Application to the Differentiation of
Micro-organisms, *Adv. Mass Spectrom, 7B*(1976)1452-1457.

135. Levsen, K. and Schulten, H.-R., Analysis of Mixtures by Collisional Activation Spectrometry; Pyrolysis Products of Deoxyribonucleic Acid, *Biomed. Mass Spectrom., 3*(1976)137-139.

136. Uden, P.C., Henderson, D.E. and Lloyd, R.J., Comprehensive Interfaced Py-GC Peak Identification System, *J. Chromatogr., 126*
(1976)225-237.

137. Oxborrow, G.S., Fields, N.D. and Puleo, J.R., Preparation of
Pure Microbiological Samples for Py-GC Studies, *Appl. Environ.
Microbiol., 32*(1976)306-309.

138. Risby, T.H. and Yergey, A.L., Identification of Bacteria Using
Linear Programmed Thermal Degradation MS. The Preliminary Investigation, *J. Phys. Chem., 80*(1976)2839-2845.

139. Jones, C.E.R. and Cramers, C.A. (Eds.), *Analytical Pyrolysis,*
Elsevier, Amsterdam, 1977.

140. May, R.W., Pearson, E.F. and Scothern, D., *Pyrolysis-Gas
Chromatography,* Chemical Society, London, 1977.

141. Hughes, J.C., Wheals, B.B. and Whitehouse, M.J., Simple Technique for the Py-MS of Polymeric Materials, *Analyst, 102*(1977)
143-144.

142. Saferstein, R. and Manura, J.J., Py-MS--A New Forensic Science
Technique, *J. Forensic Sci., 22*(1977)748-756.

143. Ohya, K. and Sano, M., Analysis of Drugs by Pyrolysis. I.
Selected Ion Monitoring Combined with a Pyrolysis Method for
the Determination of Carpronium Chloride in Biological Samples, *Biomed. Mass Spectrom., 4*(1977)241-246.

144. Schmid, P.P. and Simon, W., A Technique for Curie-point Py-MS
with a Knudsen Reactor, *Analyt. Chim. Acta, 89*(1977)1-8.

145. Tsuge, S. and Takeuchi, T., Vertical Furnace-Type Sampling
Device for Py-GC, *Analyt. Chem., 49*(1977)348-350.

146. Walker, J.Q., Py-GC Correlation Trials of the American Society
for Testing and Materials, *J. Chromatogr. Sci., 15*(1977)267-
274.

147. Eshuis, W., Kistemaker, P.G. and Meuzelaar, H.L.C., Some Numer-
ical Aspects of Reproducibility and Specificity, in C.E.R.
Jones and C.A. Cramers (Eds.), *Analytical Pyrolysis,* Elsevier,
Amsterdam, 1977, pp. 151-166.

148. MacFie, H.J.H., Gutteridge, C.S. and Norris, J.R., Use of
Canonical Variates Analysis in Differentiation of Bacteria by
Py-GC, *J. Gen. Microbiol., 104*(1978)67-74.

149. Hughes, J.C., Wheals, B.B. and Whitehouse, M.J., Py-MS. A Technique of Forensic Potential?, *Forensic Sci.*, *10*(1977)217-228.

150. Schulten, H.-R. and Gortz, W., Curie-point Pyrolysis and FIMS of Polysaccharides, *Analyt. Chem.*, *50*(1978)428-433.

151. Posthumus, M.A., Kistemaker, P.G., Meuzelaar, H.L.C. and Ten Noever de Brauw, M.C., Laser Desorption-MS of Polar Nonvolatile Bio-organic Molecules, *Analyt. Chem.*, *50*(1978)985-991.

152. Sugimura, Y., Tsuge, S. and Takeuchi, T., Characterisation of Ethylene-methacrylate Copolymers by Conventional and Stepwise Py-GC, *Analyt. Chem.*, *50*(1978)1173-1176.

153. Eustache, H., Robin, N., Daniel, J.C. and Carrega, M., Determination of Multipolymer Structure Using Py-GC, *Eur. Polym. J.*, *14*(1978)239-243.

154. Danielson, N.D. and Rogers, L.B., Determination of Tryptophan in Proteins by Py-GC, *Analyt. Chem.*, *50*(1978)1680-1683.

155. Irwin, W.J. and Slack, J.A., Detection of Sulphonamides in Urine by Py-GC-MS, *J. Chromatogr.*, *153*(1978)526-529.

156. Mitchell, A. and Needleman, M., Technique for the Prevention of Column Contamination in Py-GC, *Analyt. Chem.*, *50*(1978)668.

157. Windig, W., Kistemaker, P.G., Haverkamp, J. and Meuzelaar, H.L.C., The Effects of Sample Preparation, Pyrolysis and Pyrolysate Transfer Conditions on Py-MS, *J. Anal. Appl. Pyrol.*, *1* (1979)39-52.

158. Hickman, D.A. and Jane, I., Reproducibility of Py-MS Using Three Different Pyrolysis Systems, *Analyst*, *104*(1979)334-347.

159. Blomquist, G., Johansson, E., Soderstrom, B. and Wold, S., Reproducibility of Py-GC Analyses of the Mould Penicillium Brevi-Compactum, *J. Chromatogr.*, *173*(1979)7-17.

160. Choi, P., Criddle, W.J. and Thomas, J., Determination of Cetrimide in Typical Pharmaceutical Preparations Using Py-GC, *Analyst*, *104*(1979)451-455.

161. Schaden, G., Short-time Pyrolysis and Spectroscopy of Unstable Compounds. 9. Elucidation of Thermal Rearrangement Mechanisms by Curie-point and Flow Pyrolysis, *J. Anal. Appl. Pyrol.*, *1* (1979)159-164.

162. Andersson, E.M. and Ericsson, I., Determination of the Temperature-time Profile of the Sample in Py-GC, *J. Anal. Appl. Pyrol.*, *1*(1979)27-38.

163. Shimizu, Y. and Munson, B., Py-CIMS of Polymers, *J. Polym. Sci.*, *17*(1979)1991-2001.

164. Larter, S.R., Solli, H., Douglas, A.G., de Large, F. and de Leeuw, J.W., Occurrence and Significance of Prist-1-ene in Kerogen Pyrolysates, *Nature*, *279*(1979)405-408.

165. Saiz-Jimenez, C., Haider, K. and Meuzelaar, H.L.C., Compari-
     sons of Soil Organic Matter and its Fractions by Py-MS,
     Geoderma, 22(1979)25-37.

166. Gutteridge, C.S. and Norris, J.R., The Application of Pyroly-
     sis Techniques to the Identification of Micro-organisms, *J.
     Appl. Bacteriol., 47*(1979)5-43.

167. Reiner, E., Abbey, L.E. and Moran, T.F., Py-GC of Normal Human
     Cells and Amniotic Fluid, *J. Anal. Appl. Pyrol., 1*(1979)123-
     132.

168. Franklin, W.E., Direct Pyrolysis of Cellulose and Cellulose
     Derivatives in a MS with a Data System, *Analyt. Chem., 51*
     (1979)992-996.

169. Ulehla, J., Py-GC of Amino Acids, *Zivocisna Vyroba, 5*(1960)
     567-574. (*Chem. Abstr., 55*(1961)5242g.)

170. Winter, L.N. and Albro, P.W., Differentiation of Amino Acids
     by GLC of their Pyrolysis Products, *J. Gas Chrom., 2*(1964)1-6.

171. Simon, W. and Giacobbo, H., Thermische Fragmentierung und
     Strukturbestimmung organischer Verbindungen, *Chemie-Ing. Techn.,
     37*(1965)709-714.

172. Simon, W., Kriemler, P. and Steiner, H., Elucidation of the
     Structure of Organic Compounds by Thermal Fragmentation, *J.
     Gas Chrom., 5*(1967)53-57.

173. Lien, Y.C. and Nawar, W.W., Thermal Decomposition of Some
     Amino Acids. Valine, Leucine and Isoleucine, *J. Food Sci., 39*
     (1974)911-913.

174. Patterson, J.M., Haidar, N.F., Papadopoulos, E.P. and Smith,
     Jr., W.T., Pyrolysis of Phenylalanine, 3,6-Dibenzyl-2,5-piper-
     azinedione and Phenethylamine, *J. Org. Chem., 38*(1973)663-666.

175. Ratcliff, Jr., M.A., Medley, E.E. and Simmonds, P.G., Pyrolysis
     of Amino Acids. Mechanistic Considerations, *J. Org. Chem., 39*
     (1974)1481-1490.

176. Irwin, W.J. and Slack, J.A., Analytical Pyrolysis in Biomedical
     Studies, *Analyst, 103*(1978)673-704.

Chapter 2

Pyrolysis Methods

2.1. CONTROL OF PYROLYSIS

The essential requirement of the pyrolysis unit in analytical pyrolysis is that of reproducibility. Replicate analyses should produce
the same product profile, and pyrolysis fragments should be transmitted to the analytical device with equivalent efficiency. To
achieve this performance, quantitative control of the pyrolysis system is necessary. A consideration of the events occurring during
pyrolysis will illustrate the variables. These are collected together in Appendix 2 to provide a rapid assessment of pyrolysis
procedures.

When a sample is subjected to pyrolysis, primary-bond-fission
processes are initiated. These may proceed by several temperature-
dependent and competing reactions, which make the final product distribution highly dependent upon the pyrolysis temperature. Figure
2.1 indicates the dramatic temperature effect on polymethyl methacrylate in which methanol, a major product at lower temperatures,
almost totally disappears under more forcing conditions.

Clearly, the first concern in pyrolysis is to use a *precisely
controlled temperature* for analysis. This is often referred to as
the final pyrolysis temperature, or, more properly, the *equilibrium
temperature* (T_{eq}). Generally, as the pyrolysis temperature increases,
smaller and less characteristic fragments begin to dominate the pyrogram, and pyrolysis temperatures in the range 600 to 800°C are
preferred.

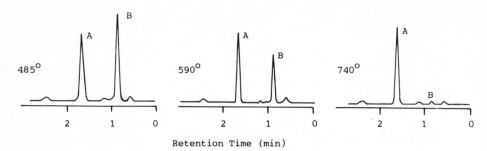

FIGURE 2.1 Variation of polymethyl methacrylate pyrograms with temperature: A, methyl methacrylate; B, methanol. (From Ref. 1, courtesy of Japan Analytical Industry Co., Ltd., Tokyo, Japan.)

The next concern is the temperature-time profile (TTP) of the
pyrolysis unit. This arises because of the very rapid degradation
reactions which occur during pyrolysis. It has been calculated that,
even for very stable materials such as Teflon (half-life 26 ms at
600°C), pyrolysis may be effectively complete before equilibrium
temperature is attained [2]. To achieve control of pyrolysis behavior, therefore, it is necessary for the sample to be reproducibly
heated, and ideally for total decomposition of the sample to occur
over the same temperature range.

The situation may be even more complex. The extent of pyrolysis during a linear temperature rise period may be defined in terms
of the characteristic temperature (T_s), which is the temperature at
which 36.8% (100/e %) of the initial sample remains unpyrolyzed.
It has been shown that T_s for polystyrene depends dramatically upon
the temperature rise time [3], with the temperature required for
complete pyrolysis of the sample falling markedly as the total heating time increases (Table 2.1). It is clear from this data that the

TABLE 2.1 Variation of Pyrolysis Temperature of Polystyrene with
Temperature Rise Time

Temperature rise time	(ms)	0.72	6.5	74	850	3 s	940 s
Complete-pyrolysis temperature	(°C)	590	530	480	420	370	310
Characteristic temperature	(°C)	550	500	450	400	350	300

Source: Ref. 3.

pyrolysis temperatures frequently quoted in literature reports (usually T_{eq}) are definitely not the temperatures at which pyrolysis occurs. In fact, the true pyrolysis temperature has been defined as
the temperature at which the rate of power consumed by the sample is
equal to the power supplied by the system [3]. With care, this temperature may be measured and the system described exactly.

The effect of the temperature rise time on the appearance of
the pyrogram from an isoprene-styrene copolymer is illustrated in
Fig. 2.2 [4]. This shows the increase in the more volatile isoprene
component as the heating rate is increased. Curves showing an ideal
and a poor temperature-time profile are depicted in Fig. 2.3. For
the equilibrium temperature to have significance, the temperature
rise time (TRT) should be rapid compared to the degradation rate of
the sample. The ideal situation would be instantaneous heating of
the pyrolyzer up to T_{eq} when energy is applied and rapid cooling
when this is removed. With a practical instrument three situations
may be envisaged, their importance being determined by the degree of
deviation from ideality. At lower equilibrium temperatures (<400°C),
pyrolysis is incomplete when T_{eq} is achieved, so that decomposition
has both nonisothermal and isothermal components. At usual temperatures (600 to 700°C), pyrolysis may be complete before T_{eq} is

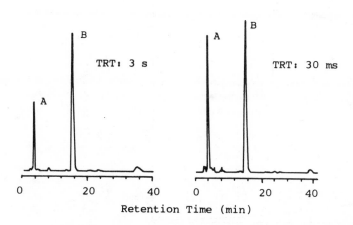

FIGURE 2.2 Pyrograms from an isoprene-styrene (68:32) copolymer.
The effect of temperature rise time. Peaks are A, isoprene, and B,
styrene. (From Ref. 4, reproduced with permission from the *Journal
of Chromatographic Science,* by permission of Preston Publications,
Inc.)

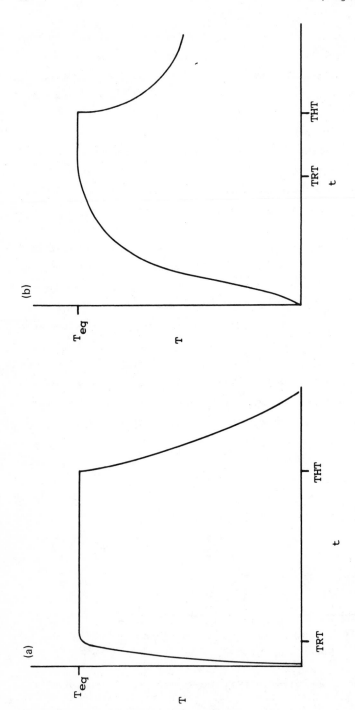

FIGURE 2.3 Pyrolysis temperature-time profile: (a) ideal and (b) poor performance.

TABLE 2.2 Curie-Point Pyrolysis of [^{14}C]Sodium Benzoate: Distribution of Radioactivity

Source of Radioactivity after Pyrolysis	Activity (%) using sodium benzoate	
	^{14}C ring	^{14}COO$^-$
Ferromagnetic conductor	7	26
Pyrolysis glass capillary	44	48
Components eluted from GC	35	23
Losses	14	3

Source: Ref. 5.

reached, and decomposition, which is essentially nonisothermal, depends upon the heating rate. With slower heating rates, higher temperatures (>800°C) may not directly influence the primary pyrolysis pathways, but changes in the appearance of the final pyrogram may result, due to either volatilization or secondary pyrolysis of products which have condensed onto surfaces within the pyrolyzer. This possibility is illustrated in Table 2.2, which shows that after pyrolysis a considerable amount of the sample may remain associated with the pyrolyzer and is not transmitted for analysis (Py-GC [5]).

The cooling rate of the pyrolyzer is also of importance in ensuring reproducibility. Slow and variable cooling will result in a variable degree of volatilization of these residues and, hence, will further decrease pyrolysis reproducibility. This problem may be reduced if the pyrolyzer can be bypassed, so that products released at this stage are not transmitted to the analytical device. To control these problems, the design of modern pyrolyzers combine a fast warm-up (8 ms to 600°C), a precisely defined equilibrium temperature which may be held for predetermined periods of time, and a well-designed heater of low thermal mass, to allow rapid cooling to be achieved. In addition, the pyrolysis products must escape rapidly from the pyrolysis zone without being subjected to further intense heating.

To complete the design of the pyrolysis unit this must be coupled to the analytical device. Ideally, the design must allow total transmission of volatile products for analysis. In pyrolysis mass

spectrometry this means that the pyrolyzer should be within the ion
source of the mass spectrometer if at all possible. With Py-GC,
gaseous diffusion must be minimized, pyrolysis should be effected as
near to the column top as possible, and the unit should be located
within the column heater zone so that the condensation onto cool
surfaces is reduced. Lack of control at this interface will also
disrupt pyrogram reproducibility.

The use of a modern pyrolysis unit effectively installed does
not, however, guarantee reproducibility. To overcome differences in
pyrolysis pathways which are caused by temperature-dependent effects
it is necessary that the whole of the sample is subjected to the same
temperature profile. The rapid heating capability of modern pyrolyzers can only be exploited if heat transfer to the sample is efficient and rapid. This demands that the sample is in intimate contact
with the pyrolysis heater, so that a temperature gradient within the
sample cannot be established. Thus, very thin films are most appropriate for a high degree of reproducibility. It has been shown that,
typically for sample weights of 1 µg or less, the temperature-time
profile of the sample is very close to that of the pyrolysis wire
[2]. With small sample-sizes, the absolute performance of the pyrolyzer is also a less critical factor in ensuring reproducibility.
Increasing sample-size will cause increased deviations from ideality,
and may introduce variations due to temperature gradient effects.
Data in Table 2.3. shows that degradation rates are not constant
until very thin sample films are used. Sample sizes are low as 10 ng
were used, and this in turn demands the use of sensitive detection
methods [6].

TABLE 2.3 Effect of Sample Thickness on the Rate of Pyrolysis of
Polymethylmethacrylate at 375°C.

Film thickness (nm)	160	130	100	96	64	28	24	16
Degradation in 10 s (%)	8	20	27	32	42	47	47	47

Source: Ref. 6.

A further consideration concerning sample size is the fate of the primary pyrolysis products. When produced from thin films, these should be transmitted rapidly to the analytical device without further reaction or degradation. When increased amounts of sample are used the primary products, or precursor radicals, may collide with other molecules and undergo a series of complex, competitive reactions which are a function of many variables, including TTP, the temperature gradient within the sample, pressure and the pyrolyzer design. In Py-GC the nature of the carrier gas and its flow rate also influence product distribution. The pyrolysis behavior of a styrene-divinylbenzene copolymer is illustrated in Fig. 2.4, and shows the effect of sample loading on secondary reactions [7]. Differences in quantitative recovery after pyrolysis are also susceptible to variations in sample load. Sodium benzoate, for example, is converted into volatile components with a 79% efficiency when 2 μg are pyrolyzed, but this falls to 49% when 100 μg are used [8].

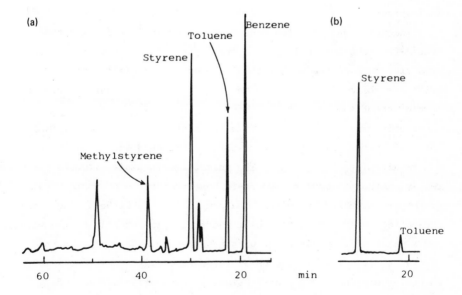

FIGURE 2.4 Pyrograms from a styrene-divinylbenzene (96:4) copolymer showing the effect of sample loading: (a) 650 μg and (b) 100 μg. Each pyrolysis at 550°C. (From Ref. 7.)

2.2. CLASSIFICATION OF PYROLYSIS SYSTEMS

Although many types of pyrolysis units have been described in the
literature, the classification of Levy [9] focuses attention on the
important variations in design. Two major groups may be identified,
depending upon the heating mechanism. Heat may be applied with a
continuous source of energy (continuous-mode pyrolyzers) or as a
pulsed source (pulse-mode pyrolyzers). The continuous-mode pyrolyzers are usually energized by electrical resistive heating, although
molten metal baths are also used. They are usually termed *microre-
actors* or *furnace pyrolyzers* and are effectively a link between classical pyrolysis and the true analytical technique. With pulse-mode
units the energy source may be resistive heating of a filament (filament or ribbon pyrolyzers), or inductive heating of a ferromagnetic
metal by means of an applied radiofrequency (Curie-point pyrolyzers).
In addition, laser pyrolysis, electrical discharge systems and
radiation-induced pyrolysis are classified as pulse-mode systems. A
consideration of the events occurring in typical systems on pyrolysis
is instructive.

The most popular pulse-mode systems available today are the
heated-filament and Curie-point pyrolyzers. In these units a wire
or a ribbon is coated with a thin film of the sample (perhaps 10
molecules thick, in a submicrogram experiment). The surface temperature of the filament is raised rapidly from ambient temperature to
the equilibrium temperature. This is maintained for a preset time
interval, after which the power source is removed and immediate cooling results (cf. Fig. 2.3a). The thin film of sample situated directly on the filament allows rapid heat transfer and ensures that
no temperature gradient within the sample is established. The sample
thus effectively duplicates the temperature-time profile of the wire,
and a pulse of thermal energy with a half-square-wave profile is de-
livered to it. This increases vibrational energy within the sample
molecules and initiates bond fission--perhaps after a delay of 10^{-5}
to 10^{-3} s, depending upon the thermal stability of the sample. Volatilization may thus precede bond fission and fragmentation may occur

when the molecule is migrating from the hot surface. Reactions to yield the final stable products should be essentially unimolecular. This is due to low sample loadings and the migration of the products from the hot surface to cooler regions of the pyrolysis unit. Both factors reduce the number of effective collisions which may lead to secondary products. The whole process probably occurs within 10 µm of the surface of the pyrolysis wire.

In contrast, a furnace pyrolyzer usually consists of a tube, the walls of which are uniformly and continuously heated externally. The sample, usually up to milligram amounts, is introduced into the tube, free from wall contact. The temperature of the sample begins to rise, with the surface being at the highest temperature. Thermal equilibrium with the gaseous surrounds will eventually be attained at the pyrolysis temperature, which will be substantially lower than that of the pyrolyzer walls. The pyrolysis products migrate, therefore, from the sample to hotter regions of the pyrolyzer. This may initiate additional fragmentation and produce secondary products by further reaction. This possibility is also increased, due to the fact that products must migrate through the residual sample before being released into the pyrolyzer.

For analytical pyrolysis, it would therefore appear more appropriate to use a pulse-mode system, with its associated advantages ensuring a higher degree of reproducibility in operation. This is not to say that good results cannot be achieved with furnace pyrolyzers, or that pulse-mode units must operate with submicrogram amounts and very rapid warm-up rates. Nevertheless, the best prospect for achieving universal reproducibility and applicability of pyrograms is through the recognition and control of pyrolysis variables. Table 2.4 compares the relative merits of pulse-mode and continuous-mode pyrolyzers.

TABLE 2.4 Properties of Typical Pulse-Mode and Continuous-Mode Pyrolyzers

CONSIDERATION	PULSE-MODE	CONTINUOUS-MODE
Temperature rise time	Fast and measurable	Generally unknown
Pyrolysis temperature	Approaches T_{eq} with small TRTs.	Generally unknown, but lower than temp. of reactor
Isothermal degradation	Approached with small TRTs.	Difficult to achieve
Heat transfer to sample	Fast with small sample sizes	Slow
Heat transfer within sample	Fast	Slow, producing a temp. gradient within sample
Flow-rate dependence	Low	High
Release of primary products	Rapidly; to cooler zone	Slowly; to hotter zone through sample matrix
Secondary reactions	Low probability	High probability
Detector sensitivity required	High	Low
Sample application	Care required	Easy
Reproducibility	High	Low
Scale	μg	mg

2.3 PYROLYSIS APPARATUS

Most current work using analytical pyrolysis is concerned with the analysis of involatile samples. The available apparatus for the pyrolysis of these materials includes a vast array of different systems. Many of these have been discussed previously [9,10]; so emphasis here will be on the types of system currently in use which enable controlled and reproducible pyrolysis to be achieved. Specific examples refer to Py-GC systems. In this mode, the pyrolyzer is interfaced to the GC column inlet, usually at the injector mounting. The carrier gas flow is directed through the pyrolyzer so that on pyrolysis the volatile products are swept onto the column. In most applications carrier gas flow is continuous, so that products are removed rapidly from the heated zone and analysis is undertaken without delay. Variations for Py-MS are dealt with in Chap. 4.

2.3.1. Heated-Filament Pyrolyzers

The use of heated-filament pyrolyzers has a long history, and the basic design requirements were recognized at an early stage [11,12]. The principle of operation is that an electric current is passed through a resistive wire or ribbon and the dissipation of power increases the temperature of the conductor. This type of unit is very easy to construct so that many systems have been described. However, such units, which operate at a fixed voltage, have extremely long temperature rise times; 10 to 30 s would not be unusual in these systems [6,9]. This time can be reduced, together with an increase in T_{eq}, by an increase in the applied voltage. To decrease the TRT substantially, say, to 1 s, however, the final temperature of the conductor becomes very high and perhaps exceeds the melting point of the metal. To allow very rapid TRTs with control of T_{eq}, it is necessary to control the power-input profile closely.

To this end, Krejci and Deml monitored the filament resistance as a function of time, and used this to produce a system capable of achieving a constant temperature with a much improved warm-up time (80 ms to 500°C) [13]. Boosted-filament pyrolyzers were also described by Lehrle and Robb, who supplied an appropriate boost-current for the first second of heating [6], and by Cogliano, who used a high-voltage pulse to achieve a short TRT [14]. A further improvement was described by Levy [3,15], who introduced the concept of voltage sweep. In this design, a high voltage is applied across the filament and the potential is rapidly reduced to a maintenance level. The use of capacitor discharge (10 mF) to effect a rapid TRT, followed by a constant current to maintain T_{eq} for a specified period thus allows high control of the temperature-time profile. Temperature rise times of 12 ms to reach 700°C are achievable [3].

The development of boosted heated-filament pyrolyzers continued with the introduction of the Pyroprobe (Fig. 2.5) [16]. This effectively extended the principle of feedback control to produce a highly predictable temperature-time profile for the filament. In this instrument, the resistive filament is used simultaneously as the heating element and as the temperature sensor. The filament is connected

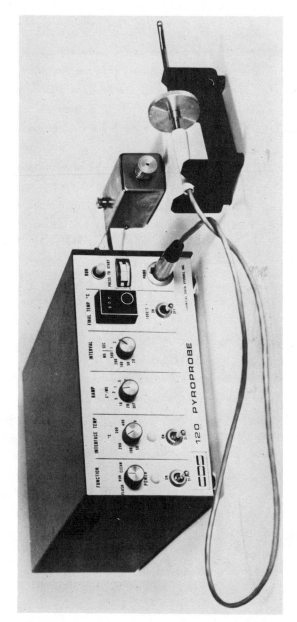

FIGURE 2.5   The Pyroprobe 120 heated-filament pyrolysis unit. (Courtesy of Chemical Data Systems, Inc., Oxford, Pa.)

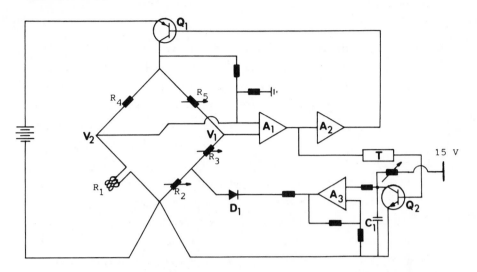

FIGURE 2.6 Basic circuit diagram of Pyroprobe pyrolysis unit: R1, heated filament (cold resistance to 0.25R); R2, final temperature control; R3, set proportional to cold resistance of R1; R4, massive resistance with a low temperature coefficient (0.2R); R5, calibrates R2; A1 and A3, operational amplifiers; A2, power amplifier; Q1 and Q2, power transistors; D1, diode; C1, capacitor; T, timer; V1 and V2, bridge voltages. (Courtesy of Chemical Data Systems, Inc., Oxford, Pa.)

to a Wheatstone bridge type of control circuit, and in the general circuit indicated in Fig. 2.6 the pnp transistor (Q_1) controls the filament current to yield a preset final temperature.

The bridge circuit is balanced ($V_1 = V_2$) when $R_1/(R_2 + R_3) = R_4/R_5$. In operation, the sample is placed on the platinum ribbon (R_1)--with typical dimensions of $38 \times 1.6 \times 0.0013$ mm--and the variable resistors are set so that bridge balance is achieved at the desired filament temperature; the resistance of R_1 increases from 0.25R at ambient temperature to about 0.7R at 500°C. When the pyrolyzer is switched on, the unbalanced bridge output is driven by the operational amplifier (A_1) through the power amplifier (A_2) to provide a base current for the power transistor (Q_1). The transistor is thus switched on, the full voltage of the power supply (~6 V) appears across the bridge, and the filament is heated rapidly (144 W) as a high current (~24 A) surges through it. This increase in tem-

perature causes an increase in the resistance of the filament, and
the bridge voltage drop decreases. This in turn reduces the bias on
the operational amplifier and the base current to Q_1 falls, reducing
the heating current through the filament. This control keeps the pyrolysis filament at a closely defined temperature after the initial
rapid temperature rise period. The timer switches off the base current after a predetermined time interval, and allows the filament to
cool rapidly (700 to 500°C in 150 ms), due to its low thermal mass.
Feedback is provided to linearise control, and a small permanent
bias to A_1 is present so that the system is self-priming.

In practice, the system is also provided with a means to vary
the heating rate linearly over the initial temperature rise period
(ramp control). This is achieved by connecting a timing circuit
across R_2. This consists of a capacitor-variable resistor network,
which provides a linearly increasing voltage to the bridge via a
further operational amplifier A_3. This voltage, rather than R_2, de-
termines the degree of imbalance of the bridge, so that the rate of
capacitor charging controls the heating rate of the filament. When
the ramp voltage through A_3 exceeds that of the bridge, the diode D_1
becomes reverse biased and control reverts to the basic bridge circuit through R_2.

This pyrolyzer design enables variation of heating rate (0.1 to
$20°C\ ms^{-1}$), from ambient temperature to 1000°C. Typical TRTs (without ramp control) are 8 ms to 600°C and 17 ms to 1000°C. For less
tractable samples, a coiled filament is also available--the sample
being held within a quartz tube or boat positioned within the coil.
The higher thermal mass of this system results in slower heating
(600°C in 600 ms) and cooling (700 to 500°C in 1.4 s). The furnace-
type pyrolysis conditions which may thus ensue make this option a
second choice unless sample-handling properties makes direct coating
of the filament impractical.

Versatility is also apparent, in that samples may be repetitively
pyrolyzed at progressively increasing temperatures (stepped pyrolysis),
the unit may be directly fitted to the column of a gas chromatograph or
designed as a direct insertion probe for mass spectrometry.

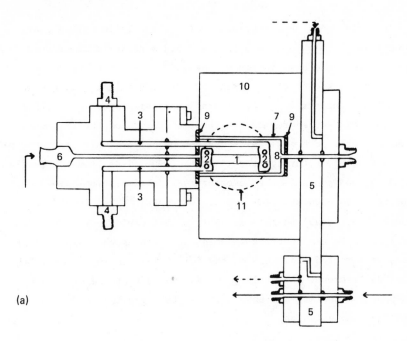

(a)

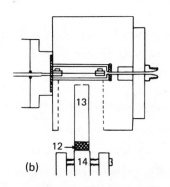

(b)

FIGURE 2.7 Heated-filament pyrolyzer with photodiode temperature
monitor: (a) above view and (b) side view. 1, Pt foil (15 × 2.6 ×
0.012 mm); 2, Pt plates; 3, Pt rod; 4, brass contact to power
supply; 5, three-way valve (up to flush pyrolyzer with carrier gas);
6, carrier gas inlet; 7, replaceable pyrolysis cell (24 × 10 × 5 mm);
8, outlet to GC; 9, Viton packing; 10, Al chamber (35 × 50 × 50 mm);
11, access hole for photodiode; 12, photodiode. (From Ref. 17.)

The performance characteristics of the pyrolysis unit are of paramount importance, and no pyrolyzer should be designed or used without an adequate description of the variables. The introduction of a further heated-filament system by Ericsson illustrates the approach [17]. The pyrolyzer configuration is displayed in Fig. 2.7, with a photodiode used to measure the temperature of the filament. Heating is achieved by two half-square-wave pulses, which induce currents of up to 60 A through the filament. The first is variable from 8 to 80 ms, and controls the temperature-rise time. The second may be varied up to a duration of 1 min. to maintain T_{eq}. This design includes a replaceable pyrolysis cell, allows variable materials and dimensions to be used for the filament and enables the sample to be located in a reproducible way on the foil. This is important in that all heated filaments are cooler towards the ends of the wire, due to end-conduction cooling and thus a temperature gradient may exist across the sample length.

2.3.2. Curie-Point Pyrolyzers

Developing in parallel with the heated-filament methods was Curie-point pyrolysis. This offered a means of rapid temperature rise to a precisely controlled equilibrium temperature, and development of this technique owes much to Simon and his group [5]. The source of heat in this pyrolysis system is the interaction of a high-frequency oscillator with a ferromagnetic metal. The sample holder may be a straight wire, a tube or a shape specifically designed to hold the sample firmly, such as a loop, a spiral, a boat, or a folded foil. The heating characteristics, however, depend upon this shape. An alternating magnetic flux is induced in the conductor; eddy currents in the conductor surface and hysteresis losses cause the temperature to increase rapidly. At a temperature specific for the material, a transition from ferro- to paramagnetism occurs (the Curie-point temperature). The energy intake by the wire falls and the temperature is held at the Curie-point. The power consumption of a ferromagnetic wire situated axially within a high-frequency induction coil has been calculated, using the expression:

$$N = 2\pi \ 2^{1/2} \ H^2 \ \rho(\tfrac{1}{S}) \ F(r,s)$$

where $S = (\dfrac{\rho}{\pi\nu\mu_o\mu_r})^{1/2}$

N = power consumption
H = magnetic field strength inside coil
r = radius of conductor
ρ = specific resistance of conductor
S = skin depth of eddy current
ν = oscillator frequency
μ_r = relative permeability (>>1 for ferromagnetic conductors
     below Curie point, ~1 for nonferromagnetic conductors)
μ_o = permeability of free space
μ = magnetic permeability ($\mu_r\mu_o$)
F = a Bessel function depending upon r and s

As $r/s \to \infty$, $F(r/s) \to 0.7$.

The relative permeability of a ferromagnetic conductor drops suddenly when the Curie point is approached. An increase in skin depth occurs, so that $1/S$ and $F(r,s)$ become smaller, thus reducing the power consumption of the conductor. The behavior of the sample holder at this point depends upon its dimensions, oscillator frequency and the magnetic field strength. With wires of more than 1 mm in diameter, power consumption above the Curie point is still evident, and temperatures above this may result. The optimal wire diameter allowing rapid warm-up and a well-defined end temperature decreases with the oscillator frequency. Typically, a frequency of 500 kHz requires a wire diameter of 0.5 mm, while with 1.2 MHz a diameter of 0.2 mm is appropriate. Only small differences are noted between the temperature-rise times between the surface and the centre of the conductor. Typically, at a field of 1170 Oe with a 1.5 kW oscillator, temperature rise times are around 30 ms (770°C), and higher powers (10 MHz with 0.05-mm-diameter wires) can reduce this to 5 ms. Smaller wires are also advantageous, in that rapid cooling is possible and thus the temperature-time profile approaches that of the ideal half square wave. Commercial units may have dramatically inferior performances [5].

The final temperature in these Curie-point systems is selected by the use of an appropriate alloy of ferromagnetic materials. It

TABLE 2.5 Curie points of Some Ferromagnetic Metals

METAL	Ni	Ni:Fe	Ni:Cr:Fe	Ni:Fe	Ni:Co	Fe	Ni:Co	Ni:Co	Co
Composition (%)	100	45:55	51:1:48	60:40	67:33	100	55:45	40:60	100
Curie-point (°C)	358	400	480	590	700	770	800	900	1128

Source: Ref. 18.

TABLE 2.6 Transfer Efficiency of Pyrolysis Products from Ferromagnetic Tubes and Wires

Source of Radioactivity after Pyrolysis	Activity (%) from labelled Benzoic Acid					
	2 µg			100 µg		
	Fe-Wire	Fe-Tube	Fe-Tube Capillary	Fe-Wire	Fe-Tube	Fe-Tube Capillary
Volatile Components	79	92	96	49	89	98
Walls of Chamber	14	4	1.5	43	4	1
Fe-Conductor (+ Capillary)	7	4	2.5	8	7	1

Source: Ref. 8.

should be noted, however, that different alloys with the same Curie-point temperature will have different temperature-rise profiles. This selection of temperature does not allow the versatility of temperature control or the possibility of stepped-pyrolysis that is available with filament units, but the reproducible profile without temperature calibration, which is almost independent of sample size, is attractive. Pyrolysis wires are available to provide a range of pyrolysis temperatures (Table 2.5) [18].

The basic arrangement of a typical Curie-point pyrolyzer is shown in Fig. 2.8a [19]. In this form, the unit is designed for Py-GC and fits directly onto the top of the GC column. A removable quartz tube is used to line the internal cavity and the pyrolysis wire, coated with sample, is placed within the tube and the induction coil. Energization of the coil from an external power supply causes rapid pyrolysis, and the carrier gas which flows through the quartz tube flushes the products onto the column. The quartz tube acts to reduce the dead volume of the system to enable chromatographic resolution to be retained, and as a disposable chamber to trap involatile residues. These may then be discarded so that background effects from earlier pyrolyses may be eliminated. The rf power of commercial units (30 W to 1.5 kW) varies considerably, and is a measure of the relative temperature rise times within these systems. Typically, a 30-W unit is an order of magnitude slower than the more powerful system, but the detailed behavior also depends upon the particular temperature chosen.

The use of ferromagnetic tubes, rather than wires, has also been advocated by Simon [8]. The extent to which undecomposed sample or involatile fragments is deposited onto the walls of the pyrolysis chamber increases with the loading of the wire, and is a possible source of poor performance. Table 2.6 illustrates this effect and shows that transfer is much more efficient with a low loading. The use of tubes as sample holders allows thermal equilibration between the sample and sample support, and a more efficient pyrolysis is achieved. This can be further increased if the sample is contained in a quartz capillary held in a ferromagnetic tube, although the temperature rise time of this system is longer. Table 2.6 compares

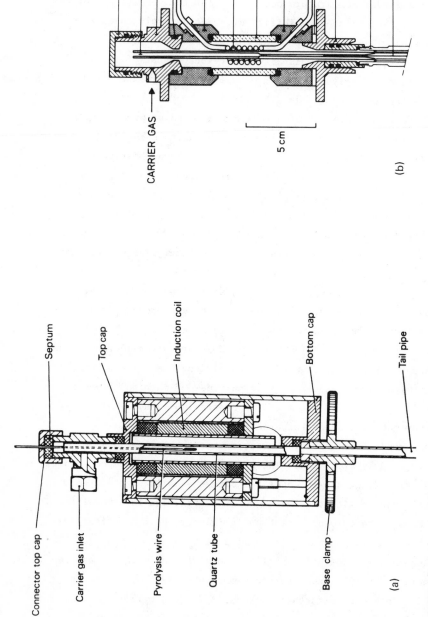

FIGURE 2.8 Curie-point pyrolysis: (a) packed-column system (from Ref. 19) and (b) capillary-column system (from Ref. 20). 1, cap; 2, quartz tube into which is placed ferromagnetic tube containing sample; 3, mounting flange; 4, Viton O-ring; 5 and 8, Delrin mounts; 6, water-cooled induction coil; 7, Pyrex tube; 9, Teflon seal; 10, mounting flange; 11, GC injector; 12, capillary column.

the efficiencies of pyrolysis of labelled sodium benzoate and shows
that for the tube sample supports, transfer is almost independent of
sample size. It would appear that the extra possibility of wall
collisions in this system may increase pyrogram complexity. For
example, a significant amount of benzene was produced in the pyrolysis of polystyrene [8].

As in the case of ferromagnetic wires, the warm-up characteristics of ferromagnetic tubes are dependent upon dimensions, radiofrequency, and oscillator power output. Iron tubes (d = 1 mm) with a
thickness equalling 20% of the radius may be heated and stabilized
at the Curie point within 30 ms (480 kHz, 1170 Oe, 1.5 kW), whereas
thicker walls take considerably longer. Thicker tubes (d = 2 mm)
may also be heated to temperatures in excess of the Curie point. In
practice, the oscillator was modified to provide an initial high-
powered pulse, which is followed by a further signal, the intensity
of which maintains the sample holder at the Curie-point for a specified period. In this way, a temperature rise time of 60 ms (from
20 to 700°C, d = 1.5 mm, t = 10%, ν = 480 kHz) was achieved [8]. An
advance pyrolysis unit capable of using a ferromagnetic tube as the
sample holder is shown in Figure 2.8b [20]. This system may be interfaced to a capillary-column GC with no loss of chromatographic
efficiency. This is achieved by gas flows being routed through and
around the pyrolysis tube before entering the capillary column.
This enables independent optimization of flow rates for both pyrolysis and separation efficiency. Curie-point pyrolyzers are also
available which can be used for both Py-GC and Py-MS applications.

2.3.3. Furnace Pyrolyzers

In contrast to heated-filament or Curie-point systems, furnace pyrolyzers are characterized by the continuous application of energy,
the sample is frequently held in a tube or a boat and large amounts
of sample may be used. The sample is usually dropped or pushed into
the heated zone and in many cases the use of such conditions provides slow (20 to 50 s) and variable pyrolysis with many secondary
products. This is particularly true if static pyrolysis conditions

operate. Here, the pyrolysis is allowed to proceed for a specific
time without removal of the products. Such procedures may also be
undertaken off-line so that trapping and sample-handling steps add
to the analytical problems. Although results using these systems
may be of practical value, they are unlikely to be reproducible in a
quantitative way. Furnace-type pyrolyzers, however, may be useful
when the lack of sample homogeneity necessitates the use of large
amounts of sample, or in cases when slow thermal interactions are
monitored. Studies such as the modelling of smoking conditions or
the effect of heat on carbohydrates to expose flammability profiles
may be usefully explored by these means.

To improve the conditions within furnace-type pyrolyzers Shulman and Simmonds introduced a design which featured an energy pulse
to give a rapid warm-up of the sample, allowed the variation and
estimation of pyrolysis temperature, and held a replaceable pyrolysis
chamber to prevent background contamination [21,22]. When this system was coupled to a capillary column GC very high resolution pyrograms were obtained.

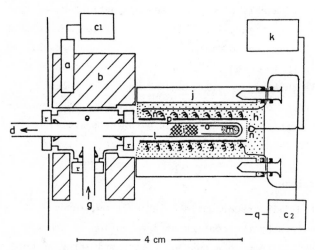

FIGURE 2.9 Controlled furnace-type pyrolyzer. a, heater; b, Al
block; c, variable transformer; d, gas outlet to column; e, Swagelok
union; f, column oven; g, gas inlet; h, cement; i, glass wool plug;
j, insulating block; k, pyrometer; l, stainless steel chamber; m,
sample; n, heater thermocouple; o, pyrolysis tube; p, ceramic tube;
q, line voltage. (From Ref. 23, reproduced from the *Journal of
Chromatographic Science*, by permission of Preston Publications, Inc.)

Figure 2.9 shows the arrangement of the unit. Accurate weighing of the sample (1 mg was initially used, but 80 µg gave better resolution) is achieved by the use of a pyrolysis tube which is held within a removable chamber. Power supplied to the pyrolysis heater is controlled by a thermocouple situated in the heater block, which has been calibrated to give a known sample temperature. Variation in the supply voltage enabled different heating rates and final temperatures to be selected. Typical temperature-rise times for this unit are 6 to 10 µs (to 600°C), and pyrolysis products migrate to hotter zones on formation. In addition to the possibility of further reactions occurring, this also results in a possibly better transfer of higher molecular weight fragments. Those which may be eluted represent a bonus, giving an increased number of characteristic peaks. Those which remain on the column cause a substantial decrease in column life [24]. Power was removed as soon as the final temperature was reached. This causes a decrease in temperature and pressure within the pyrolyzer. The back-flush of carrier gas thus reduces slow volatilization of residues. Although this unit may have certain limitations, at the time of its introduction it was a vast improvement over many contemporary designs. Furthermore, impressive results on a wide range of materials was obtained and the product distribution has been found comparable to that observed with highly refined Py-MS systems [25].

Perhaps the most advanced furnace pyrolyzer is that described by Tsuge and Takeuchi. The initial model is shown in Fig. 2.10 [26]. This is a vertical system and the sample (50 µg) is held in a platinum bucket suspended above the heating zone. This is released and falls rapidly into the heated zone, the diameter of which is progressively reduced to decrease dead volume to a minimum, and to increase the linear velocity of the carrier gas. The dimensions and flow rate are such that the carrier gas is heated to an equilibrium temperature so that no cooling of the sample occurs. It is estimated that on pyrolysis of small amounts of polymer at 600°C the fragments are effectively transferred to the column within 0.8 s-- comparable to many pulse-mode systems. Reproducibility data was

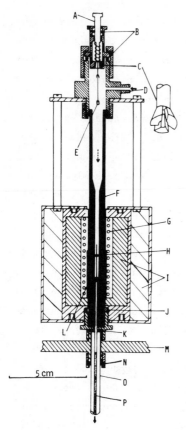

FIGURE 2.10 Controlled vertical Fur-
nace pyrolyzer. A, push button; B,
O-ring; C, fall-down device; D, carrier-
gas inlet; E, sample holder; F, quartz
tube; G, heater coil; H, hang-up gig;
I, heat insulation; J and K, O-rings;
L, silver foil; M, GC body; N, O-ring;
O, glass micropacked column; P, packing.
(From Ref. 26, reprinted with permission
from S. Tsuge and T. Takeuchi, Vertical
furnace-type sampling device for Py-GC,
Anal. Chem. *49*:348-350. Copyright 1977
American Chemical Society.)

impressive and resolution was much improved compared to a more traditional furnace system.

The simplest furnace pyrolyzer is the injection-port heater of
a gas chromatograph. This may be used effectively in such processes
as flash methylation [27,28] but cannot be recommended for the controlled pyrolysis of solid samples.

2.3.4. Laser Pyrolysis

Pyrolysis by laser [29,30] is a more specialized technique than the
methods described above, but useful results, particularly with geochemical samples, have been obtained. Indeed, the first reports of
laser pyrolysis (Py-MS) by Vastola and Pirone were concerned with

the analysis of coal samples [31]. The use of a laser beam to effect pyrolysis may have several advantages over the more traditional
methods. Sample handling, in particular, is facilitated in that the
grinding of samples to a fine powder (which may also alter constitution) is not required. Moreover, the coherent energy beam enables a
small volume (0.01 cm^3) of the sample to be selected for pyrolysis
so that specific areas of the sample may be studied. The intense,
short-duration beam enables very rapid temperature rise times (1 ms
to 3200°C), followed by rapid cooling to be achieved. The possibility of limiting reaction to surface degradation is also advantageous,
in that primary products do not migrate through the sample [32,33].

In a laser pyrolysis experiment, the laser beam from a ruby or
neodymium source is focused onto the sample, contained in a quartz
tube, perhaps using a second, low-power sighting laser, and a burst
of energy (1 ms) is released. This typically involves 2 to 10 J of
energy, which, if focused onto a spot 1 mm in diameter dissipates
0.26 to 1.27 MW cm^{-2}. A fraction of this intense energy is absorbed
by the sample, either by multiphoton absorption or electron tunneling, at the focal point of the beam and this creates rapid volatilization of fragments, causing a plume to propagate along the axis of
the beam. The absorptivity of the plume is greater than that of the
sample, so that the more energetic fragments, in particular, electrons, are subjected to even greater energies. The plume is thus at
too high a temperature (>10,000°C) to maintain structural integrity
and a plasma of electrons, atoms and perhaps stable radicals such as
C_2˙ results. The sample experiences shock waves, due to the thermal
stresses of plasma production, is subjected to intense heating due
to the presence of the plasma, and molecular fragments are mobilized
from the sample into the plasma. This also results in a crater being formed within the sample. When the energy pulse has decayed the
system returns rapidly--almost as fast as the heating process--to
ambient temperature. This is aided by radiation losses and collision with walls and carrier gas molecules (Py-GC), which also limits
plume propagation.

The events within the plasma plume may lead to three types of
product. *Plasma products* are produced by recombination of the species

produced at high temperature within the plume. They are characterized by low-molecular-weight gases such as acetylene and acetylene
derivatives (C_4H_2). *Thermal shock products* are derived from pyrolysis processes within the sample, and consist of volatile, stable
and characteristic fragments. *Plasma-tagged products* are normal pyrolysis products which have undergone further reaction with the
plasma-generated acetylene. This may be due to plasma radicals in-
vading the sample surface, or to radicals from sample pyrolysis entering the cooling plasma. In addition, further reactions, such as
hydrogen abstraction may result [34,35]. The product distribution
from the laser pyrolysis of polystyrene thus shows products such as
ethylacetylene and benzylacetylene in addition to the anticipated
monomer unit.

When ruby (694.3 nm) or neodymium (1040 nm) lasers are used,
clear or translucent materials may not absorb radiation efficiently.
This has been overcome by the addition of graphite [36,37], or metals
such as nickel [38], or by depositing the sample as a thin film on a
blue cobalt-glass rod. Initial heating of the sample is achieved by
contact with the additive or rod, and catalytic effects or gas evolution from a powdered sample may complicate the pyrogram, particularly when graphite is used [39]. Using the cobalt rod technique it
has been assessed that laser pyrolysis was similar to conventional
pyrolysis at 900-1200°C. Deliberate defocusing of the laser beam
may be used to enhance larger, more characteristic fragments [40].
Alternatively, the laser may be focused through a quartz tube [41].
The use of laser pyrolysis is complicated by the difficulty in controlling the final temperature, the development of thermal gradients
within the sample and the dependence of pyrolysis upon color. However, it is unique in that very large energies may be dissipated
very rapidly. In the field of organic geochemistry, in particular,
organic matter may be largely condensed aromatic polymers, and normal pyrolysis may give low yields of volatile products. Laser pyrolysis, which complements rather than competes with other methods,
enhances the information available from such systems and also enables specific areas of a sample to be probed [42].

An alternative approach to laser pyrolysis is the use of a laser which outputs energy at wavelengths corresponding to the vibrational energies of organic bonds (infrared: 2.5 to 16 μm). Such sources utilize HF and DF (2.8 to 4 μm), CO (5.0 to 6.0 μm) and CO_2 (9.1 to 11.0 μm). In contrast to the pulsed high-power lasers, these systems are operated so as to provide a continuous source of low-energy radiation. The pyrolysis time is controlled by a shutter. The use of lower energies reduces the possibility of electronic or photochemical fragmentation inherent with pulsed systems, and excitation is essentially vibrational or thermal in nature. Control of the temperature-time profile is still difficult, but pyrograms obtained using this technique are similar to those from conventional pyrolysis, with less involvement of complex plasma reactions [43,44]. For example, phenylalanine yielded benzene, toluene, xylene, and styrene in addition to gaseous fragments [44]. The presence of xylene indicates some complex interactions, but the remaining products are similar to those reported in Curie-point systems [18].

2.3.5. Pyrolysis of Volatile Materials

Although analytical pyrolysis has been mainly concerned with nonvolatile samples, considerable interest has also been shown in the pyrolysis of volatiles. This has led to studies which have examined the utility of Py-GC as a means of identifying lower molecular weight molecules, especially in comparison with mass spectrometry, and has developed into pyrolysis systems which are capable of identifying peaks eluted from a GC column. Pyrolyzers in this group are invariably furnace types and the design of Cramers and Keulemans [45] enables control of temperature and reaction time to be achieved. The reactor is a gold tube (1 m × 1 mm) wound around a silver core and surrounded by a heating wire (Fig. 2.11). Characteristic pyrograms from vapor-phase systems usually result only at low pyrolysis efficiencies (~20%). With higher conversions, smaller fragments predominate. A long length of narrow-bore tubing allows rapid heating and cooling of the sample, and a high carrier gas flow rate maximizes

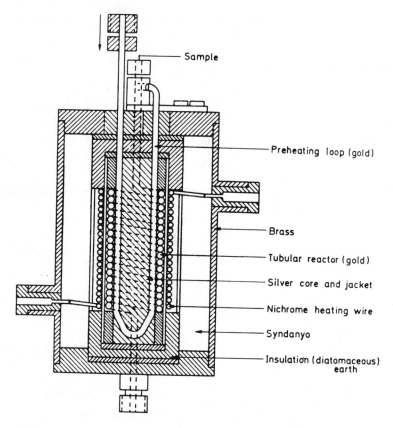

FIGURE 2.11 Furnace pyrolyzer for volatile samples. (From Ref. 45, reproduced from the *Journal of Gas Chromatography*, by permission of Preston Publications, Inc.)

efficiency. This unit has been used to demonstrate that the pyrolysis behavior of alkylbenzenes is a more useful identification parameter than mass spectral (EI) fragmentation [46]. This view has been supported for hydrocarbons [47] and for various aliphatic alcohols, aldehydes, and ketones when pyrolysis following gas chromatographic purification of samples was undertaken [48].

A typical GC–Py–GC system is shown in Fig. 2.12. Variants include in-line trapping of eluted components [49], the use of a delay coil to select peaks for Py–GC analysis [50], or interrupted elution which enables collection and further characterization of components

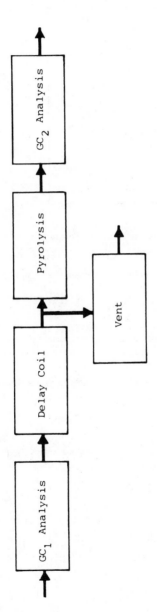

FIGURE 2.12 Typical GC-Py-GC arrangement for peak identification.

TABLE 2.7   Some Other Pyrolysis Systems

Technique	Reference
Curie-point	54-56
Dielectric breakdown	57
Electric arc	58
Furnace	59-67
Furnace - for volatiles	68-72
Photolysis	73
Pyrolysis-hydrogenation	74
Pyrolysis infrared	75
Thermal radiation	76
Thermolysis-TLC (TAS)	77

to be undertaken [51]. By using two columns, one specific for gaseous fragments and the other for larger molecules much evidence may be gained on the identity of the component under test [52,53].

2.3.6.   Other Pyrolysis Units

Table 2.7 records some of the other pyrolysis units which have been described to date.

In addition, pyrolyzers using in particular Curie-point and laser techniques have been developed for Py-MS applications. These, together with probe pyrolysis and highly controlled pyrolysis-ionization methods, such as field desorption, will be discussed in Chap. 4.

2.4.   SAMPLE-LOADING CONSIDERATIONS

A loss of quantitative precision is observable unless sample holders are scrupulously clean. Catalytic effects, increased background or reduced thermal transfer may result. Overnight heating (600 to 700°C) of solvent-washed wires, heating in a water-saturated stream of hydrogen at 550°C or chemical polishing followed by washing with

acetone and inductive heating has been used for the pretreatment of
ferromagnetic wires and tubes [20]. Direct flame cleaning may oxi-
dize the metal and change heat transfer characteristics. Heating
filament systems to 1000°C for 2 s is satisfactory. Sample con-
tainers, such as quartz tubes or boats, should also be thoroughly
cleaned and heated before use. Contamination becomes more evident
when small sample sizes are used, and solvents used for final washing should be freed from impurities such as the ubiquitous phthalates.

Although in normal operation it is unusual to monitor the
temperature-time profile of the pyrolyzer, this may be done as a
final control on standardization. This may be necessary, due to
lack of control in the production of ferromagnetic alloys, resulting
in Curie-points significantly different from nominal values [78].
Dimensions are also important in these systems. Filament systems
are subject to end-cooling effects and may also age, due to carbon
becoming dissolved in the metal [79], but with low loadings, long
lifetimes are expected. The temperature of the pyrolyzer may be estimated by using compounds of known melting points [80], calibrated
thermocouples [23], the color scale of temperature [79], and optical
pyrometry [12]. Methods have also been described using photodiodes
[3,17] which have very rapid response times, but most satisfactory
results for thin filaments or wires were obtained by the use of low-
mass thermocouples [3]. This method allows the temperature rise
time at precise positions on the filament to be measured. Such a
system enables the true pyrolysis temperature to be measured. This
shows up in a comparison between the temperature-time profiles of
the filament with and without sample. A plateau shows the true pyrolysis temperature and its duration as the sample absorbs energy
for the endothermic pyrolysis reaction.

The way in which the sample is loaded onto the pyrolyzer is
also important. A thin, uniform coating is ideal. In filament
units this should not be too near the ends, or else temperature
gradients will be experienced. In Curie-point systems the heating
rate is dramatically affected by the position of the wire within the

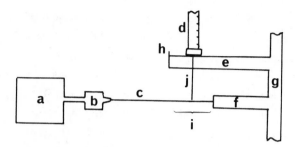

FIGURE 2.13 Loading of a rotating pyrolysis wire. a, motor; b, chuck; c, pyrolysis wire; d, syringe; e, syringe support; f, wire support; g, stand; h, syringe stop; i, loading zone; j, needle.

induction coil [3]. Coating should therefore be as controlled as possible, so that the sample is placed in the same position each time. For soluble samples the deposition from solution is usual. A microlitre syringe is used to transfer a measured volume of a standard solution (toluene, CS_2, or methanol) to the wire. Rotation of the wire during loading allows a uniform coat to be applied (Fig. 2.13 illustrates a simple laboratory set-up for this purpose). An alternative is to dip the wire into a tared solution and estimate the transfer by weight loss. This is, however, a less controllable procedure. Evaporation of the solvent is achieved with a hairdryer, an infrared lamp, or a low filament temperature, preferably in an inert atmosphere.

For insoluble samples the loading presents more difficulties. Large samples, such as may be available in geochemical work, may be reduced in size by grinding before use. Care should be taken to ensure that representative samples are taken for analysis. For Curie-point work, small samples of uniform size and shape may be held in a coil or loop of wire (Fig. 2.14), or wrapped in an iron foil [81]. The precise heating characteristics of such systems will, however, be unpredictable. Such samples are also readily handled by means of the tubes or boat sample holders available with filament pyrolyzers, and with the baskets used in furnace systems. To achieve rapid pyrolysis, the size of the sample should be kept to a minimum. Powders may be applied directly to a filament by prior wetting [82], and a

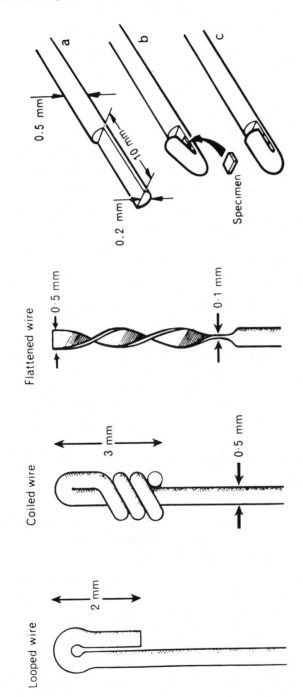

FIGURE 2.14 Looped pyrolysis wire for holding insoluble samples. (a) End of wire ground or filed flat. Specimen size is about 0.5 × 1.0 × 0.01 mm. (b) Loop formed with forceps and specimen placed in loop. (c) Loop closed with forceps to hold specimen firmly. (From Ref. 19.)

technique for the coating of Curie-point wires has been developed by
Meuzelaar and in't Veld [83]. The method is ideal for powdered and
microbiological samples. Typically, a freeze-dried culture is suspended in carbon disulphide (1 mg ml^{-1}) with ultrasonic agitation.
When a uniform dispersion is obtained, 5 to 15 µl are transferred to
a horizontal, rotating wire. CS_2 evaporates rapidly, forms stable
suspensions due to its high specific gravity and is not detected by
FID. It is thus an excellent medium, but due care must be given to
its volatility, low flash point, and toxicity. Methanol is a safer
alternative. The solvent and coating technique should be standardized, for changes will influence pyrogram appearance.

Finally, variability in sample preparation should be minimized.
This is particularly true of complex biological samples where freeze-
dried samples of viable cultures or sterilized pathogens may be
examined. The growth history and culture mediums add to the
considerations.

2.5. PYROLYZER COMPARISON

The choice of pyrolyzer is governed by several considerations--the
nature of the sample, the information required, the back-up facilities available and whether inter-laboratory comparisons are to be
made. When in-house fingerprint pyrograms are required of simple
synthetic polymers, it is probable that the majority of systems,
applied thoughtfully, will provide usable information. However, although such systems may be easy to introduce, they are difficult to
upgrade or replace because they leave a legacy of system-specific
pyrograms. With complex samples of biological origin, fingerprint
comparisons may depend upon small differences in the intensity of
fragments common to all pyrograms. Here, absolute control of the
pyrolysis conditions is essential. In all applications, control of
sample size, loading and of the heating characteristics of the pyrolyzer offers the best hope for long-term and interlaboratory reproducibility. In quantitative, mechanistic and kinetic work the occurrence of unpredictable secondary reactions must be minimized.

Here, too, the control afforded by modern design is essential. In analytical pyrolysis applications it is therefore recommended that a controlled system, such as a boosted filament or a high-power Curie-point system is used.

With the best available units there appears little difference between heated-filament, Curie-point, and furnace methods, although no stringent comparisons have been undertaken. Certainly, with current furnace systems the plethora of secondary products which have characterized this pyrolysis mode may be eliminated. Nevertheless, in general terms it is probable that at present the heated-filament and Curie-point systems are somewhat more useful, with the heated-filament being more versatile. Curie-point units may suffer from small variations, due, perhaps, to differences in alloy composition, sample loading and positioning of the conductor within the induction coil. These effects are magnified if the conductor is bent, twisted or folded to hold insoluble samples. With intractable materials, it is easier to replicate sample handling if the specimen is held in a boat within a heated-filament coil. Filament systems also enable rapid and continuous variation in pyrolysis temperature. Such an arrangement is useful in the study of mechanistic and kinetic profiles, and in optimization studies prior to quantitative work. The use of stepped pyrolysis, in which the same sample is pyrolyzed at successive, increasing temperatures is also useful in kinetic studies [84], when a complex mixture is analyzed [85], or when an organic component is trapped in an inorganic matrix [63]. In contrast, automation has essentially involved Curie-point systems [81,86].

The pyrolysis of volatile fragments should be designed so that the sample experiences a precise temperature for a predetermined period of time. Although packed tubes have been used for this purpose, reproducibility may vary and catalytic effects abound [87]. A flow-through gold tube is probably the best compromise [45,68]. Furnace pyrolyzers are best considered for applications involving large amounts of sample, in which relatively slow heating rates are involved. Such effects give useful information on thermal degradations but are characterized by the appearance of many secondary

products. The use of analytical pyrolysis is advantageous in such
cases, to illuminate initial degradation pathways [88].

Despite this assessment, analytical pyrolysis may be undertaken
satisfactorily with furnace systems [89,90], provided adequate control is applied, and it has been reported that a static furnace
method gave better differentiation of petroleum fractions than a
dynamic filament method [62].

Several comparisons of the performance of various pyrolyzers
have been described. These clearly do not enable a total understanding of pyrolyzer differences to be obtained. In many cases, limited
studies on a small range of samples without a full description of the
variables have appeared, and the pyrolyzers used have often been
superceded by modern models. Care should be exercised when assessing
data, to ensure that sample size and thermal profiles have been adequately described, and that analytical conditions are identical. The
identification of products also adds to a more thorough discussion.
Nevertheless, useful generalities emerge from some of this work.

Comparisons range from studies of whole organisms by Hall and
Bennett [91], to an examination of pure hydrocarbons by Walker and
Wolf [92]. In these specific examples, it was found that, with
cockroaches, a Curie-point method gave characteristic pyrograms, but
was easily contaminated, difficult to load and led to severe fragmentation of internal standards. With a furnace model, specificity
and reproducibility were difficult to achieve. Contamination increased with use, and this led to progressive changes in the pyrograms. Contamination was also observed with a pulsed-tube reactor
system, but here cleaning was easily accomplished, and in general,
this system was preferable [91]. With hydrocarbons, boosted-filament
and Curie-point pyrolyzers gave different product distributions with
low efficiency (7 to 30% and 4% respectively). A vapor-phase reactor
pyrolyzed 13 to 94% of the sample, and the product distribution was
more predictable by theoretical considerations [92]. Considerably
greater efficiency in the pyrolysis of less volatile compounds would
be expected (cf. Table 2.5). In contrast, a Curie-point system was

TABLE 2.8 Feature Comparison of Pyrolysis Units

	Heated Filament	Curie-point	FURNACE		Laser
			Solids	Volatiles	
Sample size range (µg)	0.1-500	0.1-500	50-5000	0.001-10	500 +
Sample state	solid	solid	solid	volatile	solid
Temperature variability	continuous	discrete	continuous	continuous	uncontrolled
Temperature range (°C)	1000	980	1500	800	10^9
TRT	10ms	70ms-2s	0.2s-1min.	-	10µs
Additives	-	-	-	+ [a]	+++
Catalytic reactions	+	++	+	+	-
Contamination [b]	+++	+++	++	+++	+++++
Cost	+	+++++	+++	+++	+++
Dead volume	+++++	+	+++++	+++	+
Ease of use	+	+++++	+	+	+++++
Hazards	++++	+	+++++		+++++
Insoluble materials	+++++	+++	+++++	++	++
Kinetic uses	+++++	+++	+++		++
Py-GC	+++++	+++++	+++++	+++	+++++
Py-MS	+++++	+++++	-		++++
Quantitative uses	+++++	+++++	+++		+++
Reproducibility	+++++	++++	++	++	++
Secondary products	++	++	+++++	++	++++
Stepped pyrolysis	+++++	-	+++		-
Taxonomic uses	+++++	+++++	-		-
Temperature gradients	+	+	+++++	+	++++

[a] Much higher in packed tubes.
[b] Depends upon detailed design (little with removable lining to pyrolysis chamber).
Source: Refs. 95 and 96.

TABLE 2.9   Variation in the Pyrolysis Products of Polystyrene

PRODUCT	Py-GC						Py-MS		
	Laser	Laser	Furnace	Fila-ment	Curie-Point	Curie-Point	Laser	Curie-Point	F.I.
CH$_4$		1							
C$_2$H$_2$	16.7	7							
C$_2$H$_4$		7							
HC≡CEt	0.4								
HC≡C-CH=CH$_2$	0.5								
HC≡C-C≡CH	10.6								
HC≡C-CH$_2$-C≡CH	0.3								
Ph-H	17.0	12				21			
HC≡C-CH=CH-C≡CH	1.1								
Ph-Me	2.5	18	1.6		0.9	13			
Ph-Et	1.0					2			
Ph-CH=CH$_2$	48.0	54	81.7	74	99.0	14	100	60	59
Ph-C(Me)=CH$_2$			1.6			24			3
Ph-CH=CHMe			0.6						
Ph-CH$_2$-CH=CH$_2$						6			
Ph-CH(Me)CH=CH$_2$	0.3								
Ph-CH$_2$-C≡CH	1.4								
Ph-CH=CH-CH$_2$Ph			4.0						2
Ph-CH$_2$-CH=C(Me)Ph			9.4						
Ph-CH=CH-CH=CH-Ph			1.0						
Ph-CH$_2$-CH$_2$-C(=CH$_2$)Ph				12		+		34	11
Phenylnaphthalene									4
Ph-CH$_2$-CH$_2$-CH(-Ph); CH$_2$=C(Ph)-CH$_2$				14		+		6	18
C$_{24}$H$_{18}$									4
REFERENCE	34	38	97	89	98	8	99	101	102

more effective than a furnace system in the elucidation of the micro-structure of polymers [93]. Note has also been made that results equivalent to those from a Curie-point method may require lower temperatures in filament pyrolysis [94]. Assessments of various pyrolyzers have been made, and Table 2.8 has been adapted from these sources [95,96].

Table 2.9 records some results from the pyrolysis of polystyrene which have been recorded in the literature [97-102]. Significant differences are observed in the pyrolysis device, the analysis method and the products. Such differences may reflect a true difference in the polymer, depending upon the molecular weight of the sample, which indeed does have a real but small effect [97]. It is certain, however, that the technique is by far the major source of variation.

Some differences are clearly due to the pyrolysis events. The plasma-tagged products detected in the pulsed-laser Py-GC study [34], for instance, are unique to that system, and compare dramatically to results using a low-power laser in a Py-MS study [99]. In the other Py-GC work, the Curie-point results [8], using ferromagnetic tubes, deviate from the simple pyrogram of the filament method [98], and reveal the presence of dimeric and trimeric fragments [102], while one furnace system detects other complex products [97], but under more controlled conditions monomer, dimer and trimer are obtained [89]. In addition to control of the pyrolysis, however, reproducibility depends upon the method of analysis of the fragments. This demands efficient transfer from the pyrolysis zone to the analytical device, and the effective analysis of these products. The most common analytical devices are gas chromatography (Py-GC), in which variations are due to different chromatographic conditions, and mass spectrometry (Py-MS) in which variations may be caused by differences in the pyrolysis-ionization method and by the mass range of the instrument. Integrated Py-GC and Py-MS techniques are the subjects of the following chapters.

REFERENCES

1. Fujita, K. and Mizuki, K., Analysis of Polymers by Py-GC, Japan Analytical Instrument Bulletin, 1975, pp. 38-54.

2. Farre-Ruis, F. and Guiochon, G., On the Conditions of Flash Pyrolysis of Polymers as used in Py-GC, *Analyt. Chem.*, 40(1968) 998-1000.

3. Levy, R.L., Fanter, D.L. and Wolf, C.J., Temperature Rise-Time and True Pyrolysis Temperature in Pulse Mode Py-GC, *Analyt. Chem.*, 44(1972)38-42.

4. Walker, J.Q., Py-GC Correlation Trials of the American Society for Testing and Materials, *J. Chromatogr. Sci.*, 15(1977)267-274.

5. Buhler, Ch. and Simon, W., Curie-point Py-GC, J. Chromatogr.
    Sci., 8(1970)323-329.

6. Lehrle, R.S. and Robb, J.C., The Quantitative Study of Polymer
    Degradation by GC, J. Gas Chromatogr., 5(1967)89-95.

7. Sellier, N., Jones, C.E.R. and Guiochon, G., Py-GC as a Means of
    Determining the Quality of Porous Polymers of Styrene cross-
    linked with Divinylbenzene, in C.E.R. Jones and C.A. Cramers
    (Eds.), Analytical Pyrolysis, Elsevier, Amsterdam, 1977, pp.
    309-318.

8. Oertli, Ch., Buhler, Ch. and Simon, W., Curie-point Py-GC using
    Ferromagnetic Tubes as Sample Supports, Chromatographia, 6(1973)
    499-502.

9. Levy, R.L., Py-GC. A Review of the Technique, Chromatogr.
    Revs., 8(1966)48-89.

10. May, R.W., Pearson, E.F. and Scothern, D., Pyrolysis-Gas Chrom-
    atography, Chemical Society, London, 1977.

11. Jones, C.E.R. and Moyles, A.F., Rapid Identification of High
    Polymers using a Simple Pyrolysis Unit with a GLC, Nature, 189
    (1961)222-223.

12. Jones, C.E.R. and Moyles, A.F., Pyrolysis and GLC on the Micro-
    gram Scale, Nature, 191(1961)663-665.

13. Krejci, M. and Deml, M., Methode und Vorrichtung fur die Unter-
    suchung der Pyrolytischen Zersetzung von Substanzen bei Konstan-
    ten Temperatur, Coll. Czech. Chem. Comm., 30(1965)3071-3079.

14. Cogliano, J.A., Apparatus for Rapid Pyrolysis under Controlled
    Conditions, Rev. Sci. Instr., 34(1963)439-440.

15. Levy, R.L., Trends and Advances in Design of Pyrolysis Units
    for GC, J. Gas Chromatogr., 5(1967)107-113.

16. Martin, A.J., Sarner, S.F., Averitt, O.R., Pruder, G.D. and
    Levy, E.J., Pyroprobe 100 Series, Chemical Data Systems Inc.,
    Oxford, Pennsylvania.

17. Tyden-Ericsson, I., A New Pyrolyser with Improved Control of
    Pyrolysis Conditions, Chromatographia, 6(1973)353-358.

18. Simon, W. and Giacobbo, H., Thermal Fragmentation and the De-
    termination of Structure of Organic Compounds, Angew. Chem.
    (Int. Ed.), 4(1965)938-943.

19. Coupe, N.B. and McKeown, M.C., Py-GC using a Curie-point System,
    Column, 2(1968)8-12. Pye Unicam Ltd., Cambridge, England.

20. Schmid, J.P., Schmid, P.P. and Simon, W., Instrumentation for
    Curie-point Py-GC using High-Resolution Glass Open-tubular
    Columns, Chromatographia, 9(1976)597-600.

21. Shulman, G.P. and Simmonds, P.G., Thermal Decomposition of
    Aromatic and Heteroaromatic Amino acids, Chem. Communs. (1968)
    1040-1042.

22. Simmonds, P.G., Shulman, G.P. and Stembridge, C.H., Organic
     Analysis by Py-GC-MS. A Candidate Experiment for the Biologi-
     cal Exploration of Mars, *J. Chromatogr. Sci., 7*(1969)36-41.

23. Quinn, P.A., Development of High-resolution Py-GC for the Iden-
     tification of Micro-organisms, *J. Chromatogr. Sci., 12*(1974)
     796-806.

24. Quinn, P.A., Identification of Micro-organisms by Pyrolysis:
     The State of the Art, in H.H. Johnston and S.W.B. Newsom (Eds.),
     Rapid Methods and Automation in Microbiology, Learned Informa-
     tion (Europe) Ltd., Oxford, 1976, pp. 178-186.

25. Schulten, H-R, Beckey, H.D., Meuzelaar, H.L.C. and Boerboom, A.
     J.H., High-Resolution FIMS of Bacterial Products, *Analyt. Chem.,
     45*(1973)191-195.

26. Tsuge, S. and Takeuchi, T., Vertical Furnace-Type Sampling De-
     vice for Py-GC, *Analyt. Chem., 49*(1977)348-350.

27. Blau, K. and King, G.S. (Eds.), *Handbook of Derivatives for
     Chromatography,* Heyden, London, 1977.

28. Nicholson, J.D., Derivative Formation in the Quantitative GC
     Analysis of Pharmaceuticals, *Analyst, 103*(1978)1-28, 193-222.

29. Kimelt, S. and Speiser, S., Lasers and Chemistry, *Chem. Revs.,
     77*(1977)437-472.

30. Charschan, S.S., *Lasers for Industry,* Van Nostrand Reinhold,
     New York, 1972.

31. Vastola, F.J. and Pirone, A.J., The Use of the Laser Micro-Py-
     MS in Studying the Pyrolysis of Coal, *Amer. Chem. Soc. Div.
     Fuel Chem. Preprints, 10*(1966)C53-58.

32. Vanderborgh, N.E. and Ristau, N.T., Laser Py-GC for the Charac-
     terisation of Soil Materials, *Amer. Lab., 5*(1973)41-48.

33. Vanderborgh, N.E., Laser-Induced Pyrolysis Techniques, in C.E.
     R. Jones and C.A. Cramers (Eds.), Analytical Pyrolysis, Else-
     vier, Amsterdam, 1977, pp. 235-248.

34. Vanderborgh, N.E., Fletcher, M.A. and Jones, C.E.R., Laser Py-
     rolysis of Carbonaceous Rocks, *J. Anal. Appl. Pyrol., 1*(1979)
     177-186.

35. Jones, C.E.R. and Vanderborgh, N.E., Elucidation of Geomatrices
     by Laser Py-GC and Py-MS, *J. Chromatogr., 186*(1979)831-841.

36. Falmer, O.F. Jr. and Azarraga, L.V., A Laser Pyrolysis Apparatus
     for GC, *J. Chromatogr. Sci., 7*(1969)665-670.

37. Ristau, W.T. and Vanderborgh, N.E., Effect of Carbon Loading
     upon Product Distribution of Laser-Induced Degradations, *Analyt.
     Chem., 44*(1972)359-362.

38. Fanter, D.L., Levy, R.L. and Wolf, C.J., Laser Pyrolysis of
     Polymers, *Analyt. Chem., 44*(1972)43-48.

39. Falmer, O.F., Laser Py-GC Application to Polymers, *Analyt. Chem.*, *43*(1971)1057-1065.

40. Guran, B.T., O'Brien, R.T. and Anderson, D.H., Design, Construction and Use of a Laser Fragmentation Source for GC, *Analyt. Chem.*, *42*(1970)115-117.

41. Ristau, W.T. and Vanderborgh, N.E., Simplified Laser Degradation Inlet System for GC, *Analyt. Chem.*, *42*(1970)1848-1849.

42. Merritt Jr. C., Sacher, R.E. and Petersen, P.A., Laser Pyrolysis-GC-MS Analysis of Polymeric Materials, *J. Chromatogr.*, *99* (1974)301-308.

43. Coloff, S.G. and Vanderborgh, N.E., Time-Resolved Laser-Induced Degradation of Polystyrene, *Analyt. Chem.*, *45*(1973)1507-1511.

44. Means, J.C. and Perkins, E.G., Laser Py-GC-MS of Membrane Components in C.E.R. Jones and C.A. Cramers (Eds.), *Analytical Pyrolysis*, Elsevier, Amsterdam, 1977, pp. 249-260.

45. Cramers, C.A.M. and Keulemans, A.I.M., Pyrolysis of Volatile Substances (Kinetics and Product Studies), *J. Gas Chromatogr.*, *5*(1967)58-64.

46. Svob, V. and Deur-Siftar, D., Comparison of Degradation Products and Processes in MS and Py-GC Applied to the Identification of Alkylbenzenes, *J. Chromatogr.*, *135*(1977)85-92.

47. Kelley, J.D. and Wolf, C.J., A Comparison of Py-GC and MS, *J. Chromatogr. Sci.*, *8*(1970)583-585.

48. Boss, B.D. and Hazlett, R.N., Application of Py-GC and MS for Identification of Alcohol and Carbonyl Isomers, *Analyt. Chem.*, *48*(1976)417-420.

49. Goforth, R.R. and Harris, W.E., A System for the Identification of GC Peaks by Py-GC with High-Speed Temperature Programming, in C.L.A. Harbourn and L. Stock (Eds.), *Gas Chromatography 1968*, Institute of Petroleum, London, 1968, pp. 261-275.

50. Levy, E.J. and Paul, D.G., The Application of Controlled Partial Gas-Phase Thermolytic Dissociation to the Identification of GC Effluents, *J. Gas Chromatogr.*, *5*(1967)136-145.

51. Walker, J.Q. and Wolf, C.J., Complete Identification of Chromatographic Effluents Using Interrupted Elution and Py-GC, *Analyt. Chem.*, *40*(1968)711-714.

52. Groenendyk, H., Levy, E.J. and Sarner, S.F., Controlled Thermolytic Dissociation of Hexadecane and Methyl Decanoate, *J. Chromatogr. Sci.*, *8*(1970)115-121, 599-601.

53. Merritt, C. Jr. and DiPietro, C., Qualitative Analysis of GC Eluates by Means of Vapour-Phase Pyrolysis, *Analyt. Chem.*, *44* (1972)57-59.

54. Schaden, G., Short-time Pyrolysis and Spectroscopy of Unstable Compounds, V. Improvement in Curie-point Py-GC, *J. Chromatogr.*, *136*(1977)420-422.

55. Luce, C.C., Humphrey, E.F., Guild, C.V., Norrish, H.H., Coull, J. and Castor, W.W., Analysis of Polyester Resins by GC, *Analyt. Chem.*, *36*(1964)482-486.

56. Thompson, W.C., Device for Pyrolysing Insoluble Solids Using the Curie-point Principle, *Lab. Pract.*, *18*(1969)1074.

57. Barlow, A., Lehrle, R.S. and Robb, J.C., Direct Examination of Polymer Degradation by GC. I. Applications to Polymer Analysis and Characterisation, *Polymer*, *2*(1961)27-40.

58. Sternberg, J.C. and Litle, R.L., Electrical Discharge Pyrolyser for GC, *Analyt. Chem.*, *38*(1966)321-330.

59. Hewitt, G.C. and Witham, B.T., The Identification of Substances of Low Volatility by Py-GC, *Analyst*, *86*(1961)643-652.

60. Ettre, K. and Varaldi, P.F., Py-GC Technique. Effect of Temperature on Thermal Degradation of Polymers, *Analyt. Chem.*, *35* (1963)69-73.

61. Deur-Siftar, D., Bistricki, T. and Tandi, T., An Improved Pyrolytic Device Suitable for the Study of Polymer Microstructure by Py-GC, *J. Chromatogr.*, *24*(1966)404-406.

62. LePlat, P., Application of Py-GC to the Study of the Nonvolatile Petroleum Fractions, *J. Gas Chromatogr.*, *5*(1967)128-135.

63. Levy, R.L. and Wolf, C.J., A GC Method for Characterisation of the Organic Content Present in an Inorganic Matrix, *J. Chromatogr. Sci.*, *8*(1970)524-526.

64. Armitage, F., The Determination of the Polybutadiene Concentration in High-Impact Polystyrene by Py-GC, *J. Chromatogr. Sci.*, *9*(1971)245-248.

65. Polak, R.L. and Molenaar, P.C., A Method for Determination of Acetylcholine by Slow Pyrolysis Combined with Mass Fragmentography on a Packed Capillary Column, *J. Neurochem.*, *32*(1979)407-412.

66. Higman, E.B., Severson, R.F., Arrendale, R.F. and Chortyk, O.T., Simulation of Smoking Conditions by Pyrolysis, *J. Agric. Food Chem.*, *25*(1977)1201-1207.

67. Shafizadeh, F., Furneaux, R.H., Stevenson, T.T. and Cochran, T.G., Acid-catalysed Pyrolytic Synthesis and Decomposition of 1,4:3,6-Dianhydro-α-D-Glucopyranose, *Carb. Res.*, *61*(1978)519-528.

68. Dhont, J.H., Identification of Aliphatic Alcohols by Pyrolysis, *Analyst.*, *89*(1964)71-74.

69. Gough, T.A. and Walker, E.A., The Use of Pyrolysis and Hydrogenation in the GC of Aliphatic Hydrocarbons, *J. Chromatogr. Sci.*, *8*(1970)134-138.

70. Tinkelenberg, A., Reaction Kinetics in a Microreactor-GC-Combination, *J. Chromatogr. Sci.*, *8*(1970)721-723.

71. Walker, J.Q. and Maynard, J.B., Analysis of Vapour-Phase Pyrolysis Products of the Four Trimethylpentane Isomers, *Analyt. Chem., 43*(1971)1548-1557.

72. Lysyj, I., Nelson, K.H. and Webb, S.R., Analysis of Multi-Component Organic Mixtures in Aqueous Media by Pyrolysis, *Water Res., 4*(1970)157-163.

73. Juvet, R.S. Jr., Smith, J.L.S. and Li, K-P, Polymer Identification and Quantitative Determination of Additives by Photolysis-GC, *Analyt. Chem., 44*(1972)49-56.

74. Dimbat, M., Pyrolysis-Hydrogenation-Capillary GC. A Method for the Determination of the Isotacticity and Isotactic and Syndiotactic Block Lenth of Polypropylene, in R. Stock (Ed.), *Gas Chromatography 1970,* Institute of Petroleum, London, 1971, pp. 237-246.

75. Truett, W.L., Py-IR Analysis of Polymeric Materials, *Amer. Lab., 9*(1977)33-38.

76. Martin, S.B. and Ramstad, R.W., Compact Two-stage GC for Flash Pyrolysis Studies, *Analyt. Chem., 33*(1961)982-985.

77. Stahl, E., Advances in the Field of Thermal Procedures in Direct Combination with TLC, *Acc. Chem. Res., 9*(1976)75-80.

78. Jackson, M.T. Jr. and Walker, J.Q., Py-GC of Phenyl Polymers and Phenyl Ether, *Analyt. Chem., 43*(1971)74-78.

79. Jennings, E.C. and Dimick, K.P., GC of Pyrolytic Products of Purines and Pyrimidines, *Analyt. Chem., 34*(1962)1543-1547.

80. Zulaica, J. and Guiochon, G., Fast Qualitative and Quantitative Microanalysis of Plasticisers in Plastics by Py-GC, *Analyt. Chem., 35*(1963)1724-1728.

81. Coulter, G.L. and Thompson, W.C., Automatic Analysis of Tyre Rubber Blends by Computer-Linked Py-GC in C.E.R. Jones and C.A. Cramers (Eds.), *Analytical Pyrolysis,* Elsevier, Amsterdam, 1977, pp. 1-15.

82. Levy, R.L., Gesser, H., Halevi, E.A. and Saidman, S., Py-GC of Porphyrins, *J. Gas Chromatogr., 2*(1964)254-255.

83. Muezelaar, H.L.C. and in't Veld, R.A., A Technique for Curie-Point Py-GC of Complex Biological Samples, *J. Chromatogr. Sci., 10*(1972)213-216.

84. Ericsson, I., Sequential Py-GC Study of the Decomposition Kinetics of cis-1,4-Polybutadiene, *J. Chromatogr. Sci., 16*(1978) 340-344.

85. Hu, J.C-A., Py-GC Analysis of Rubbers and Other High Polymers, *Analyt. Chem., 49*(1977)537-540.

86. Meuzelaar, H.L.C., Ficke, H.G., and den Harink, H.C., Fully Automated Curie-Point Py-GC, *J. Chromatogr. Sci., 13*(1975)12-17.

87. Verzele, M., Van Cauwenberghe, K. and Bouche, J., Experiments with GC Support Materials. Observations on Pyrolysis, *J. Gas Chromatogr.*, 5(1967)114-118.

88. Ohnishi, A., Kato, K. and Takagi, E., Curie-Point Pyrolysis of Cellulose, *Polymer, J.*, 7(1975)431-437.

89. Sugimura, Y. and Tsuge, S., Fundamental Splitting Conditions for Pyrogram Measurements with Glass Capillary GC, *Analyt. Chem.*, 50(1978)1968-1972.

90. Tsuge, S., Kobayashi, T., Nagaya, T. and Takeuchi, T., Py-GC Determination of Run Numbers in Methyl Methacrylate-Styrene Copolymers Using the Boundary Effect, *J. Anal. Appl. Pyrol.*, 1 (1979)133-141.

91. Hall, R.C. and Bennett, G.W., Py-GC of Several Cockroach Species, *J. Chromatogr. Sci.*, 11(1973)439-443.

92. Walker, J.Q. and Wolf, C.J., Py-GC: A Comparative Study of Different Pyrolysers, *J. Chromatogr. Sci.*, 8(1970)513-518.

93. Alekseeva, K. and Khramova, L.P., Comparison of Various Pyrolysers in the Study of Polymer Compositions, *J. Chromatogr.*, 69 (1972)65-70.

94. Galin-Vacherot, M., Pyrolyse Eclair de Polyisoprenes. Essai de Correlation Entre les Produits de Degradation et la Microstructure des Polymeres, *Eur. Polym. J.*, 7(1971)1455-1471.

95. Walker, J.Q., A Comparison of Pyrolysers for Polymer Characterisation, *Chromatographia*, 5(1972)547-552.

96. Crighton, J.S., Characterisation of Textile Materials by Thermal Degradation: A Critique of Py-GC and Thermogravimetry, in C.E.R. Jones and C.A. Cramers (Eds.), *Analytical Pyrolysis*, Elsevier, Amsterdam, 1977, pp. 337-349.

97. Tsuge, S., Okumoto, T. and Takeuchi, T., Study of Thermal Degradation of Fractioned Polystyrenes by Py-GC, *J. Chromatogr. Sci.*, 7(1969)250-252.

98. Ericsson, I., Py-GC--A Potential Technique for Characterisation of Complex Organic Material, *Acta Path. Microbiol. Scand.*, Sect. B. (Suppl.) 259(1977)37-42.

99. Kistemaker, P.G., Boerboom, A.J.H. and Meuzelaar, H.L.C., Laser Py-MS: Some Aspects and Applications to Technical Polymers, *Dynamic Mass Spectrom.*, 4(1975)139-152.

100. Schmid, P.P. and Simon, W., A Technique for Curie-Point Py-MS with a Knudsen Reactor, *Analyt. Chim. Acta*, 89(1977)1-8.

101. Schudemagge, H.D.R. and Hummel, D.O., Characterisation of High Polymers by Pyrolysis within the FIMS, *Adv. Mass Spectrom.*, 4 (1968)857-866.

102. Schmid, J.P., Schmid, P.P. and Simon, W., Application of Curie-Point Py-GC Using High-Resolution Glass Open-Tubular Columns, in C.E.R. Jones and C.A. Cramers (Eds.), *Analytical Pyrolysis*, Elsevier, Amsterdam, 1977, pp. 99-105.

Chapter 3

Pyrolysis Gas Chromatography

3.1. GAS CHROMATOGRAPHY

Gas chromatographic separation of pyrolysis products involves a
plethora of variables [1-8]. For the universal application of analytical pyrolysis, each should be defined, and, where possible, optimized
to yield an effective and reproducible system. Consideration must be
given to the overall design, so that pyrolysis, separation, detection,
identification, and quantification are undertaken in an integrated and
efficient way to enable a full and repeatable profile of the sample to
be obtained. The almost universal availability of gas chromatographs
and the ease with which these may be interfaced with readily available
pyrolysis units means that Py-GC analyses may be undertaken with relatively little difficulty. Before such a course of action is initiated, however, a consideration of the variables involved and a preliminary setting up of the system, in a standardized way, will rapidly
repay the small investment of time required.

3.1.1. The Column

The GC column is the heart of the separation process, and its design
and operation must be adequately controlled to ensure maximum efficiency. The most important factor is whether a packed or a capillary
column is used. This will determine the degree of resolution observed, and the overall appearance of the pyrogram. Many of the
operating parameters, and the nature of chromatograph-pyrolyzer
interface, also depend on the type of column used.

Packed Columns

Packed columns have been used extensively in pyrolysis gas chromatography. This in part stems from the relative ease with which the
columns may be prepared and connected to a pyrolysis unit. Two variants are common: the column, typically 1 to 3 m × 2 to 4 mm, may be
packed with an inert support, which has been coated or impregnated
with a chromatographically active liquid stationary phase, or else
porous polymer spheres may be used.

The inert support should be

Of *uniform particle size* and shape. This allows the column to
be packed uniformly, maximizes efficiency and reduces the
pressure drop across the column. Usual sizes range from
125 to 250 µm (120 to 60 mesh).

Of *large specific surface area,* so that a high-capacity column
may be obtained with a thin coating of stationary phase.
This speeds up mass transfer and results in a high-
efficiency column. Typically, specific surface areas of
1 to 4 m^2g^{-1} are used.

Of *controlled pore size.* Mass transfer rates within the particles should be of the same order as transfer rates between the mobile and stationary phases. Thin coatings and
small regular particles thus aid efficiency.

Chemically inert to prevent adsorptive losses of components.
Diatomaceous supports are common. White materials (e.g.,
Chromosorb W) are less active but are more fragile and
have a lower specific surface area than the pink grades
(e.g., Chromosorb P). Such materials may be acid washed
(AW) to decrease the activity of the support (perhaps
through the removal of metal ions), and residual surface
silanol groups ($\equiv$Si-OH) may be silylated [e.g., with di-
methylchlorosilane (DMCS) or with hexamethyldilsilazane
(HMDS)]. High-performance (HP) supports are an improved
class of AW-DMCS materials.

Mechanically and thermally stable to allow coating, packing and
use without degradation of the support.

The choice of stationary phase is almost limitless, with dis-
tributers being able to offer over 300 stock items [9,10]. The
choice of material is critical to achieving a good separation of the
components. Considerations include the following:

Selectivity. This implies a matching of the partition characteristics of the components to that of the stationary phase (i.e., "likes like") and is generally considered as a polarity factor. Thus, polar stationary phases (e.g., polyethylene glycols, Carbowax) will increase the retention of more polar molecules in a mixture, and should be chosen for compounds such as alcohols. This phase, with the incorporation of KOH, is also suitable for amine components. Conversely, for nonpolar products, stationary phases such as the OV series of silicones are available. The Rohrschneider or McReynolds' numbers are the commonest means of quantifying stationary phase polarity [11]. The McReynolds system uses the Kovats retention index (I):

$$I = 100 \left[\frac{\log t' - \log t_n'}{\log t_{n+1}' - \log t_n'} \right] + 100n$$

where t' is the net retention time of the component under test; t_n' and t_{n+1}' are the net retention times of the normal paraffins with n and n + 1 carbon atoms, which are eluted on either side of the component peak.

For alkanes the bracketed expression is zero; hence these compounds have a Kovats index of 100 times the carbon number. Division of I by 100 gives the number of carbon atoms in an alkane which is equivalent in retention behavior to the component under consideration.

Polarity indexes are calculated by measuring the Kovats index for a range of solutes (10 in McReynolds initial paper [12]), which are selected to introduce various solvent-solute interactions. Thus proton-donation or acceptance (butanol, pyridine), dipole-dipole interactions (1-nitropropane, pentan-2-one), and induction interactions (benzene) are modeled. The difference in the retention index (ΔI), obtained from a non-polar squalane column and that from the stationary phase under test, is determined for each of the standard components. Direct comparison, summation or calculation of the mean value provides a polarity index for the stationary phase--the higher the index, the greater the polarity. Many stationary phases have equivalent McReynolds' numbers [12] and therefore may be substituted

for each other with little chromatographic loss. This is important
when attempting to repeat literature work in which rare or unobtain-
able stationary phases are used. With care, a direct equivalent
should be available. Further, highly selective phases to aid a
special separation may be identified by the variation of McReynolds
number with the structure of the solute. Figure 3.1 illustrates the
effects of stationary phase polarity on the appearance of a pyrogram
[17].

Stability. The stationary phase must be chemically stable to prevent
thermal degradation during chromatography. The more usual problem
with such phases, however, is that many exert significant vapor pressures at elevated temperatures. This results in the volatilization
of the stationary phase (column bleed), a high background signal,
leading to lack of sensitivity and linearity (and which may be
totally unacceptable if temperature-programming or GC-MS work is in
progress) and to progressive degeneration of the column. Stationary
phases thus have a recommended temperature range to ensure useful
performance over a substantial period of time. If all other factors
are consistent, it is appropriate to choose the phase with the
greatest thermal stability.

Viscosity. To achieve adequate mass transfer rates, the stationary
phase should have a low viscosity. This state is enhanced by an increase in temperature, but the normal lower operating temperature
for a column is usually specified by the melting point of the phase
in question. For instance, the common phase Carbowax 20M has a useful temperature range of 60 to 250°C, and at ambient temperatures
this material is in the solid phase. If a temperature-programmed
run is initiated, considerable loss of efficiency will result until
a liquid phase at equilibrium temperature is attained.

Solubility. In order to be effectively coated onto the support
material, the stationary phase must be adequately soluble in a volatile solvent. The stability, viscosity, and solubility requirements
mean that stationary phases are a compromise between conflicting

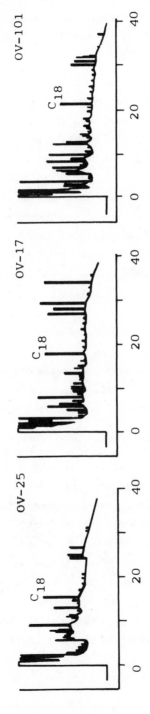

FIGURE 3.1 Influence of column polarity on the pyrogram of *Periplaneta Australasiae* (female). (From Ref. 17, reproduced from the *Journal of Chromatographic Science*, by permission of Preston Publications, Inc.)

requirements. Nevertheless, satisfactory phases may be found for
most applications. A range of silicones (OV) and Carbowax 20M cover
many Py-GC applications, and details of these phases are presented
in Table 3.1.

Loading. This is usually expressed as the weight of stationary phase
per 100 g of support, and is typically in the range 3 to 20%. The
loading depends upon the capacity of the support material. For
example, in the Chromosorb series $P(S.S.A. = 4 \text{ m}^2\text{g}^{-1})$ may be loaded
to 30%, $W(S.S.A. = 1 \text{ m}^2\text{g}^{-1})$ to 25% and $G(S.S.A. = 0.5 \text{ m}^2\text{g}^{-1})$ to 5%.
The minimum loading should be such that a total covering of the sup-
port is obtained. Heavy coatings of stationary phase will slow mass
transfer rates and may occlude the column, thus resulting in poor
chromatographic performance. It must be noted also that the loading
itself is not an adequate description of the column. For example, a
3% loading on Chromosorb A is equivalent to a 2.5% loading on Chromosorb G and to a 5.3% loading on Chromosorb W. The optimal loading
occurs when the stationary phase occupies 5 to 15% of the volume of
gas in the column.

        To coat the support, several procedures are available. One ef-
fective method to produce a 5% OV17 column is to dissolve the stationary phase (1 g) in a purified solvent (toluene 40 ml)—manufacturers recommend an appropriate solvent for each phase (chloroform,
methanol, acetone or toluene are common)—and to this the appropriate
amount of washed and deactivated support (Chromosorb W HP, 20 g) is
carefully added. The slurry is evaporated under reduced pressure
(rotary evaporator and a slowly rotating fluted flask to prevent
fractionation on drying and damage to the particles). Final drying
is best achieved with a fluidized-bed drier, in which nitrogen is
passed from below through a bed of the coated support. The gas is
distributed through a fine wire gauze at such a rate to suspend the
particles. The vessel is then heated to remove final traces of
solvent. This procedure enables a high quality packing material to
be obtained, due to the prevention of fractionation or mechanical
damage.

TABLE 3.1 Properties of Some GC Liquid Stationary Phases

PHASE	STRUCTURE	McREYNOLDS' NUMBER FOR SOLUTE:[a]										REL. POLARITY	TEMP. RANGE °C	SOLVENT
		1	2	3	4	5	6	7	8	9	10			
OV 1	Me-silicone	-1	23	4	16	32	44	45	55	42	65	33	100-350	T
OV 101	Me-silicone	-2	23	4	17	33	45	46	57	43	67	33	0-350	C
OV 3	10% Ph-silicone	17	46	39	44	55	81	84	86	88	124	66	0-350	T
OV 7	20% Ph-silicone	35	66	68	69	77	111	120	113	128	171	96	0-350	T
OV 11	35% Ph-silicone	59	92	103	102	100	145	164	142	178	219	130	0-350	T
OV 17	50% Ph-silicone	69	105	119	119	112	162	184	158	202	243	147	0-350	A
OV 22	65% Ph-silicone	99	132	152	160	133	191	228	188	253	283	182	0-350	A
OV 25	75% Ph-silicone	113	147	169	178	144	208	251	204	280	305	200	0-350	A
OV 210	50% trifluoropropyl Me	60	56	139	146	206	358	283	238	310	468	226	0-275	A
OV 225	25% Ph:25% cyanopropyl Me.	117	150	226	228	282	338	342	369	386	492	293	0-275	A
CARBO-WAX 20M	Polyethyleneglycol (20,000)	148	221	282	322	387	368	434	536	510	572	378	60-250	C
DEGS	Diethyleneglycol succinate	237	321	418	492	579	581	705	733	791	833	569	20-200	A
OV 275	Ph:cyanopropyl Me	-	-	-	629	-	763	-	872	849	1106	ca.700	0-275	C

[a]1, cis-Hydrindene; 2, oct-2-yne; 3, 1-iodobutane; 4, benzene; 5, 2-methylpentan-2-ol; 6, pentan-2-one; 7, 1,4-dioxan; 8, butan-1-ol; 9, pyridine; 10, 1-nitropropane; A, acetone; C, chloroform; T, toluene.

Source: Adapted from Ref. 9.

Column packing may be readily effected by attaching a vacuum pump to the column outlet, carefully adding small amounts of the packing material and *gently* vibrating the column with an electric vibrator. At all stages, care must be taken to retain the integrity of the coated support. Metal columns have adsorption sites which frequently cause tailing, or prevent the transmission of polar materials. Glass columns are generally preferred, and these are usually further deactivated by silanisation before use. The column should now be *conditioned* by gradually increasing the temperature to somewhat less than the maximum recommended (~20°C), while a carrier gas flow is maintained (~20 ml min^{-1}). As much as 24 to 48 h may be required, particularly for polyester phases which may contain low-molecular-weight impurities. During this stage, the column must not be connected to the detector, or else contamination will result. The column should never be left heated without a flow of carrier gas, unless specifically recommended.

The procedures outlined here manifest many variables, each of which may lead to differences in column behavior, with consequent lack of reproducibility in Py-GC. To make standardization of the column easier, it has been recommended that *porous polymer supports* be used, particularly for the forensic investigation of paint and polymer samples [13-15]. These materials are highly porous polymers, based upon styrene-divinylbenzene, and are capable of direct separation of a wide variety of compounds, including permanent gases, polar compounds, and high-boiling materials. One advantage of these packing materials is the elimination of the nonlinear adsorption isotherms, which may reduce chromatographic resolution with coated materials. Reproducibility of performance is also enhanced because many of the variables of coated phases have been eliminated, and overall control depends upon the manufacturer's quality assurance, rather than at each point of application, although variations in quality still occur [16]. The main types are the Chromosorb 100 series and the Poropaks, most of which may safely be operated at 250°C isothermally and up to 300°C on a programmed run. Others, such as

TABLE 3.2   Retention Indexes of Solutes on Chromosorb Porous Polymers

SOLUTE	B.Pt.(^{o}C)	Mol.wt.	RETENTION INDEX (I) OF CHROMOSORB:				
			101	102	103	104	105
Methanol	64.7	32.00	440	360	420	625	365
Ethanol	78.4	46.10	495	425	495	690	435
Propanol	97.2	60.10	595	510	595	795	535
Butanol	118.0	74.12	700	615	705	905	655
Acetone	56.5	50.10	555	480	530	755	465
Ethylene glycol	196.2	62.10	780	625	-	-	-
Propionaldehyde	49.5	58.08	550	475	535	725	480
Acetic acid	118.1	60.50	610	435	-	985	555
Butyric acid	164.0	88.10	805	610	-	1155	765
Ethyl acetate	77.1	88.10	655	580	630	785	585
Butyl acetate	126.1	116.16	540	780	825	990	790
Acetonitrile	81.6	41.05	580	460	565	855	480
Ethylamine	16.6	45.08	-	-	470	-	-
Butylamine	77.8	73.14	-	-	670	-	-
Diethylamine	56.0	73.14	-	-	590	-	-
Triethylamine	89.0	101.19	-	-	710	-	-
Ethylene diamine	117.2	60.10	-	-	740	-	-
Dioxan			775	675	770	935	-
Benzene	80.1	78.10	745	650	720	835	635
Aniline	184.4	93.13	-	-	1140	-	-
Pyridine	115.5	79.10	850	705	820	1025	720
Phenol	182.0	94.11	1070	905	-	-	-
R.T. MeOH (s)			28	26	28	63	30

Tenax (2,6-diphenyl-*p*-phenylene oxide) are stable up to 375°C [18].
The retention indexes for a variety of solutes on a selection of
Chromosorb phases are displayed in Table 3.2.   Porapak Q has been
particularly recommended for Py-GC work, because of its discriminating power, its long, useful life and intercolumn reproducibility.
Silylated phases (e.g., Porapak QS) are also available to eliminate
residual tailing with very polar components.   The retention behavior

of some solutes on Porapak columns is presented in Table 3.3. It
should be noted that, in common with traditional packed columns, the
substitution of one phase by another may change the order of elution
of the components. This may have a significant bearing upon the interpretation of pyrograms, particularly when pyrolysis products have
not been identified adequately. In comparison with coated supports,
porous polymers have a lower capacity, are thus more readily overloaded, and they frequently require higher temperatures for equivalent analyses.

Other phases, such as porous silica beads with high specific
surface areas (e.g., Porasil, Spherosil; S.S.A. = 2 to 500 m^2g^{-1})
and controlled particle sizes (40 to 200 μm) are also available [19],
as are chemically bonded supports, which range from octadecyl hydrocarbons to Carbowax 20M coatings. Graphitized carbon black (Carbo-
pak) [20], and porous polymers coated with a liquid stationary phase
[21] have also been used.

Column Performance

Figure 3.2 illustrates a typical chromatogram obtained from a two
component mixture. Under a fixed set of operating conditions, the
void volume of the column is V_N and the volume of carrier gas required to elute the components are V_A and V_B at a carrier gas flow
rate of F_C ml min^{-1}. The *retention time* (t) is given by

$$t_A = \frac{V_A}{F_C} \qquad t_B = \frac{V_B}{F_C}$$

the *net retention time* (t_A') by

$$t_A' = t_A - t_0 \qquad t_B' = t_B - t_0$$

and the retention behavior of each component may be expressed as the
retention factor (R)

$$R_A = \frac{V_N}{V_A} = \frac{t_0}{t_A} \qquad R_B = \frac{V_N}{V_B} = \frac{t_0}{t_B}$$

TABLE 3.3   Relative Retention Times of Solutes on Porapak Porous Polymers

PORAPAK	R.T. MeOH (s)	MeOH	MeCHO	EtOH	MeCN	AcMe	AcOH	EtOAc	BuOH	PhH	PhMe
P	31	1.00	1.00	1.23	1.72	1.57	1.90	2.25	2.92	3.43	5.87
Q	38	1.00	1.33	1.72	2.26	2.70	3.30	6.39	8.43	9.13	21.34
QS	44	1.00	1.27	1.72	2.13	2.60	2.83	6.36	7.99	8.66	20.07
R	50	1.00	1.06	1.71	1.99	2.17	7.28	4.80	8.17	6.89	13.50
S	50	1.00	1.11	1.73	2.07	2.33	11.37	5.13	8.69	7.44	16.96

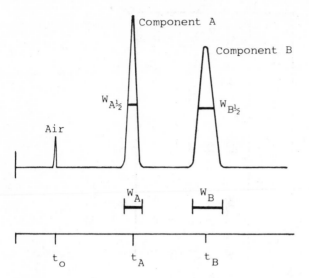

FIGURE 3.2 Typical two-component chromatogram (t_0, gas hold-up time; t_A and t_B, retention times of components A and B; W_A and W_B are the base widths; and $W_{A\frac{1}{2}}$ and $W_{B\frac{1}{2}}$ are the half-height widths).

or as the *capacity factor* (k)

$$k_A = \frac{t_A - t_0}{t_0} \qquad k_B = \frac{t_B - t_0}{t_0}$$

where $R_A = \dfrac{1}{(k_A + 1)}$

$$k_A = \frac{(1 - R_A)}{R_A}$$

The capacity factor is effectively the volume of the mobile phase, measured in column void volumes, necessary to effect elution of a component. A range of 1 to 10 is optimum. Larger volumes should be avoided, as these are generally characterized by broad peaks and lengthy analysis times. The retention factor is the fraction of the solute in the mobile phase, and thus R has values in the range 0 to 1. The *relative retention factor* (α), which is also corrected for column dead volume, is given by

$$\alpha = \frac{k_B}{k_A}$$

The *column efficiency* (N) is estimated by

$$N = 16\left(\frac{t_A}{W_A}\right)^2 = 5.54\left(\frac{t_A}{W_{A\frac{1}{2}}}\right)^2$$

which is the number of theoretical plates in the column. Typical values are 1500 to 3000 m^{-1}. This may alternatively be expressed as the *height equivalent of a theoretical plate* (H)

$$H = \frac{L}{N}$$

where L is the column length. The separation between two adjacent components may be quantified by the *resolution* (R_S) (ideally, with peaks of equal area)

$$R_S = \frac{2(t_B - t_A)}{W_A + W_B}$$

With symmetrical peaks of value $R_S = 1$, a usable separation (2% overlap) is obtained, and almost total separation (~0.3% overlap) is indicated when $R_S = 1.5$. For partial separations ($R_S < 0.8$) a change in chromatographic conditions are required [22]. To increase resolution several options are available. These may be illustrated by an alternative expression for R_S:

$$R_S = \frac{1}{4} \frac{(\alpha - 1)}{\alpha} \frac{k_B}{k_B + 1} N_B^{\frac{1}{2}}$$

This indicates that the resolution depends on

α: The relative positions of the two components. This is a function of the selectivity of the stationary phase.

k: The number of column dead volumes of carrier gas required for elution of the components. This is determined by the partition coefficient of the solute into the stationary phase, the loading of the column and the temperature of operation.

N: The efficiency of the column. This parameter may be described by the simplified van Deemter equation:

$$H = A + \frac{B}{U} + CU$$

where A, B, and C are constants, and U is the true linear gas velocity in the column.

For an *open tube* this depends on the flow-rate and the column dimensions:

$$U = \frac{LF_c}{V_c} = \frac{F_c}{d_c}$$

where V_c = column volume

d_c = column diameter

In a *packed column*, U is the interstitial velocity and is given by

$$U = \frac{LF_c}{V_0} = \frac{L}{t_0}$$

The constant A is an eddy diffusion term:

$$A = 2\lambda d_p \qquad \lambda \approx 1$$

It is directly related to particle diameter (d_p), and also depends on the uniformity of column packing, and represents the loss of efficiency due to different path lengths available for flow through the column. The contribution is minimized by the regular packing of small, uniform particles. Gaps, voids, or channels in the column decrease efficiency dramatically. The term B/U models the longitudinal diffusion of solute molecules in the gas phase.

$$B = 2\gamma D_g \qquad 0.5 \leq \gamma \geq 0.7$$

This depends upon the tortuosity of the packing material (γ) and the diffusivity of the solute in the carrier gas (D_g). In practical terms, the CU term predominates and accounts for the rate of mass transfer between the liquid and gas phases.

$$C = C_g + C_1 = \frac{Kd_p^2}{D_g} + \frac{2kd_f^2}{3(k + 1)^2 \cdot D_1}$$

when C_g and C_1 are the resistance to mass transfer in the gaseous and liquid phases, K is a constant depending upon the capacity factor and the packing efficiency, and D_1 is the solute diffusivity in the stationary phase of thickness d_f. C thus increases with particle size and also with the thickness and viscosity of the stationary

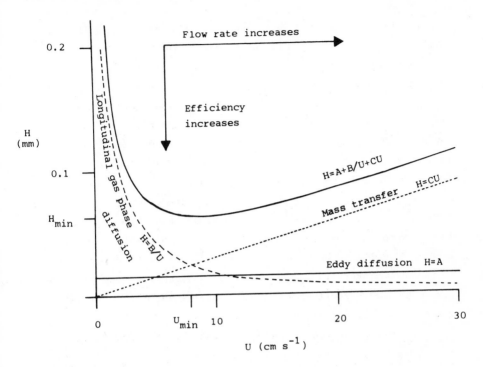

FIGURE 3.3 Column efficiency (H) as a function of carrier gas velocity (U) (A, 0.015 cm; B, 0.2 cm^2s^{-1}; C, 0.003 s; U_{min}, 8.16 cm s^{-1}; H_{min}, 0.064 mm).

phase. Low values for the capacity factor (k) also increase the value of C. Typical ranges are A = 0.01 to 0.02 cm, B = 0.1 to 0.2 cm^2s^{-1} and C = 0.002 to 0.004 s. For a given column, it is apparent that an optimum flow-rate exists for maximum efficiency. This is illustrated in Fig. 3.3.

The minima are found when B/U = CU. Differentiation yields

$$U_{min} = \left(\frac{B}{C}\right)^{\frac{1}{2}} \qquad H_{min} = 2(BC)^{\frac{1}{2}} + A$$

In practice, A is usually small, so that B (intercept) and C (slope) may be evaluated from a plot of HU versus U^2.

It is important when developing a separation to ensure that the flow rate is optimized. When repeating literature work, it must be noted that the flow rate through the column is NOT the controlling

factor. This is mediated by the internal diameter of the column, the
particle size, and the packing efficiency. The parameter to standardize is U, the interstitial linear gas flow *velocity*. In practice,
carrier gas flows faster than optimum may be adopted to speed analysis. The shallow hyperbola (Fig. 3.3) means that this does not dras-
tically reduce column efficiency. Optimization also depends on the
nature of the carrier gas. Many Py-GC applications are described
with nitrogen as the carrier gas, whereas helium is the gas of choice
for Py-GC-MS work. Typical flow velocities are 3 to 6 cm s^{-1} for
nitrogen and 10 to 20 cm s^{-1} for helium. Dependence of product distribution upon the nature of the carrier gas has also been noted.
It was found that nitrogen produced more low molecular weight components than helium when the pyrolysis of an isoprene-styrene copolymer was undertaken [23], and reduction was observed when iron tubes
were used for the Curie-point pyrolysis of stearic acid, with hydrogen as the carrier gas [24].

Improvement in resolution is achieved by variation of α, k, or
N, which may either increase the separation between the peaks $t_B - t_A$
without affecting peak width, or by reducing peak width $W_A + W_B$ without increasing separation. In practice, both are varied together,
and optimization between separation and band broadening is required.
Figure 3.4 illustrates the effects of varying each factor in turn on
the resolution between adjacent peaks. It is seen that when α or k
are small, significant improvements are readily obtained, whereas
diminishing returns result if these parameters are already large.
The column efficiency N is a square root term, and therefore substantial increases are necessary for an observable effect on resolution.

Selectivity (α) is mainly dependent upon the choice of stationary phase. Guidance may be obtained from related separations or
McReynolds numbers. The capacity factor (k) is determined by the
partition coefficient and the loading of the stationary phase. This
parameter thus increases with loading, and decreases with temperature,
and is thus a ready source of variation. Changes, however, also effect the column efficiency (N), which, in contrast to α and k

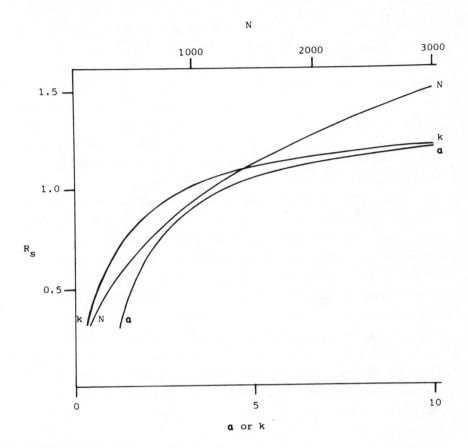

FIGURE 3.4   Effect of independent changes in α, k, or N on resolution (α: N = 500, k = 0.3; k: N = 500; α = 1.3; N: α = 1.5, k = 0.5).

(thermodynamic control), is essentially a kinetic phenomenon. This arises because of the dependence of C in the van Deemter equation upon k. Columns are most efficient if run at the lowest possible temperature, but at the cost of increased analysis time. Thus, a compromise is sought, and in practice, maximum speed for acceptable resolution is the guiding principle. One further important effect of temperature should be noted. The viscosity of a gas increases with temperature, and the flow rate through a column will drop if the temperature is raised. This may be avoided by the use of mass flow controllers. If these are not available, flow should be monitored when

thermal equilibrium is attained. Increased loading also reduces N
by increasing the time required for mass transfer between the mobile
phase and the thicker liquid layer. Small, uniform particles with a
large surface area and a small loading of stationary phase enable
optimization of N to be achieved. An increase in the column length
is also an effective way to increase efficiency. If adequate resolution cannot be achieved by temperature or flow-rate optimization,
then an increase in column length (1.5- to 2-fold) is indicated.
This is frequently satisfactory, but length cannot be increased in-
definitely. This arises because of the drop in pressure across the
column; ideally the inlet pressure should be no more than double the
outlet pressure. Although most of the pressure drop appears near
the outlet, flow rates will differ along the column, and significant
deviations from optimum efficiency will result.

    For many applications of Py-GC, packed columns have acceptable
resolution capabilities. It has been observed, however, that complex samples, such as microorganisms, may yield more than 200 pyrolysis products [25], whereas packed columns fail to separate more than
one-third of these. This reduces the effectiveness of such pyrograms
because small chromatographic changes may alter the profile of overlapping peaks significantly. To allow full separation and identification of pyrolysis products, the use of capillary columns has been
advocated.

Capillary Columns

Capillary columns are long, open tubes (10 to 150 m × 0.25 to 0.75
mm), which have stationary phase coated onto the walls [26,27]. In
contrast to packed columns, the permeability of these columns is
very high, and long lengths may be used without detriment. Thus
very high column efficiencies are available. Two types of capillary
columns are generally available. These are wall-coated (WCOT) and
support-coated (SCOT) open-tubular columns. WCOT columns are characterized by the deposition of a continuous thin film (0.2 to 0.4 μm)
of stationary phase on the inside wall of the column. In contrast,

the inside wall of a SCOT column is coated with a porous support layer, which is impregnated with stationary phase (1 to 5 μm). These columns do not have efficiencies vastly superior to those of packed columns (N, 1000 to 5000 m^{-1}) but are effective because much longer columns may be used. Open-tubular columns require smaller flow rates than packed columns, and they also have a much lower capacity. The support-coated columns may be considered as intermediate between the WCOT type and packed columns, in having a higher capacity, but with lower overall column efficiency. Representative data is presented in Table 3.4

The performance of capillary columns is outstanding, so that selectivity is less important than with conventional packed columns. This means that only a few phases (perhaps OV-17 and carbowax high polymer ≡ carbowax 20M) may be suitable for many separations, although newer systems, such as graphitized columns show promise [28]. The capacity factor is also significantly less in capillary columns, so the main influence upon the resolution capabilities is column efficiency. This may be obtained by calculation of the minimum plate height (cf. packed columns, p. 105, A = 0 for WCOT)

$$H_{min} = 2 \ (BC)^{\frac{1}{2}}$$

For capillary columns

$$C = C_g + C_1$$

$$= \frac{1 + 6k + 11k^2}{24(1 + k)^2} \ \frac{r^2}{D_g} + \frac{k}{6(1 + k)^2} \ \frac{d_f^2}{D_1}$$

with the assumption that $C_g >> C_1$, the minimum plate height (H_{min}) is approximated by

$$H_{min} = r \left(\frac{1 + 6k + 11k^2}{3(1 + k)^2}\right)^{\frac{1}{2}}$$

The coating efficiency (E) may then be calculated:

$$E = \frac{H}{H_{min}} \times 100\%$$

TABLE 3.4 Wall-Coated and Support-Coated Open-Tubular Columns

COLUMN	I.D.	S.S.A.	Solid Support	Liquid Phase	Phase Ratio	F_C -N_2	U-N_2	F_C -He	U-He	Capacity per Peak	N
	mm	cm^2/m	mg/m	mg/m	β	ml/min	cm/s	ml/min	cm/s	ng	m^{-1}
WCOT	0.5	15.7	-	0.15	320	2-4	10-15	4-10	20-35	5-100	1500-2500
SCOT	0.5	15.7	8.75	4.1 2.6 0.51	50 67 360	2-4	10-15	4-10	20-35	30-300	600-2000

[a]The gas-liquid phase ratio is $\beta = r/2d_f$, where r is the radius of column bore and d_f is the film thickness. For a packed column, $\beta \approx 20$).

where H is the plate height measured from the chromatogram. For
capillary columns, which are characterized by large dead volumes, it
is more appropriate to calculate H from the effective plate number
(N_{eff}), which is corrected for column dead volume:

$$N_{eff} = 5.54 \left(\frac{t - t_0}{W_{\frac{1}{2}}}\right)^2$$

This correction has little effect for a packed column, but is significant for open tubular systems. E should be close to 100%, but
in practice 60 to 80% is obtained. Lower efficiencies herald poor
coating and rapid degeneration of the column. One recommended scheme
of coating glass capillaries [29] involves leaching the glass surface
(20% HCl, 160°C, 16 h) to remove inorganic impurities and etching the
silica surface with barium carbonate (0.5 to 0.05%) to reduce catalytic effects further, and to provide a key for polar stationary phases.
Coating of the column with polar [30] or apolar [31] stationary phases
is achieved by forcing a plug of stationary phase, dissolved in a
volatile solvent (methylene dichloride, pentane, 1%) at a controlled
rate through the cleaned, dried column. An additional piece of tubing may be added to ensure a uniform coat throughout the column
length. Stainless steel columns have also been used [25]. It must
be noted that nonpolar phases, such as OV-101 have little ability to
cover adsorption sites, although useful columns with a high degree of
thermal stability are possible. Careful deactivation of the column
is necessary with these phases [32,33]. A static coating procedure,
in which the capillary is filled with a solution of the stationary
phase, or else a suspension of a finely divided solid in the dissolved stationary phase, is normally used for SCOT columns. One end
of the tube is closed and the other end is slowly drawn through an
oven. The volatile solvent is removed and a porous layer, impregnated with stationary phase, is slowly deposited onto the wall.

Under normal conditions of use, the capacity of capillary columns demands that very small amounts of sample are transmitted to
the column. This has been achieved in Py-GC by using very small

sample sizes and pyrolyzers designed with low dead volumes [34-36].
This approach, however, demands low flow rates of carrier gas for the
optimal operation of the column. In turn, the flow through the pyrolyzer may be too slow for the rapid removal of products, and secondary reactions may be involved. An alternative approach is to split
the flow onto the column so that only a fraction of the initial
charge is analyzed. This approach is traditionally used in normal
GC analyses [37] and has been studied in detail by Sugimura and Tsuge
[38]. Problems with this mode of operation are that the product distribution fed to the column may not be representative of the sample,
depending upon relative volatilities; high efficiency involves the
optimization of many variables, including sample size, splitting
ratio, carrier gas flow rate, dead volume, and temperature. A make-
up flow of gas is usually supplied after the column outlet, for proper functioning of the detector system. A modified version of a
furnace pyrolyzer, with independent heating of the splitter and inlet
system, was used with polyethylene and polystyrene to explore the
effect of these variables. The system is illustrated in Fig. 3.5.

The splitter was designed to be used empty, or packed with a
support coated with stationary phase, and was provided with a heater
for temperature control. Variations in peak shape, product distribution, and reproducibility were observed, and results were obtained
with a heated or lagged splitter, with a split ratio of 50, or 100:1.
A comparison of data is recorded in Table 3.5.

The appearance of tarry products, which result in the serious
deterioration of capillary columns, has been noted [25,34]. These
contaminants build up after a period of time, and prevent the establishment of long-term reproducibility. Removal of the initial section of the column has been employed, but a technique which overcomes
the problem has been proposed by Mitchell and Needleman [39], and
combines a disposable precolumn with a back flushing arrangement.

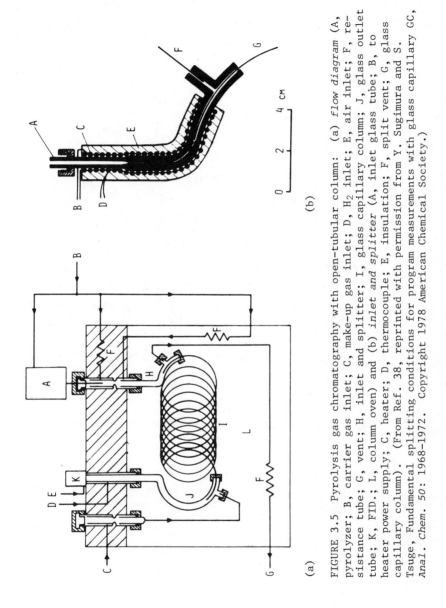

FIGURE 3.5 Pyrolysis gas chromatography with open-tubular column: (a) *flow diagram* (A, pyrolyzer; B, carrier gas inlet; C, make-up gas inlet; D, H_2 inlet; E, air inlet; F, resistance tube; G, vent; H, inlet and splitter; I, glass capillary column; J, glass outlet tube; K, FID.; L, column oven) and (b) *inlet and splitter* (A, inlet glass tube; B, to heater power supply; C, heater; D, thermocouple; E, insulation; F, split vent; G, glass capillary column). (From Ref. 38, reprinted with permission from Y. Sugimura and S. Tsuge, Fundamental splitting conditions for program measurements with glass capillary GC, *Anal. Chem. 50:* 1968-1972. Copyright 1978 American Chemical Society.)

TABLE 3.5   Reproducibility of Polystyrene Pyrograms

PRODUCT	PRODUCT YIELD (%)					COEFFICIENT OF VARIATION (%)				
	A	B	C	E	P	A	B	C	E	P
MONOMER	81.9	77.8	80.3	77.1	77.0	1.1	0.7	2.1	0.4	0.5
DIMER	7.6	8.2	6.3	8.3	8.7	6.2	2.4	9.0	4.8	5.1
TRIMER	10.4	14.0	13.4	14.6	14.3	12.7	2.1	10.1	2.7	3.8

[a]Capillary column, split ratio 100:1; A, empty inlet tube (no heating); B, empty inlet tube at 250°C; C, packed inlet tube (no heating); E, packed inlet tube at 250°C; P, packed column results; C.V. = $\sigma/\bar{x}$ %.

Source: Adapted from Ref. 38, reprinted with permission from Y. Sugimura and S. Tsuge, Fundamental splitting conditions for pyrogram measurements with glass capillary GC, Anal. Chem. 50: 1968–1972. Copyright 1978 American Chemical Society.

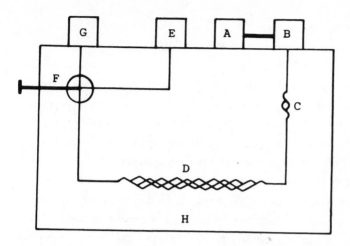

FIGURE 3.6 Prevention of column contamination by back-flushing. In
normal operation, the inlet supply (A) passes through the pyrolyzer
(B) and flushes the products through a precolumn (C), and onto the
analytical column (D). The carrier gas flow is made up from inlet
(E) and is swept into the detector through valve (F). Back-flushing
is achieved by closing valve (F) and disconnecting the inlet line
(A,B). (Adapted from Ref. 39, adapted with permission from A.
Mitchell and M. Needleman, Technique for the prevention of column
contamination in Py-GC, *Anal. Chem. 50*: 668. Copyright 1978 American
Chemical Society.)

The experimental setup is depicted in Fig. 3.6. The splitter system
of Tsuge [38] may also be used as a precolumn to reduce column contamination without a decrease in performance. Moreover, the system
may be used to effect pyrolysis-hydrogenation sequences [40]. This
is achieved by the incorporation of a catalyst tube between the pyrolyzer and the inlet system, and the use of hydrogen as carrier gas.
The triplets characteristic of polyethylene (up to C_{31}) are replaced
by sharp singlets of the saturated hydrocarbons.

Temperature Programming

In many applications of Py-GC, the range of products obtained is
such that no single set of operating conditions is suitable for the
resolution of all components. Separation of the more volatile products may mean that later peaks are broad, tail badly, or are simply

not transmitted by the system. Conversely, to achieve efficient separation of these compounds, conditions which cause overlap or superimposition of early peaks are necessary. Such problems may be largely overcome by changing the chromatographic conditions throughout a run. One possibility is to change the carrier gas flow rate (flow programming [41]), but most applications now involve temperature programming [42]. In those mode of operation, an initial isothermal period (temperature and duration programmed) is followed by a period in which the temperature is increased linearly, to the final column temperature (rate of increase to final temperature programmed), and this is then maintained for a preset interval, before the column is rapidly cooled to the initial isothermal temperature. This results in products being essentially fixed at the beginning of the column. These are mobilized when appropriate conditions are reached. Analysis time is reduced, peaks may be evenly spaced throughout the chromatogram, and the sensitivity for late peaks is substantially improved.

This clearly introduces several more variables, which must be controlled before reproducible pyrograms are obtained. Retention times and peak heights depend dramatically on the program used, and reproducibility demands that this is well defined. Not only should columns behave identically at one temperature, but the temperature profiles must also be similar. The greatest variation then may be due to the flow rate of the carrier gas through the column, which decreases with increasing temperature, e.g., 2 ml min^{-1} at 75°C fell to 1.5 ml min^{-1} at 165°C [43]. Thus, the temperature at which the flow is set must be established. Modern gas chromatographs, however, are equipped with mass flow controllers to maintain a constant flow of carrier gas. Temperature programming also increases the degree of column bleed throughout the run. This manifests itself as a rise in base line at some point in the chromatogram. The use of a second identical column reduces this effect (dual column operation). Here, two columns are operated in tandem; one for analysis and the second as a reference. The flow rates are adjusted so that retention times are equivalent, and the detectors are adjusted for optimal response. With the detectors wired in parallel, so that continuous changes in both columns (provided these are in fact identical) are not transmitted to the recorder, the detected signal is essentially due to the pyrolysis products.

The column performance parameters in temperature-programmed GC are somewhat different to those calculated in isothermal work. This arises because fundamental properties, such as partition coefficient vary throughout the run. The retention temperature (T_R) is the most characteristic parameter in the programmed mode. This is equivalent to retention time in an isothermal chromatogram and is the temperature at which the peak maximum elutes. The efficiency of the column (N) may be calculated by determining the retention time (ti) of a component, measured from an isothermal run at the retention temperature, and the base width (W) in the temperature programmed mode:

$$N = 16 \left(\frac{ti}{W}\right)^{\frac{1}{2}}$$

Adequate description of chromatographic resolution also demands that an isothermal run, in addition to the programmed analysis, is made [40]. One definition of *resolution* is

$$R_S = \frac{T_{R,B} - T_{R,A}}{ti_{,A} + ti_{,B}} \quad \frac{N^{\frac{1}{2}}}{2W_L \, \Delta}$$

where W_L is the weight of liquid phase on the column, Δ is the heating rate, T_R is the programmed retention temperatures of components A and B, and ti is the isothermal retention time of components A and B at the retention temperature.

It should be noted that all column performance characteristics are dependent upon symmetrical, gaussian peak shapes. Where this is not so, significant errors may result [44,45]. Despite this difficulty, and the problems associated with the adequate description of parameters in temperature programming work, it is essential that the behavior of the column is measured numerically so that changes in performance may be readily identified.

3.1.2. The Detector

Many detectors have been used in gas chromatography [46]. Detectors
may exhibit

> Integral or differential responses
>
> Response to concentration (TCD) or to mass flow (FID)
>
> Selectivity (ECD) or universal response (TCD)
>
> Destructive (FID) or nondestructive (TCD)

Requirements are

> Stability to exclude short-term or long-term drift
>
> Reproducibility of response and operating conditions
>
> Selectivity to respond to analyte
>
> Absence of background response
>
> Fast response time to retain resolution of eluted peaks
>
> High sensitivity to low amounts of analyte for efficient
> chromatography
>
> Linear response for quantitative analysis

The response of the detector (σ_d) may be expressed as

$$\sigma_d = \frac{A_A' \sigma_r F_c}{V \, w_A}$$

where A_A' is the area of the peak produced by w_A mg of component A,
σ_r is the recorder sensitivity, and V is the chart speed. A more
useful parameter is the minimum detectable quantity

$$E_u = \frac{2N_B}{\sigma_d}$$

assuming that a signal is detectable if it exceeds the background
noise (N_B) by a factor of two. This parameter is in terms of flow
through the detector. For an FID detector the response is dependent
upon the rate of flow of a component, rather than the concentration
in the carrier gas. Thus, the minimum amount of sample on the column
(Q) to yield a detectable peak may be estimated from

$$Q = E_u F_c W_{\frac{1}{2}}$$

A comparison of detector properties is given in Table 3.6. Flame ionization detectors (FID) are most commonly used in Py-GC because of their high sensitivity, wide linear range, and robustness. The sensitivity depends somewhat upon the gas flows (H_2, 30 ml min^{-1}, and air, 400 ml min^{-1} are typical), and these should be optimized. The detector responds to most organic molecules, but it is less useful if volatile gases, such as water or carbon dioxide are to be monitored [47]. The use of alkali halide beads suspended above the flame increases the specificity of the detector by yielding high fluxes of ions when characteristic species are present (e.g., CN from many nitrogenous compounds). Such a detector has been used to enhance components from protein pyrolysis [48]. Thermal conductivity detectors are now less popular but respond to both organic and inorganic components. They are less sensitive than FIDs and are therefore difficult to use with low sample loadings, but are useful if eluate collection is required, although FID with an outlet splitter is also applicable. Thermal conductivity detectors have thus been used as the basis for tandem GC systems [49,50]. The electron-capture detector, which has found much use in general analytical work, has found little application in Py-GC. This detector, although very sensitive to those compounds which can capture electrons, has little affinity for many of the products normally found in pyrograms. Although ECD would appear to be ideal for the study of halogenated polymers, the operating care which these systems require (e.g., they are easily overloaded and contaminated), mediate against their routine use. One further point is the linear range of these detectors. The intensity of important pyrolysis products may differ by orders of magnitude. For effective control of the pyrogram, the detector should have a high dynamic range. This is achieved with FID, but ECD (even with the modern pulsed-modulated systems) is substantially poorer.

Detectors such as the Argon ionization detector, which enabled the realization of high-efficiency analytical pyrolysis [51] have now generally been superceded; but novel systems for special purposes, e.g., the beta-induced luminescence detector (BILD) for

TABLE 3.6   Common GC Detectors

DETECTOR	AVERAGE SENSITIVITY	LINEARITY RANGE	APPLICATION
ALKALI BEAD-FLAME IONISATION (ND)	100pg	10^4	Determination of organo-nitrogen compounds. Normally used isothermally.
ELECTRON CAPTURE (ECD)	2×10^{-14} g/s (1pg)	normal:10^2 pulsed:10^3	Analyte must capture electrons. Useful for halogenated compounds. Hydrocarbons not detected. Easily saturated and sensitive to impurities and temperature changes.
FLAME IONISATION (FID)	2×10^{-2} C/g (1ng)	10^7	Suitable for most organic compounds. Responds poorly or not at all to water, non-combustible gases and polyhalogenated compounds.
FLAME PHOTOMETRIC (FPD)	8×10^{-13} g/s (P: 1pg) 1×10^{-11} g/s (S: 10pg)	10^3	Determination of organophosphorus and organo-sulphur compounds.
THERMAL CONDUCT-IVITY (TCD)	1.9×10^{-9} g/ml (100ng)	10^4	Sensitive to a wide range of organic and inorganic components. Products must not corrode filaments.

volatile products insensitive to FID [47], continue to appear. One
such detector with general availability is the flame photometric detector, which has a highly selective response to phosphorous and
sulfur compounds. The gas density balance detector has been used
for molecular weight determinations [52].

One other detector has high sensitivity and a high degree of
specificity, combined with general applicability. This detector is
the mass spectrometer, and its use has revolutionized GC analyses in
many areas, particularly when combined with online computing
facilities.

3.1.3. GC-MS: Component Identification

The identification of pyrolysis fragments is an essential control
feature in analytical pyrolysis. It makes a full description of the
pyrogram available, it allows intercolumn comparisons to be made,
e.g., when overlap or reversal of eluted components is observed, and
assists in the choice of model compounds for GC optimization studies.
Furthermore, a significant amount of structural evidence may be obtained from a detailed study of product distribution.

In the case of simple polymers, a comparison of the retention
times of standard compounds may be adequate; but with more complex
samples the procedure becomes unreliable. Retention times may match
but structures may not. The tandem GC principle [49,50] has been extended by Uden to provide a comprehensive peak identification system
[52]. This system is a combination of seven instruments, and allows
infrared spectra, elemental analysis, molecular weight determination,
and pyrolysis gas chromatography on eluted components to be undertaken. The advantage of this system is that a series of online
analyses may be undertaken so that sufficient data for characterization is available.

A system which undertakes pyrolysis molecular-weight chromatog-
raphy-IR analyses of polymers has also been described by Kiran and
Gillham [53]. The mass chromatograph uses two gas chromatographs,
each with a different carrier gas, to analyze the pyrolysis products.

Detection is by means of gas-density detectors, and the molecular
weight of the eluted components may be calculated from the response
ratio. Quantification and IR analysis is also available. Notwith-
standing the utility of these concepts, however, an interfaced mass
spectrometer is the commonest system for the routine identification
of pyrolysis products. Such systems are widely available and are of
sufficient sensitivity to be used with capillary columns [54-56].

The combination of GC and MS produces a powerful analytical tool
which has greater potential in many areas than either instrument used
alone. The usual MS operating conditions involve electron impact
ionization with low-resolution magnetic sector analysis (see Chap.
4). However, quadrupole instruments now have adequate resolution,
and can withstand higher source pressures, so that many are now in
use. The greatest drawback is that the operating conditions required
for each are incompatible. GC uses carrier gas flow rates of 20 to
40 ml min^{-1}, whereas MS needs high vacuum conditions (10^{-6} torr).
The interface must therefore remove the carrier gas without significant loss of analyte before entry into the mass spectrometer. This
is achieved by the use of a separator which allows differential pump-
ing of the carrier gas. Such processes are much more efficient if
highly diffusable gases are used, so that in general, helium is preferred for Py-GC-MS work. The commonest separators are those which
operate on the single jet principle, but Ryhage two-stage jet separators and the Watson-Biemann glass-frit separator are also in use.
Membrane separators, in which enrichment is achieved due to the
greater diffusion of organic molecules through a silicone film, have
also been described. The resolution and product distribution detected by the mass spectrometer will be mediated by the efficiency
of the separator; the temperature of which should be carefully
monitored.

The ion flux within the mass spectrometer source rises and falls
as components are eluted from the GC and are transmitted through the
separator into the mass spectrometer. The chromographic record is
presented as a trace of this total ion current. The pyrogram

obtained from Py-GC-MS analysis may thus differ from that in a normal Py-GC run due to

> Different response factors of the MS and FID detectors for
> pyrolysis products
>
> Pyrolysis and analysis differences mediated by a change in
> carrier gas
>
> Differential transmission of products through the separator

In a typical run, a mass spectrum can be initiated as peaks of interest are eluted from the column. Spectra are generally characteristic of each component and may be interpreted by inspection of
fragmentation pathways [57,58], by comparison with tabulated data
[59,60], or with computer assistance [61-64]. Modern mass spectrometers have rapid scan and recycling times, so that many mass spectra
may be run in a very short period. With computer control this may
be developed into a very powerful system [65]. Typically, repetitive
spectra are determined throughout the course of the GC run, at intervals of about 10 s, and are stored on disk. The record may be manip-
ulated in several ways [66]. The data may be summed to provide a
total ion current GC trace (mass chromatogram), identical to that
obtained in the real-time analysis. Such traces, however, record
the coelution of background contamination which, if large, may ob-
literate peaks of interest. This may be suppressed by the recon-
struction of the chromatographic record from only a portion of the
data. Typically, drug metabolic work [67] would use the summed current from ions m/z 200-700 to exclude phthalate contaminants (base
peak m/z 149), whereas studies of the martian surface summed the intensities above m/z 47, to eliminate water and carbon dioxide contributions [68]. This process may be extended to produce a com-
ponent-specific detector. In this mode, the variation of the intensity of one or more selected masses is monitored. This process
selects for display only those components whose mass spectra contain
the chosen fragment. This greatly simplifies the mass chromatogram,
reveals overlapping fragments, and indicates those components which
have a similar structure. The recall of any selected mass spectrum

(corrected for background effects) enables the rapid characterization of the eluted fractions [69]. Algorithms also exist to enable
resolution enhancement, by considering only ions which increase or
decrease together when peaks are detected.

Clearly, such a volume of data (500 mass spectra per run) requires an efficient data-handling system. An alternative approach,
which increases detector specificity with adequate sensitivity
(picogram-to-nanogram level) and may be used in real time without
computer assistance, is mass fragmentography [70-72]. In this mode,
the mass spectrometer is tuned to allow specific ions only through
to the detector. This is illustrated in Fig. 3.7 with the pyrolysis
of an acrylate polymer, which yields methyl methacrylate ($M^{+\cdot}$100,
base peak m/z 41) and ethyl acrylate ($M^{+\cdot}$100, base peak m/z 55). This
may be achieved by programming the accelerating voltages required for
transmission of the ions in question. Rapid switching between these
preset windows allows the simultaneous detection of upto eight fragment ions. Each is monitored independently and provides a recorder
trace as the compounds containing the specific ions are eluted. The
availability of this multiple-ion detection system means that internal standards may be incorporated to improve quantitative reproducibility. Deuterated samples or isomers are most effective. Although
mass fragmentography is usually performed under low-resolution conditions, high-resolution spectra, using an isomer as internal standard,
enable even greater selectivity to be achieved, together with an increase in sensitivity. Mass fragmentography is usually employed to
increase the detector selectivity, so that low levels of products
may be quantified reliably with full utilization of instrument sensitivity. It has been applied extensively in the Py-GC-MS analysis
of quaternary ammonium compounds, such as acetylcholine [73] and
drugs [74], using either electron-impact or chemical-ionization
methods.

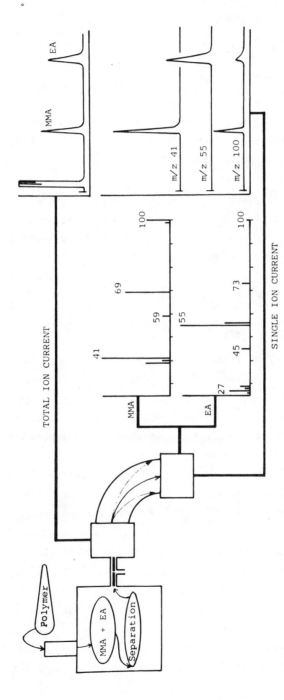

FIGURE 3.7 Mass fragmentography of an acrylate polymer (MMA, methyl methacrylate; EA, ethyl acrylate).

3.2. PYROLYSIS GAS CHROMATOGRAPHY

The attachment of the column to the pyrolyzer and to the detector
should be made so that minimum dead volume remains within the system. Such voids dramatically reduce resolution and may also trap
polar or more volatile fragments. To avoid such loss, the interface
temperatures should also be controlled so that the pyrolysis products
are not subjected to cold spots along the analysis pathway [38].
When the system is assembled it should be tested for background effects by firing the pyrolyzer without sample. This ensures that contamination from earlier runs [17], or from other sources such as the
pyrolyzer itself [25] is absent. It is now necessary to optimize the
analytical conditions to provide the information required, together
with sufficient definition of the variables to achieve reproducibility. This ensures that analyses are repeatable by different opera-
tors after various intervals of time.

3.2.1. Correlation Trials

Comparative interlaboratory tests of Py-GC reproducibility have given
an insight into variations in technique [75] and the problems associated with ensuring precision [76-80]. Trials were organized by the
Pyrolysis Sub-group of the Gas Chromatography Discussion Group (UK),
and initially three samples: polystyrene, styrene-butadiene block
copolymer and methyl methacrylate-styrene-butadiene terpolymer were
analyzed [76]. Two stationary phases were specified, but the pyrolysis and analytical conditions were selected by participants and both
quantitative and qualitative (fingerprint) information was sought.
The variation in results was remarkable: the recovery of styrene
from polystyrene ranged from 26% to 102%, and 9 to 41% from the copolymer. Although no evidence was presented as to how quantification
was achieved, or on the statistical viability of calibration lines or
replicate analyses, it is clear that variations of such magnitude
necessitated a much greater control of future trials. Fingerprint
comparisons were also poor and involved comparisons of isothermal and
temperature-programmed runs, but recognizable patterns were evident.

A second trial was designed using fingerprint assessment only [77].
Here, variables were more closely defined. Sample handling procedures and pyrolysis temperature and duration were specified. The
recommended analytical conditions included the stationary phase and
support material, the column temperature and efficiency (N > 3000
for styrene), chart attenuation, and flame-ionization detection. A
useful step also was the recommendation that chart speed and flow
rate be adjusted to give the styrene peak a retention distance of
10 cm, relative to methane.

These specifications ensured a considerable improvement, and
for easily handled copolymers 80 to 85% of the pyrograms returned
were comparable. Progressively poorer results were obtained with
more complex samples. Failure to analyze for a sufficient length of
time caused loss of information with a phenol-formaldehyde condensate, and extremely poor reproducibility was observed between pyrograms of a gloss paint film. Sample handling techniques were largely
responsible for this variation [78]. Pyrograms were altered significantly (1) when residual solvent was present, (2) when different
pyrolysis temperatures were used (it is particularly important to
ensure *all* of the sample achieves the same temperature, thus cool
spots on filaments or Curie-point wires outside the induction zone
of the pyrolyzer must be avoided), (3) when contamination from earlier analyses was evident, or (4) when large differences in sample
size occurred.

A further correlation trial also specified that sample sizes
should be below 100 µg, that as large a surface area as possible
should be used for sample coating, and that cold spots and contamination were minimized. The column efficiency was also specified to
be 3000 plates at the styrene peak obtained by pyrolyzing polystyrene.
This parameter now includes pyrolyzer performance and dead volume,
and is clearly a much more useful index. Four groups of widely differing polymers (15 in all) were analyzed, and 72% of all returned
pyrograms were acceptable, considering the ease of matching, chromatographic resolution and the absence of spurious peaks. Of the 26%

rejected pyrograms, over half were due to one participant with poor
column performance. Cross-contamination, column overload, and gross
peak tailing were responsible for other rejected pyrograms. The
necessity for control in sample handling, pyrolysis, and gas chromatography was stressed. The possibility of using a standard pyrogram
to calibrate a Py-GC system prior to analysis was also mentioned.

These trials indicated that reproducibility was more difficult
to achieve with complex samples. The fourth and final correlation
trial in this series, therefore, attempted to define conditions
which would allow a satisfactory analysis of a difficult sample to
be achieved [79]. A styrenated alkyd-base resin and an aged paint
flake prepared from the resin were used as samples. The parent
resin was easily handled in a solution, the paint flake however, demanded the pyrolysis of less than 100 µg as a solid. Despite this
difference, the two samples compared closely in all pyrogram pairs
submitted. Of these, 53% were superimposable, or nearly so, while
16% failed to give a definitive pyrogram--mainly due to the use of
excess sample. Quantitative reproducibility was still low. The
peak height ratio of the two main products (styrene and vinyltoluene)
showed coefficients of variation of 13% (resin) and 10.4% (paint)--
rather large for quantitative assessment.

This series of trials showed that readily identifiable pyrograms could be achieved on an inter-laboratory basis, providing
analytical variables were defined and adopted by all laboratories.
However, sufficient latitude remained to allow the submission of a
significant proportion of rogue results. In an attempt to overcome
these problems, a further set of correlation trials, organized by
the American Society for Testing and Materials [23], were devised.

In common with the UK trials a range of pyrolyzers were used,
but temperature and duration were again specified. In addition, the
column cleaning, packing, conditioning and operating procedures were
detailed, together with the recommendation that helium be used as
the carrier gas. Results similar to the UK trials were obtained,
with 55% of laboratories submitting pyrograms with a high degree of

similarity, while 9% were classified as not comparable. Peak-height data however, varied substantially, and was too poor for quantitative assessment. Retention-time data were more satisfactory, with coefficients of variation of 3.2 to 3.5% (polybutadiene) and of 6.7 to 7.3% (isoprene-styrene).

A second trial specified Carbowax, rather than a silicone stationary phase, to provide greater GC resolution, and isobutene-isoprene and isoprene-styrene copolymers were used as the test substances. Now, 83% of laboratories provided first class pyrograms and none were rejected outright. Nevertheless, quantitative reproducibility was still poor. To overcome this difficulty, the concept of calibrating the GC response to a standard system before pyrolysis, was introduced.

In the third correlation trial, the alkyd samples previously used by the UK group [79] were analyzed [23]. Major pyrolysis products are styrene and vinyltoluene. Participants were, therefore, provided with a quantitative liquid mixture of both products, so that GC conditions prior to pyrolysis could be standardized. This was achieved by adjusting the carrier gas flow-rate to yield specific retention times for styrene (7.5 ± 0.25 min) and vinyltoluene (13.5 ± 0.25 min). The chromatographic conditions were such that the presence of one extra product was revealed, and only one set of pyrograms were discarded, again due to large sample sizes. Results for both intensity and retention time data paralleled the UK results closely, and showed that with care, interlaboratory comparisons of data are possible. A substantial part of the variation in the results could be due to temperature differences. Curie-point systems were operated at 770°C, whereas filaments were run at 700°C (perhaps sufficient to cause problems). Within a single laboratory, quantitative reproducibility is excellent (Table 3.7) [14,23,80].

The use of Porapak Q was also reported by one participant. Temperature programming allowed better resolution of early peaks, together with a shorter overall analysis time. The easier standardization of this packing material compared to coated phases, has made

TABLE 3.7 Coefficient of Variation for Intensity of Vinyltoluene
Peak on Pyrolysis of an Alkyd Paint

	Resin	Aged paint
Programmed run	1.6	3.9
Isothermal run	8.4	6.5

Source: Ref. 23.

it a good candidate for a standardized system [13,15]. With this in
mind, a fourth correlation trial has been undertaken, which involves
for the first time, a temperature-programmed GC analysis [81]. Although results are not as yet available for this venture, recommendations for a tuned approach to Py-GC, to ensure a high degree of
reproducibility may be made.

3.2.2. A Tuned Approach

The checklists in Appendix 2 focus attention on the variables encountered in analytical pyrolysis (Fig. 1.8). A tuned approach to
Py-GC involves the adjustment of GC conditions to yield a standard
chromatogram, followed by adjustment of the pyrolysis conditions to
yield a standard pyrogram from a calibration sample.

For polymer analysis, Porapak Q (50 to 80 or 80 to 100 mesh)
has been recommended [13-15,23,81]. This material does not crush
readily and may therefore be packed into columns (1 to 1.5 m × 2 to
4 mm) by suction and vibration. With tightly packed beds, efficiencies of 1500 to 2500 plates per meter are obtained. Its use removes
many of the variables associated with packed columns. Conditioning
may be achieved by programmed heating up to 200°C for 12 h, followed
by a further 12 h at 250°C using a carrier gas flow rate of 30 ml
min^{-1} throughout. A typical analytical run will involve a run from
100 to 250°C at 5 to 8°C min^{-1}, and this should be checked without
sample to ensure that a usable baseline is obtained.

The column performance is then tuned by adjustment of the flow
rate (20 to 30 ml min^{-1}), to reproduce the retention times of

calibration standards. This has been achieved by injecting solutions into the column in isothermal [23] or temperature programmed [15] modes. Typically, a methanol-propanol mixture was injected and the flow-rate of nitrogen was adjusted to give retention times of 2.6 min and 9.1 min when the column was programmed from 100 to 200°C at 8°C min^{-1} [15]. The upper temperature was then adjusted, so that a retention time of 14 min was obtained for cyclohexane when analyzed under the same conditions. A more direct method is to use a polymer which, on pyrolysis, yields components suitable for GC calibration. One such material is Kraton 1107 [23], which is an isoprene-styrene (86:14) copolymer. This sample (20 μg) is pyrolyzed with the column programmed from 200 to 250°C at 8°C min^{-1}, and then maintained at 250°C, and the helium flow-rate is adjusted to elute styrene in 6.5 ± 0.3 min and dipentene in 10.3 ± 0.5 min [81]. This procedure also enables a check on the efficiency of the Py-GC system to be made, rather than the GC component alone.

Finally, tuning of the pyrolysis unit is required. This is necessary as pyrolyzers will differ in their temperature profiles with possible product variation, in addition to other problems, such as large dead volumes or cold regions decreasing transfer efficiency. This tuning may be achieved by the use of a polymer, the fragmentation of which is highly sensitive to pyrolysis temperature, a procedure suggested by E. J. Levy [82]. Kraton 1107 is again useful. Table 3.8 summarizes the fate of the four major products as a function

TABLE 3.8 Effect of Pyrolysis Temperature on Product Distribution (%) from Kraton 1107

Product	Pyrolysis Temperature (°C)					
	400	500	600	700	800	900
Isoprene	29	46	50	63	71	75
Styrene	0	16	24	21	19	19
Dipentene	68	32	21	13	7	4
Dimethyl-vinyl cyclohexane	3	6	5	4	3	2

Source: Ref. 23.

of pyrolysis temperature, with the pyrolysis of 15 μg of polymer and
a TRT of 15 ms to 600°C [23]. Under the prescribed conditions, the
isoprene:dipentene ratio (y) varies linearly with the pyrolysis
temperature:

$$y = 0.01393T - 7.21$$

and thus variations in pyrolysis temperature may be evaluated and
corrected by quantitative assessment of the pyrogram. Clearly, such
correction is far easier with a heated filament pyrolyzer than with
Curie-point systems.

The use of dual standardization is at present the best approach
to ensure interlaboratory reproducibility of pyrograms. It is anticipated that with this procedure an atlas of polymer pyrograms, which
also contains tuning instructions, could be made available. There is
little doubt that if authors were to define their systems by these
procedures, the pyrolysis literature would be a much more fruitful
source of data. This is particularly true of reports which use pyrograms for fingerprint comparisons, without the identification of
components. In these instances, in particular, calibration with
standards is essential to define the analytical system.

Interlaboratory correlation trials have effectively been limited
to synthetic polymers. It would appear that reproducibility deteri-
orates as sample complexity increases. It is perhaps not surprising,
therefore, that materials of geological origin, or microorganisms,
have not been studied seriously from this viewpoint. The problems
of sample-handling to ensure reproducible loading of a representative
and repeatable sample, together with the large number of products obtained in these analyses, makes standardization in these areas even
more difficult. Although results from one laboratory are generally
satisfactory, the broader tasks of, for instance, the production of
an atlas of pyrograms for microorganism taxonomy are yet to be accomplished. Nevertheless, some pointers are available to suggest a
possible approach to making data more universally applicable.

The correlation trials with polymers have shown that reproducibility is improved substantially if resolution of adjacent peaks is

achieved. For complex samples of biological origin, capillary columns are therefore essential. The tuning of the GC system may be
achieved in several ways. The use of a polymer such as Kraton 1107
as already described--attractive since this serves a dual purpose
and allows both retention time and pyrolyzer performance to be adjusted. In addition, long-chain hydrocarbons (C_{18} and C_{32}) were
found to have satisfactory retention times; so they could be used as
reference substances for the pyrolysis of cockroaches [17]. Moreover, under the conditions used (600°C, TRT 3s), volatilization
rather than pyrolysis was observed, so that the alkanes could be
used as internal standards to continuously monitor the chromatographic performance. An alternative approach is to pyrolyze hydrocarbons or polyethylene, to provide a homologous series of fragments,
mainly terminal olefins [83]. This procedure was developed to aid
GC-MS interpretation by the provision of appropriate mass markers.
The same system would appear to be useful for GC calibration purposes. Perhaps the simplest approach here is to use polyethylene.
On pyrolysis, this material fragments to yield a homologous series
of triplets, composed of the α,ω-diolefin, the α-olefin and the
n-alkane. An excellent pyrogram of this material, showing fragments
up to C_{31} has been reported [38], and is reproduced here as a bar
graph in Fig. 3.8. The generation of the triplets also allows a
check on resolution, in addition to retention-time tuning, to be
made. The use of solid material again allows the performance of the
system, rather than the GC alone, to be checked. Alternatively,
compounds which have been identified in samples may also be used.
These include indole, cresols, acetamide, and pyrrole (microorganisms [84]) and phenol, naphthalene, acetic acid, and toluene (soils
[85]). Although available evidence suggests that pyrograms of organisms are but little influenced by temperature variations above
500°C [17,25], the heating rate is important [25]. Final standardization, therefore, demands that, when retention times have been
adjusted, the pyrolyzer performance is checked by the use of a
temperature sensitive standard, such as Kraton 1107.

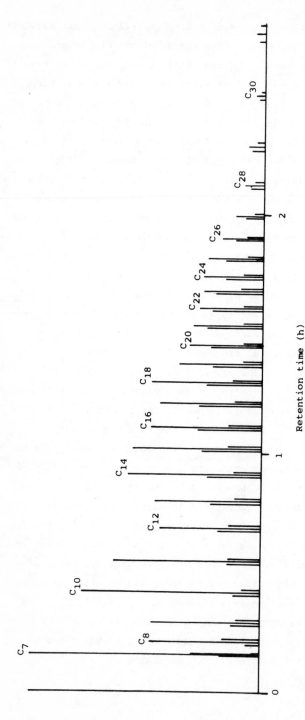

FIGURE 3.8 Bar graph of high-density polyethylene pyrogram. Pyrolysis at 650°C. Chromatography on 50-m OV 101 capillary column, programmed from 40 to 250°C at 2°C min⁻¹. The first peak in each triplet is the α,ω-diene, followed by the α-olefin and the n-alkane. (From Ref. 38, reprinted with permission from Y. Sugimura and S. Tsuge, Fundamental splitting conditions for pyrogram measurements with glass capillary GC, *Anal. Chem. 50*: 1968–1972. Copyright 1978 American Chemical Society.)

An alternative approach to the precise tuning of the GC conditions, which may be adopted to overcome small variations in operating parameters and sample size, is to normalize the retention-time data [86]. Here, the retention times of readily identifiable marker peaks are used to calculate corrected retention times for other peaks in the pyrogram (see Chap. 5). Normalization of peak area is also undertaken. This manipulation of the crude data allows a file of standardized data to be obtained, which is easier to compare or assess than the original pyrograms.

3.2.3. Quantitative Analysis

Many applications of analytical pyrolysis depend upon the quantitative assessment of pyrograms. Although some polymers [15] and drug series [87-89] may be characterized by the production of a unique pyrolysis fragment which enables identification to be achieved rapidly by visual inspection, the qualitative identification of an unknown frequently requires that the relative intensities of pyrolysis fragments are assessed. Good chromatographic performance is essential so that retention times and peak intensities may be measured accurately. For sharp peaks, peak heights may be recorded and are more precise than peak area measurements [14]. Peak areas may be determined by hand, with the width at half-height multiplied by the peak height giving the least biased estimate. Integrators are satisfactory with well-resolved pyrograms [86], but automatic data logging is prone to errors, due to variations in sample loading, retention times, base-line estimation, and the treatment of fused peaks. It has been estimated that peak areas determined by such systems are 2 to 3 times less reliable than peak heights measured directly from the chart [34]. Sophisticated and programmable data-logging systems are better able to accommodate pyrogram variables, but chromatographic separation should be maximized before quantitative measurements are made.

The assessment of reproducibility should also be undertaken by comparing normalized peak intensities from replicate analyses.

Table 3.9 records data from replicate results from the pyrolysis of
a paint sample, and illustrates what can be achieved [14]. Table
3.10 illustrates data from the pyrolysis of a microorganism [34].
Both data sets are from programmed runs. Some peaks will vary more
than others, but excessive variations indicate that the system is
not adequately controlled. If chromatographic performance is satisfactory, the commonest causes of poor performance are excessive
amounts of sample, poor heating of the sample due to its position or
to the heating characteristics of the pyrolyzer, contamination and
residual solvent. For many purposes, particularly those of a taxonomic nature, it is essential that the variation between replicates
(the inner variance) is minimized. The subroutine MEAN in program
STAT (Appendix 3) allows these calculations to be made easily.

The minimization of intersample variation is also essential
when quantitative analyses are undertaken. In this mode, the intensities of pyrogram fragments are used to estimate the proportion
(e.g., copolymer composition, or absolute amount, viz assay of a
biomolecule in a tissue extract) of a component. The principles of
quantitative analysis by GC [8] are applicable to such analyses, but
care must be taken to ensure satisfactory control of the pyrolysis.
A well-resolved, unique fragment which is detected as an intense and
symmetrical peak should be chosen for measurement. Further, this
fragment should be produced through one reaction pathway only via a
unimolecular decomposition process. This is to ensure that the intensity of the peak is linearly dependent upon the amount, or the
proportion, of the analyte in the sample to be analyzed. Clearly,
information concerning pyrolysis mechanisms is highly desirable in
the design of quantitative applications. Small sample sizes are
essential to avoid competing secondary reactions which cause a loss
of linearity. Calibration curves should be constructed and analyzed
statistically to confirm that adequate reproducibility and linearity
are available for assay purposes. Program STAT (Appendix 3) has options which allow simple least squares regression or analysis of
variance calculations to be performed, together with the calculation

TABLE 3.9 Reproducibility of Py-GC Data from a Paint Sample

Measurement	PRODUCT						
	1	2	3	4	5	6	7
Mean peak height ($\bar{x}$%)	23.6	12.3	8.9	38.4	7.9	3.7	5.4
Standard deviation (σ)	0.21	0.27	0.15	0.27	0.08	0.08	0.11
Coefficient of variation ($\sigma/\bar{x}$%)	0.89	2.19	1.68	0.70	1.01	2.16	2.04

Source: Ref. 14.

TABLE 3.10 Reproducibility of Py-GC Data from a Microorganism

Measurement	PRODUCT						
	1	2	3	4	5	6	7
Mean peak height ($\bar{x}$%)	57.7	62.2	82.8	61.9	55.4	79.3	50.3
Standard deviation (σ)	2.87	1.56	2.97	1.6	2.53	3.31	1.08
Coefficient of variation ($\sigma/\bar{x}$%)	4.98	2.51	3.62	2.59	4.58	4.17	2.16

Source: Ref. 34.

of errors about an interpolated assay result. Where possible, calibration curves should be constructed from pyrolysis data rather
than by the injection of pure components. This ensures that correc-
tions for the efficiency of pyrolytic fragmentation and for chromatographic differences are made. If pure components are used for
calibration purposes, peak area measurements are preferred. The
addition of a substance of generate an internal standard (an isomer
or a homolog) improves precision, provided no pyrolytic or chromatographic interference is observed.

One particularly important quantitative application of analytical pyrolysis to biological problems is the determination of the
neurotransmitter acetylcholine and related compounds. These substances are quaternary ammonium derivatives and do not exert sufficient vapor pressure for direct GC analysis. On-column thermolysis
of choline hydroxide yielded trimethylamine and acetaldehyde via
Hofmann elimination:

$$(CH_3)_3 N^+ -CH_2 -CH-OH \quad \xrightarrow{\Delta} \quad (CH_3)_3 N + CH_3 -CHO$$
$$\overline{OH} \qquad H$$
Choline

but under these conditions acetylcholine failed to fragment reproducibly [90]. Pyrolysis of the halides of acetylcholine, however,
initiated an alternative degradation:

$$(CH_3)_2 N^+ -CH_2 -CH_2 O-COCH_3 \quad \xrightarrow{\Delta} \quad (CH_3)_2 N-CH_2 -CH_2 O-COCH_3 + CH_3 X$$
$$CH_3 \quad X$$
Acetycholine

which yielded a characteristic tertiary amine fragment through loss
of a methyl halide. Other quaternary halides decomposed similarly
[91]. Transacylation between choline and acetylcholine may cause
scrambling, unless derivitization of choline is undertaken [92].
Conversion of choline to the propyl ester and the use of butyryl-
choline as internal standard, however, enables simultaneous estimation of both choline and acetylcholine by Py-GC, to a level of 25
ng [93,94]. The detection limit can be reduced substantially if

mass fragmentography is used. The copyrolysis of tetramethylammonium iodide--used as a coprecipitant, but its presence also increases the sensitivity of the analysis [92,95,96]--followed by chemical ionization with isobutane allows detection in the 10^{-9} to 10^{-13} mol range. Deuterated compounds are used as internal standards and the M + 1 ions are measured [96,97].

A technique using slow pyrolysis has also been described [73]. This fragments the acetylcholine, deposited on a pyrex rod, by heating at $250°C$ for 70 s. The volatile products are swept out and trapped on the cooled initial portion of the GC column. This technique was used to produce a small injection plug to retain column efficiency; a similar method has also been used to overcome dead-volume problems when using filament pyrolyzers with capillary columns [86]. GC analysis and electron-impact mass fragmentography followed. The advantage of this procedure is that the addition of tetramethylammonium iodide is not required to aid product transfer. This compound, which interferes with mass fragmentography through exchange reactions, and because both it and acetylcholine generate the same ions (m/z 59, 58) may safely be omitted.

The analysis of quaternary ammonium compounds by analytical pyrolysis is particularly rewarding, due to the simplicity of many of the pyrograms [98]. Herbicides, such as paraquat and diquat, yield the corresponding dipyridyls and can be estimated rapidly in pond and river waters [99] and in urine [100]. In contrast, Cetrimide, an antiseptic agent, is a mixture of components:

$$(CH_3)_3\overset{+}{N}(CH_2)_nCH_3 \quad \overset{\Delta}{\longrightarrow} \quad \begin{matrix} CH_3(CH_2)_n-Br + (CH_3)_3N \\ (CH_3)_2N(CH_2)_nCH_3 + CH_3Br \end{matrix} \quad n = 11,13,15$$

Br⁻

Cetrimide

The most intense peak is that due to NN-dimethyltetradecylamine (n = 13), and with the addition of 4-methoxyanilinium chloride (to generate 4-methoxyaniline as internal standard) quantification in various pharmaceuticals is easily achieved [101]. In this application, a heated injection port (at 450°C) was found satisfactory.

Other involatile drugs which have been assayed by pyrolysis include barbiturates [102] and penicillins [103]. Although good linearity over a vast range of amounts (10 ng to 100 µg) was reported,
the log-log plots were necessary for linearization.

Other quantitative analyses using pyrolysis include the estimation of alkaloids [104,105], surfactants [106,107], hexane in vegetable oils (but at 150°C) [108], mercury in fish [109], pollution in
waterways [110-113], organic oxygen [114,115], and atomic ratios by
pyrolytic sulfurization [116,117]. A large number of GC analyses
also include incidental thermal or pyrolytic conversions [118-122].

The quantitative analysis of synthetic polymers is an important
and informative application which is almost as old as the technique
itself [123]. Examples include the determination of copolymer composition [124], the elucidation of microstructure [125], including
stereochemistry [126], and the measurement of the proportion of
polymers in either a polymer mixture or within a matrix [127], a
technique of wide applicability in the analysis of rubbers. As with
simpler molecules, a mechanistic appreciation of the pyrolysis processes aids the choice of the analytical peaks and is essential to
many of the more interesting applications of this technique.

The pyrolysis of styrene-methacrylate copolymers is characterized by the production of the monomeric species (styrene and methacrylate ester). The yield of each depends upon the composition of
the polymer, and calibration may be achieved by the pyrolysis of
polymers of known composition [124]. The coefficient of variation
for the analysis of unknown samples using this procedure is 1 to 2%
and results were comparable to those obtained by NMR or carbon analysis. Small scale work is essential (20 µg used) and reproducible
product distribution depends upon adequate temperature control
(Table 3.11). Similar techniques may be used for rubbers [128].
Pyrograms from natural (cis-polyisoprene) rubber (NR), styrene-
butadiene rubber (SBR), butadiene rubber (BR), and a cured blend of
these components are shown in Fig. 3.9. Isoprene (NR) and styrene
(SBR) allow the direct quantification of natural and styrene-butadiene

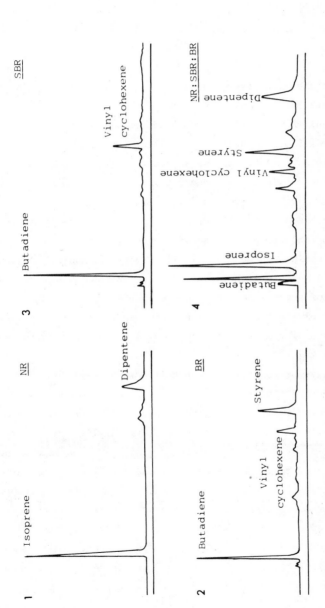

FIGURE 3.9 Pyrograms from cured rubbers: 1, natural rubber (NR); 2, styrene-butadiene rubber (23:77) (SBR); 3, butadiene rubber (BR); 4, a cured blend of NR:SBR:BR. (From Ref. 128.)

TABLE 3.11 Pyrolysis of Styrene-Methyl and Styrene-Butyl Copolymers. Effect of Pyrolysis Temperature

Pyrolysis Temperature (^{o}C)	350	450	550	650	750	950
Methyl Methacrylate (%)	45.7	44.1	43.1	42.8	42.5	31.4
Butyl Methacrylate (%)	30.5	24.7	23.7	20.5	11.1	1.7

Source: Ref. 124.

rubbers. Butadiene, however, is not a unique fragment, but is produced from both SBR and butadiene rubber. Estimation of this third
component requires correction for SBR content. Contamination by
other elastomers, e.g., butyl rubber yields isobutylene, is also
readily observed. An alternative procedure for the analysis of rubbers containing ethylene-propylene polymers has also been described.
The monomeric units here are not unique, but are produced by pyrolysis of most rubbers. Analysis of the more volatile fragments (17
components up to isoprene were separated) however, led to satisfactory quantification of the blend [127]. Additives such as carbon
black do not interfere with the validity of the assay. Copolymers
and mixtures of homopolymers may be distinguished by the different
effects of pyrolysis temperature, and because the monomer recovery
from the homopolymer mixture is essentially independent of polymer
composition [129] (Fig. 3.10).

Such assays of composition generally use high-yield monomeric
fragments. Further information on polymer microstructure may be obtained if larger fragments are considered [130]. Table 3.12 compares the pyrolysis products from polymethyl acrylate, polystyrene,
and a styrene-methyl acrylate copolymer [131]. Hybrid dimers and
trimers are observed in the pyrogram of the copolymer, which are absent in pyrograms from homopolymer mixtures. Such products, therefore, are characteristic of the sequence of monomers in the polymer,
and may be used to probe this microstructure. In this example, the
relative dimer yields from both polymeric units were similar, and
good correlations between calculated and predicted diad concentrations were obtained. Deviations result if the yield of a dimer or

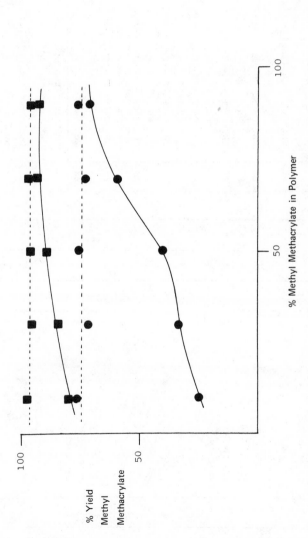

FIGURE 3.10 Pyrolysis of styrene-methacrylate copolymer and homopolymer mixture: recovery of methyl methacrylate (——— Copolymer; ----- Homopolymer mixture; ●, 510°C; ■, 610°C). (From Ref. 110.)

TABLE 3.12 Pyrolysis Products from Polymethyl Acrylate (PMA), Polystyrene (PSt), and a Styrene-Methyl Acrylate Copolymer (CoP, 55.6:44.4)

No	PRODUCTS IDENTIFIED	PMA	PSt	CoP
1	MeOH	*		*
2	$CH_2=CH-COOMe$	*		*
3	$CH_2=C(Me)-COOMe$	*		*
4	Ph-H		*	*
5	Ph-Me		*	*
6	Ph-Et		*	*
7	$Ph-CH=CH_2$		*	*
8	$Ph-C(Me)=CH_2$		*	*
9	Ph-CH=CH-Me		*	
10	$MeOOC-(CH_2)_3-COOMe$	*		*
11	(MeO-substituted 2-pyranone ring structure)	*		*
12	$CH_2=C(COOMe)-CH_2-COOMe$	*		*
13	$CH_2=C(COOMe)-CH_2-C(COOMe)=CH_2$	*		*
14	$Ph-(CH_2)_3-COOMe$			*
15	$Ph-(CH_2)_2-C(COOMe)=CH_2$			*
16	$Ph-CH=C(COOMe)-Me$			*
17	$Ph-CH=CH-CH_2-Ph$		*	*
18	$Ph-CH=CH-CH(Ph)-Me$		*	*
19	$CH_2=C(COOMe)-CH_2-CH(COOMe)-(CH_2)_2-COOMe$	*		*
20	$CH_2=C(Ph)-CH_2-CH(Ph)-Me$		*	*
21	$CH_2=C(Ph)-CH_2-CH(COOMe)-(CH_2)_2-COOMe$			*
22	$CH_2=C(COOMe)-CH_2-CH(Ph)-(CH_2)_2-COOMe$			*
23	$CH_2=C(Ph)-CH_2-CH(COOMe)-(CH_2)_2-Ph$			*
24	$CH_2=C(Ph)-CH_2-CH(Ph)-(CH_2)_2-COOMe$			*
25	$CH_2=C(Ph)-CH_2-CH(Ph)-(CH_2)_2-Ph$		*	*
26	Tetramer	*		

* Indicates the presence of a component in the chromatogram.

No Indicates the order of elution of the components.

Source: Ref. 112.

trimer fragment is influenced by its position in the polymer chain.
These discrepancies may be attributed to variations in the monomer
units adjacent to the diad unit giving rise to the dimer fragment
[132]. Calculations to account for these "boundary effects" allow
sequence distributions to be firmly proposed [133,134]. Even when
the copolymer, by analogy with the constituent homopolymer, might be
expected to yield only monomers, e.g., as polymethyl methacrylate or
poly-α-methylstyrene do, boundary effects direct fragmentation and
allow useful information to be obtained [133]. Such data may be
used to characterize a polymer by calculation of a run number [133,
135]. This parameter indicates the average number of sequences per
100 monomer units in a copolymer chain, and has a value of unity for
a homopolymer and 100 for a completely alternating copolymer [136].

Intramolecular reactions, too, may provide evidence concerning
microstructures [125]. On pyrolysis of vinyl chloride-methyl methacrylate polymers, lactonization may occur between adjacent residues:

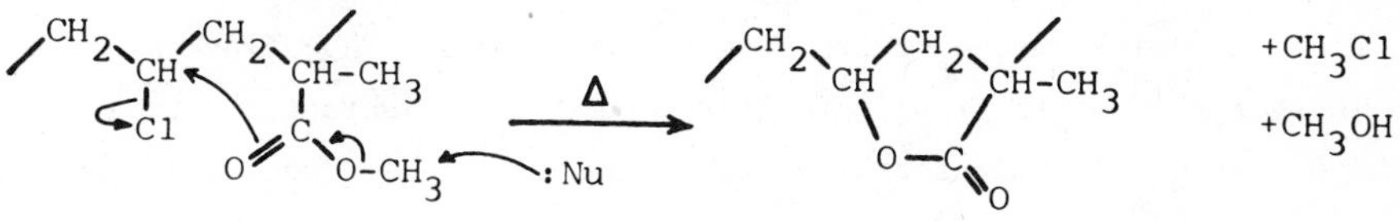

Thus an alternating polymer gives negligible amounts of methyl
methacrylate, but a high yield of methyl chloride. Elimination reactions may also be used with effect to determine the distribution
of subunits in chloropolymers [137,138]. Here, triad units yield
substituted benzene derivatives which may be used to indicate microstructure. Chloropolypropylene, for example, eliminates HCl and
yields mesitylene:

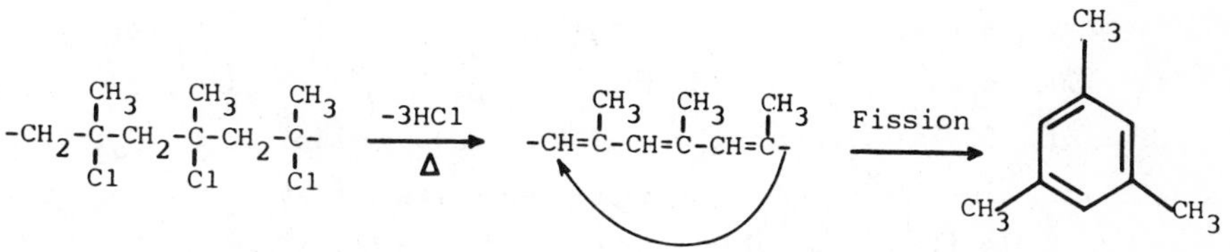

whereas the presence of head-to-head or tail-to-tail units causes
1,2,4-trimethylbenzene to result:

$$-CH_2-\underset{Cl}{\overset{CH_3}{C}}-CH_2-\underset{Cl}{\overset{CH_3}{C}}-\underset{Cl}{\overset{CH_3}{C}}-CH_2- \xrightarrow[\Delta]{-3HCl} -CH=\overset{CH_3}{C}-CH=\overset{CH_3}{C}-\overset{CH_3}{C}=CH- \xrightarrow{Fission}$$

The relative proportions of these products enables the estimation of
inverse monomer to be made [137].

The high-resolution capabilities of capillary column Py-GC allows a further imaginative use of analytical pyrolysis: the determination of tacticity and block length in polypropylene [126]. Pyrolysis and hydrogenation yields three compounds which possess two
asymmetric carbon atoms (see Fig. 3.11): 4,6-dimethylnonane (i),
2,4,6-trimethylnonane (ii), and 2,4,6,8-tetramethylnonane (iii).
Each may exist in one of two geometrical isomers, depending on
whether the methyl groups are on the same side (isotactic) or on opposite sides (syndiotactic) of the backbone. These isomers are resolved by capillary column chromatography, and hence three independent assessments of the tacticity may be made. The block length may
also be estimated by the analysis of products with three asymmetric
carbon atoms. Although assuming that all C-C bonds have the same
lability, this method illustrates how a knowledge of polymer pyrolysis pathways may be used to advantage.

The many applications of quantitative analytical pyrolysis in
polymer analysis also includes the determination of monomers in
fluorinated ethylene-propylene copolymers [139], acrylate copolymers
[140-144], styrene copolymers [145,146], rubbers [147,148], nylon
[149,150], phenol-formaldehyde resins [151,152], ethylene oxide
polymers [153,154], polyurethanes [155], and plasticizers [149,156].
The carbon loading of chromatographic supports may also be assessed
by pyrolysis [157] and molecular weight determinations of

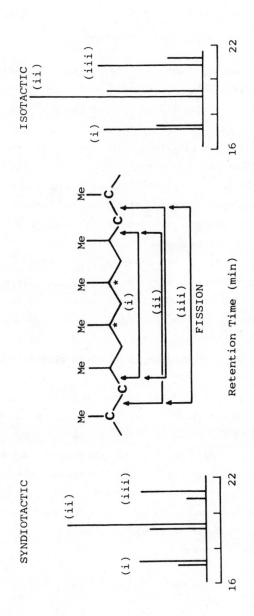

FIGURE 3.11 Pyrolysis-hydrogenation of syndiotactic and isotactic polypropylene products with two asymmetric centers. (From Ref. 126.)

polycarbonates by end-group analysis have also been undertaken [158].
Here, the intensity of *tert*-butylphenol (a chain-terminating unit)
is used for quantification:

$$-C_6H_4-O-CO-O-C_6H_4-C(Me_2)-C_6H_4-O-CO-O-C_6H_4-CMe_3 \xrightarrow{\Delta} R-C_6H_4OH$$

$$R = H, Me, Et, iso\text{-}Pr, tert\text{-}Bu, CH_2=\underset{|}{C}\text{-}Me$$

In comparison, applications involving natural polymers are
rare, but data are available on derivatives of starch [159], cellulose [160-162] and porphyrins [163]. Recent progress has shown that
pyrolysis may have a further role to play (in the structure determination of proteins). In particular, the amino acid tryptophan is
difficult to determine in such molecules due to its relative ease of
degradation. Pyrolysis of tryptophan yields indole and skatole (3-
methylindole). Although fragments which have the same retention
time as indole may be produced by other amino acids (e.g., phenylalanine), skatole is a unique fragment and is also produced by tryptophan residues in proteins. Measurement of the intensity of this
peak, produced on pyrolysis of various enzymes, and interpolation
onto a calibration line obtained from the pyrolysis of known weights
of tryptophan enabled useful estimations of the tryptophan component
to be made [164]. Quantitative analysis to assess oil-bearing shales
has also been reported and promises to be of further importance in
the future [47,165,166].

Advantages of analytical pyrolysis include simple sample preparation and sensitive and specific detection. The speed of the
analysis, however, is dependent upon the chromatographic separation.
With polymers such as some acrylates, the GC run may be complete
within minutes. In other cases, e.g., polyethylene, which yields a
long series of homologues, an analysis may take 3h. The pyrolysis
of microorganisms takes perhaps 30 to 60 min for each analysis. To
achieve the optimum turnover of results for analyses which require
long GC runs automated systems have been developed.

3.2.4. Automation

An automated Py-GC procedure requires provision for (1) remote-controlled sample loading, (2) initiation of pyrolysis, (3) control of the GC operating conditions, including recycling of the temperature program, and (4) the recording of pyrograms in digital or analogue format. In addition, fail-safe devices to operate in the event of a failure must be present. One such system has been described by Meuzelaar and co-workers [34]. This system uses a specially designed Curie-point pyrolyzer (Fig. 3.12) which allows direct

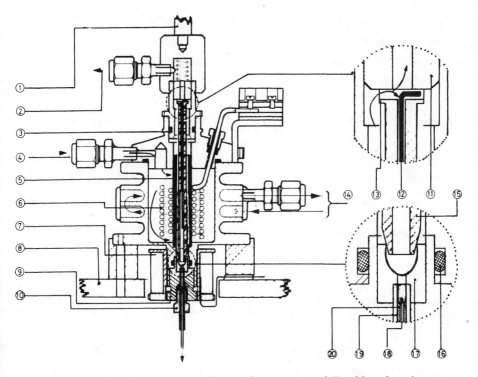

FIGURE 3.12 Curie-point pyrolyzer for automated Py-GC: 1, piston-lift; 2, carrier-gas outlet; 3, O-ring; 4, carrier gas inlet; 5, guide tube; 6, induction coil; 7, heater; 8, oven; 9, removable base; 10, LDV coupling; 11, piston; 12, pyrolysis wire; 13, flange; 14, heating or cooling flow; 15, reaction tube; 16, O-ring; 17, PTFE seal; 18, quartz wool plug; 19, PTFE shrink tubing; 20, capillary column. (From Ref. 34, reproduced from the *Journal of Chromatographic Science,* by permission of Preston Publications, Inc.)

splitless coupling to capillary columns and accurate positioning of
the sample within the induction coil. Provision for cooling of the
pyrolyzer is also made. This was found to enhance baseline stability
and quantitative reproducibility, without detectable loss of fragments. Samples are preloaded onto Curie-point wires which are then
inserted into flanged reaction-tubes, which act as low-dead-volume
removable liners in the pyrolysis chamber. The reaction tube containing the wire and sample is loaded into the pyrolyzer from the
top, through a pneumatically driven removable cover. This cover also
contains a carrier gas vent to allow purging of the reactor with carrier gas to remove volatile contaminants and atmospheric gases, and
to select a low flow rate (4 ml min^{-1}) through the capillary column.
The reaction tube is interfaced to the GC column by means of a
glass-filled Teflon seal and adequate fit is obtained simply by
dropping the tube into the pyrolyzer and closing the reactor top.
Automatic sample loading of the pyrolyzer is achieved by means of a
pneumatic device which selects a sample tube from a turntable, ro-
tates to position this over the open pyrolyzer and finally drops the
reaction tube into position (Fig. 3.13). A pressure gauge monitors
gas flow into the pyrolyzer and initiates program abortion if preset
limits are exceeded. The initiation of analysis-pyrolysis, GC cycle,
data collection via data logger and analogue and digital recorders is
controlled by an external program controller. In this application a
temperature programmer with a recycling capability was used. Cir-
cuits to provide control of extra external functions were added to
synchronize all components (Fig. 3.14). A typical cycle involves

1. Start: Pyrolysis (0.1 to 4 s) initiated and data logger
   and recorders on.

2. Programmed GC run.

3. Final column pressure printed, data logger, and recorder
   off.

4. Temperature increase for column reconditioning.

5. Cooling.

6. Column pressure monitor disengaged (2 min), sample changer
   activated (10 s), and initial column temperature reselected.

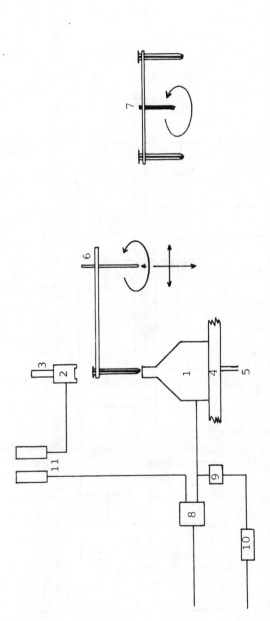

FIGURE 3.13 Arrangement of units in an automated Py-GC system: 1, pyrolyzer; 2, pyrolyzer top; 3, pneumatic piston; 4, oven; 5, column; 6, sample loader; 7, rotary sample holder; 8, carrier gas inlet valve; 9, pressure transducer; 10 LED display; 11, valves and flowmeters. (Adapted from Ref. 34, from the *Journal of Chromatographic Science*, by permission of Preston Publications, Inc.)

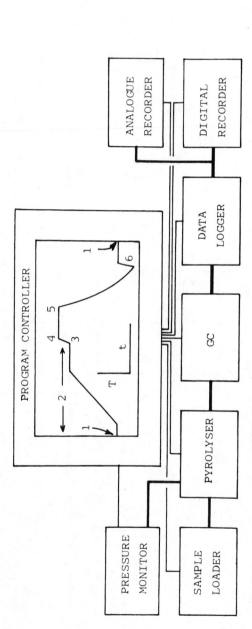

FIGURE 3.14 Control of Automated Py-GC system. (Adapted from Ref. 34, from the *Journal of Chromatographic Science*, by permission of Preston Publications, Inc.)

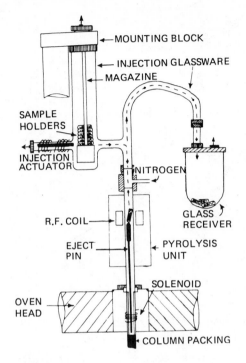

FIGURE 3.15   Automated Py-GC system.   (From Ref. 167.)

This system also has manual override and allows unattended analysis of up to 24 samples. Peak height variations were less than 5% and pressure variations of up to ±1.5% per day were noted.

An alternative procedure, based upon commercially available units has been described by Coulter and Thompson, and is used routinely in the analysis of tyre blends [167]. The basis of this system is a GC automatic solids injector [168], in which samples are held in glass holders and are sequentially ejected from a magazine into the pyrolyzer by the action of a solenoid. On pyrolysis of the sample, a further solenoid ejects the spent holder. Pyrolysis samples are prepared by enclosing polymer strips within creased iron foils (14 × 4 × 0.1 mm), which are then held in glass tubes inserted into the magazine. Such a system may be used for the continuous analysis of up to 35 samples (Fig. 3.15). A more effective system uses the control

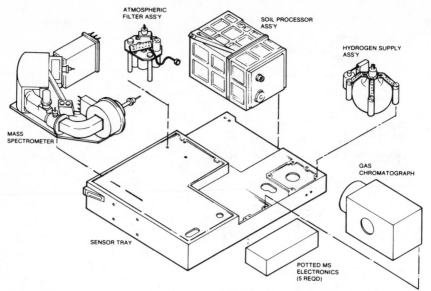

FIGURE 3.16   Components of Viking Py-GC-MS system.   (From Ref. 169.)

afforded by an online computer. This addition allows the system to
be monitored for injection, ejection, and other failures. The computer may also undertake the calculations required for the analysis
of the various samples.

One other example of automated Py-GC is worthy of note: the
Viking Py-GC-MS system used to explore the martian atmosphere and
surface [169]. The overall dimensions of this unit were 27.5 × 33
× 25 cm with a mass of approximately 20 kg [69]. The heart of this
masterpiece of miniaturization was a double-focusing mass spectrometer with a mass range of 12 to 215 amu. The component arrangement
is shown in Fig. 3.16, and their location in the Viking Lander in
Fig. 3.17. The unit was capable of two modes of operation: the determination of atmospheric gases by direct MS analysis (after removal
of CO, CO_2 and H_2O) or the GC-MS analysis of soil samples (Fig. 3.18).
Here, soil samples retrieved by the surface sampler are transmitted
to the soil processor, which grinds and measures the sample and
passes a cut (<300 μm) to the soil loader. This subassembly loads
the ovens of the pyrolyzer (three are available), arranges the gas
flow valves appropriately and initiates oven heating to a preselected

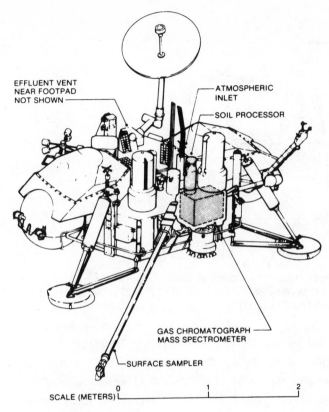

EFFLUENT VENT
NEAR FOOTPAD
NOT SHOWN

ATMOSPHERIC
INLET

SOIL PROCESSOR

GAS CHROMATOGRAPH
MASS SPECTROMETER

SURFACE SAMPLER

SCALE (METERS) 0 1 2

FIGURE 3.17   The Viking Lander.   (From Ref. 169.)

temperature (50, 200, 350, and 500°C). This causes evaporation of
volatile components and pyrolysis of nonvolatile organic residues.
The ovens were flushed with $^{13}CO_2$ because hydrogen, used in development studies and as the GC carrier gas, caused reduction of some
components. The isotopically labelled material was chosen to distinguish it from CO_2 produced by pyrolysis, and to detect other side
reactions, such as insertion during pyrolysis. The pyrolysis products were swept into the column [2,6-diphenyl-*p*-phenylene oxide
(Tenax-GC 60 to 80 mesh)] coated with polymetaphenoxylene. This system was specifically developed for this application [21], after earlier systems had revealed limitations in the separation of very polar
substances [69]. Analysis was achieved with a program consisting of

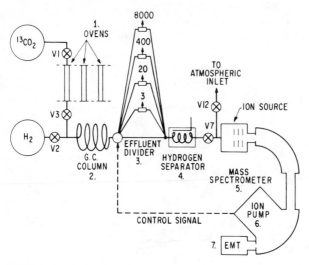

FIGURE 3.18 Plan of Viking Py-GC-MS system. (From Ref. 68, K.
Biemann et al., *J. Geophys. Res.* *82*:4641-4658, 1977, copyrighted by
the American Geophysical Union.)

an initial isothermal period at 50°C (10 min), followed by an increase of 8.3°C min^{-1} up to 200°C. The final isothermal period was
then maintained for 18, 36, or 54 min. The GC eluate is passed
through an effluent divider (held at 218°C) to control sample flow
into the mass spectrometer. This is controlled by the MS ion pump
to prevent overload as, for example, would occur with large amounts
of water slowly eluting from the system. The hydrogen separator is
an electrochemical device, and consists of a silver-palladium tube,
operating as the anode, surrounded by a NaOH-H$_2$O eutectic at 220°C
[170]. The cathode is a larger tube in contact with the electrolyte,
but open to the atmosphere. Hydrogen selectively permeates the tube
wall into the electrolyte, while the sample continues into the MS
source. Hydrogen is transported through the electrolyte, permeates
the cathode, and is expelled from the system. Such a system removes
more than 99.99995% of the carrier gas and enables source pressures
of less than 10^{-8} torr to be achieved.

     Mass spectra (500 scans) were determined continuously at 10-24
s intervals throughout the analysis. Each scan was recorded as 3840

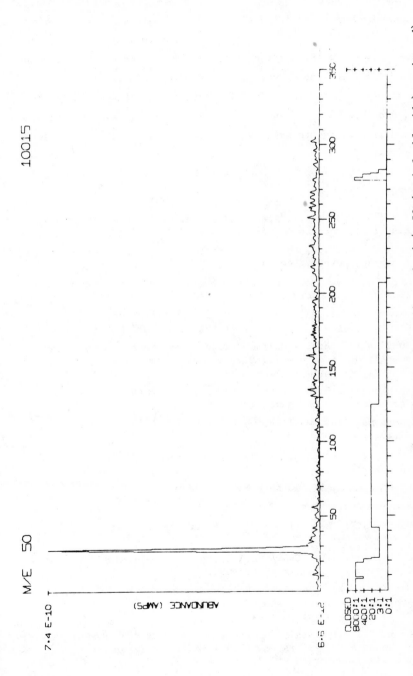

FIGURE 3.19 Mass chromatogram from Viking Lander soil sample (m/z 50 (methyl chloride) monitored). The lower trace plots the status of the effluent divider. (From Ref. 68, K. Biemann et al., *J. Geo-phys. Res. 82*:4641-4658, 1977, copyrighted by the American Geophysical Union.)

intensity cuts, each being encoded to 9 bits. Thus, about 17 M bits
per analysis were transmitted to earth at 4K bits s^{-1} via the orbiter
(16K bits per second), over a period of about 70 min. Data processing for mass scale calibration, noise rejection and mass spectra and
chromatogram regeneration was performed on earth. Figure 3.19 records a mass chromatogram (m/z 50) from the first Viking Lander analysis, and shows the presence of methyl chloride (a system contaminant) and illustrates the status of the effluent divider throughout
the analysis.

REFERENCES

1. Tranchant, J. (Ed.), Practical Manual of Gas Chromatography,
    Elsevier, Amsterdam, 1969.

2. Pattison, J.B., A Programmed Introduction to Gas-Liquid Chroma-
    tography, Heyden, London, 1973.

3. Mitruka, B.M. (Ed.), Gas Chromatographic Applications in Micro-
    biology and Medicine, John Wiley and Sons, New York, 1975.

4. Berezkin, V.G., Alishoyev, V.R. and Nemirovskaya, I.B., Gas
    Chromatography of Polymers, Elsevier, Amsterdam, 1977.

5. Gudzinowicz, B.J., Gudzinowicz, M.J. and Martin, H.F., Funda-
    mentals of Integrated GC-MS. I. Gas Chromatography, Marcel
    Dekker, New York, 1976.

6. Baiulescu, G.E. and Ilie, V.A., Stationary Phases in Gas
    Chromatography, Pergamon Press, Oxford, 1975.

7. Leathard, D.A., Qualitative Analysis by GC, Adv. Chromatogr.,
    13(1975)265-303.

8. Novak, J., Quantitative Analysis by GC, Adv. Chromatogr., 11
    (1974)1-71.

9. Chrompack UK, 114 Norbury Hill, London.

10. Supelco Inc., Supelco Park, Bellefonte, Pennsylvania.

11. Haken, J.K., Retention Indices in GC, Adv. Chromatogr., 14
    (1976)367-407.

12. McReynolds, W.O., Characterisation of Some Liquid Phases, J.
    Chromatogr. Sci., 8(1970)685-691.

13. May, R.W., Pearson, E.F., Porter, J. and Scothern, M.D., A
    Reproducible Py-GC System for the Analysis of Paints and Plas-
    tics, Analyst, 98(1973)364-371.

14. Levy, E.J., The Analysis of Automotive Paints by Py-GC, in C.E.
    R. Jones and C.A. Cramers (Eds.), Analytical Pyrolysis,
    Elsevier, Amsterdam, 1977, pp. 319-335.

15. May, R.W., Pearson, E.F. and Scothern, D., *Pyrolysis Gas Chromatography*, Chemical Society, London, 1977.

16. Sellier, N., Jones, C.E.R. and Guiochon, G., Py-GC as a means of Determining the Quality of Porous Polymers of Styrene Cross-linked with Divinylbenzene, in C.E.R. Jones and C.A. Cramers (Eds.), *Analytical Pyrolysis*, Elsevier, Amsterdam, 1977, pp. 309-318.

17. Hall, R.C. and Bennett, G.W., Py-GC of Several Cockroach Species, *J. Chromatogr. Sci., 11*(1973)439-443.

18. Freedman, A.N., Gaseous Degradation Products from the Pyrolysis of Insulating Materials Used in Large Electricity Generators, *J. Chromatogr., 157*(1978)85-96.

19. Cirendini, S., Vermont, J., Gressin, J.C. and Guillemin, C.L., Rapid GC Analysis on Spherosil, *J. Chromatogr., 84*(1973)21-36.

20. Lloyd, J.B.F., Hadley, K. and Roberts, B.R.G., Py-GC over Hydrogenated Graphitised Carbon Black, *J. Chromatogr., 101* (1974)417-423.

21. Novotny, M., Hayes, J.M., Bruner, F. and Simmonds, P.G., GC Column for the Viking 1975 Molecular Analysis Experiment, *Science, 189*(1975)215-216.

22. Scott, R.P.W., Determination of the Optimum Conditions to Effect a Separation by GC, *Adv. Chromatogr., 9*(1970)193-214.

23. Walker, J.Q., Py-GC Correlation Trials of the American Society for Testing and Materials, *J. Chromatogr. Sci., 15*(1977)267-274.

24. Schmid, J.P., Schmid, P.P. and Simon, W., Application of Curie-Point Py-GC Using High Resolution Glass Open-Tubular Columns, in C.E.R. Jones and C.A. Cramers (Eds.), *Analytical Pyrolysis*, Elsevier, Amsterdam, 1977, pp. 99-105.

25. Quinn, P.A., Development of High-Resolution Py-GC for the Identification of Micro-Organisms, *J. Chromatogr. Sci., 12*(1974) 796-806.

26. Ettre, L.S., Open Tubular Columns in Gas Chromatography, Plenum Press, New York, 1965.

27. Ettre, L.S. and Purcell, J.E., Porous-Layer Open-Tubular Columns. Theory, Practice and Applications, *Adv. Chromatogr., 10*(1974)1-97.

28. Goretti, G., Liberti, A. and Pili, G., Evaluation of Graphitised Glass Capillary Columns, *J. High Resoln. Chromatogr., 1*(1978) 143-148.

29. Grob, K. and Grob, G., A New Generally Applicable Procedure for the Preparation of Glass Capillary Columns, *J. Chromatogr. 125* (1976)471-485.

30. Grob, K., Grob, G. and Grob, K. Jr., The Barium Carbonate Procedure for the Preparation of Glass Capillary Columns. Further Information and Development, *Chromatographia, 10*(1977)181-187.

31. Grob, K. Jr., Grob, G. and Grob, K., Preparation of Apolar Glass
    Capillary Columns by the Barium Carbonate Procedure, *J. High
    Resoln. Chromatogr.*, *1*(1978)149-155.

32. McKeag, R.G. and Hougen, F.W., Dynamic Coating of Glass Capil-
    laries with Polar Phases and Silanox, *J. Chromatogr.* *136*(1977)
    308-310.

33. Grob, K., Grob, G. and Grob, K. Jr., Deactivation of Glass Cap-
    illary Columns by Silylation, *J. High Resoln. Chromatogr.*, *2*
    (1979)31-35.

34. Meuzelaar, H.L.C., Ficke, H.G., den Harink, H.C., Fully Auto-
    mated Curie-Point Py-GC, *J. Chromatogr. Sci.*, *13*(1975)12-17.

35. Schmid, J.P., Schmid, P.P. and Simon, W., Instrumentation for
    Curie-Point Py-GC Using High-Resolution Glass Open-Tubular
    Columns, *Chromatographia,* *9*(1970)597-600.

36. de Leeuw, J.W., Maters, W.L., Meent, v.d. D. and Boon, J.J.,
    Solvent-Free and Splitless Injection Method for Open-Tubular
    Columns, *Analyt. Chem.*, *49*(1977)1881-1884.

37. Schomberg, G., Dielmann, R., Husmann, H. and Weeke, F., GC
    Analysis with Glass Capillary Columns, *J. Chromatogr.*, *122*
    (1976)55-72.

38. Sugimura, Y. and Tsuge, S., Fundamental Splitting Conditions
    for Pyrogram Measurements with Glass Capillary GC, *Analyt.
    Chem.*, *50*(1978)1968-1972.

39. Mitchell, A. and Needleman, M., Technique for the Prevention of
    Column Contamination in Py-GC, *Analyt. Chem.*, *50*(1978)668.

40. Kolb, B. and Kaiser, K.H., Hydrogenation as a Technique of
    Identification in Py-GC, *J. Gas Chromatogr.*, *2*(1964)233-234.

41. Ettre, L.S., Mazor, L. and Takacs, J., Pressure (Flow) Program-
    ming in GC, *Adv. Chromatogr.*, *8*(1969)271-325.

42. Harris, W.E. and Habgood, H.W., *Programmed Temperature Gas
    Chromatography,* John Wiley and Sons Inc., New York, 1966.

43. Meuzelaar, H.L.C. and in't Veld, R.A., A Technique for Curie-
    Point Py-GC of Complex Biological Samples, *J. Chromatogr. Sci.*,
    10(1972)213-216.

44. Kaiser, R.E., Systematically Wrong Evaluation of Columns, Pack-
    ings, Instrument Data in Chromatography, *J. High Resoln.
    Chromatogr.*, *2*(1979)91-92.

45. Said, A.S., Variance of Non-Gaussian and Asymmetric Peaks in
    Chromatography, *J. High Resoln. Chromatogr.*, *2*(1979)93-94.

46. David, D.J., *Gas Chromatographic Detectors,* John Wiley and Sons,
    New York, 1974.

47. Hanson, R.L. and Vanderborgh, N.E., Characterisation of Coals
    Using Laser Py-GC, in C. Karr, Jr. (Ed.), *Analytical Methods*

for Coal and Coal Products, Volume III, Academic Press, New York, 1979, pp. 73-103.

48. Myers, A. and Smith, R.N.L., Application of Py-GC to Biological Materials, *Chromatographia, 5*(1972)521-524.

49. Levy, E.L. and Paul, D.G., The Application of Controlled Partial Gas-Phase Thermolytic Dissociation to the Identification of GC Effluents, *J. Gas Chromatogr., 5*(1967)136-145.

50. Walker, J.Q. and Wolf, C.J., Complete Identification of Chromatographic Effluents Using Interrupted Elution and Py-GC, *Analyt. Chem., 40*(1968)711-714.

51. Jones, C.E.R. and Reynolds, G.E.J., The Contribution of Integrated Py-GC to the Elucidation of Polymer Microstructure, *J. Gas Chromatogr., 5*(1967)25-29.

52. Uden, P.C., Henderson, D.E. and Lloyd, R.J., Comprehensive Interfaced Py-GC Peak Identification System, *J. Chromatogr., 126* (1976)225-237.

53. Kiran, E. and Gillham, J.K., Pyrolysis-Molecular Weight Chromatography-Vapour Phase IR Spectroscopy. An On-line System for the Analysis of Polymers, Office of Naval Research (NR 356-504) Technical Report No. 16 (1979); to appear in *Development in Polymer Degradation-2,* N. Grassie (Ed.), Applied Science, London.

54. Gudzinowicz, B.J., Gudzinowicz, M.J. and Martin, H.F., *Fundamentals of Integrated GC-MS. II. Mass Spectrometry, III. The Integrated GC-MS Analytical System,* Marcel Dekker, New York, 1977.

55. McFadden, W.H., *Techniques of Combined GC-MS: Applications in Organic Analysis,* John Wiley and Sons, New York, 1973.

56. Gross, M.L. (Ed.), High-Performance MS. Chemical Applications, in *American Chemical Society Symposium Series No. 70,* American Chemical Society, Washington, D.C., 1978.

57. Budzikiewicz, H., Djerassi, C. and Williams, D.H., Mass Spectrometry of Organic Compounds, Holden-Day, London, 1967.

58. Waller, G.R. (Ed.), Biochemical Applications of Mass Spectrometry, John Wiley and Sons, New York, 1972.

59. *Eight Peak Index of Mass Spectra,* 2nd ed., Mass Spectrometry Data Center, HMSO, London, 1974.

60. Stenhagen, E., Abrahamsson, S. and McLafferty, F.W., *Registry of Mass Spectral Data,* Wiley-Interscience, New York, 1974.

61. Kwok, K-S., Venkataraghavan, R. and McLafferty, F.W., Computer-Aided Interpretation of MS. III. A Self-Training Interpretive and Retrieval System, *J. Amer. Chem. Soc., 95*(1973)4185-4194.

62. Dayringer, H.E. and McLafferty, F.W., Computer-Aided Interpretation of MS. STIRS Prediction of Rings-Plus-Double Bonds Values, *Org. Mass Spectrom., 12*(1977)53-54.

63. Smith, D.H. (Ed.), *Computer Assisted Structure Elucidation*, American Chemical Society, Washington, D.C., 1977.

64. Hertz, H.S., Hites, R.A. and Biemann, K., Identification of MS by Computer-Searching a File of Known Spectra, *Analyt. Chem.*, *43*(1971)681-691.

65. Chapman, J.R., *Computers in Mass Spectrometry*, Academic Press, London, 1978.

66. Biller, J.E. and Biemann, K., Reconstructed MS, a Novel Approach to the Utilisation of GC-MS Data, *Analyt. Lett.*, *7*(1974)515-528.

67. Harvey, D.J., Martin, B.R. and Paton, W.D.M., Identification of di- and tri-substituted hydroxy and ketone Metabolites of Δ'-Tetrahydrocannabinol in Mouse Liver, *J. Pharm. Pharmacol.*, *29* (1977)482-486.

68. Biemann, K., Oro, J., Toulmin III, P., Orgel, L.E., Nier, A.O., Anderson, D.M., Simmonds, P.G., Flory, D., Diaz, A.V., Rushneck, D.R., Biller, J.E. and Lafleur, A.L., The Search for Organic Substances and Inorganic Volatile Compounds in the Surface of Mars, *J. Geophys. Res.*, *82*(1977)4641-4658.

69. Biemann, K., Test Results on the Viking GC-MS Experiment, *Origins Life*, *5*(1974)417-430.

70. Palmer, L. and Holmstedt, B., Mass Fragmentography--the Use of MS as a Selective and Sensitive Detector in GC, *Sci. Tools*, *22* (1975)25-31, 38-39.

71. Millard, B.J., *Quantitative MS*, Heyden, London, 1978.

72. Reinhold, V.N. and Costello, C.E., Mass Spectral Approaches to the Study of Drug Metabolism, in C. Merritt Jr. and C.N. McEwen (Eds.), *Mass Spectrometry*, Part B, Marcel Dekker, New York, 1979.

73. Polak, R.L. and Molenaar, P.C., A Method for the Determination of Acetylcholine by Slow Pyrolysis combined with Mass Fragmentography on a Packed Capillary Column, *J. Neurochem.*, *32*(1979) 407-412.

74. Ohya, K. and Sano, M., Analysis of Drugs by Pyrolysis. I. Selected Ion Monitoring Combined with a Pyrolysis Method for the Determination of Carpronium Chloride in Biological Samples, *Biomed. Mass Spectrom.*, *4*(1977)241-246.

75. Perry, S.G., Pyrolysis Sub-Group of the GC Discussion Group, *J. Chromatogr. Sci.*, *7*(1969)193-194.

76. Coupe, N.B., Jones, C.E.R. and Perry, S.G., Precision of Py-GC of Polymers. A Progress Report, *J. Chromatogr.*, *47*(1970)291-296.

77. Jones, C.E.R., Perry, S.G. and Coupe, N.B., Precision of Py-GC of Polymers. Part II. The Standardisation of Fingerprinting. Preliminary Assessment of Second Correlation Trial, in B. Stock

(Ed.), *Gas Chromatography*, Institute of Petroleum, London, 1971, pp. 399–406.

78. Coupe, N.B., Jones, C.E.R. and Stockwell, P.B., Precision of Py–GC of Polymers. Part III. The Standardisation of Finger-printing--Assessment of the Third Correlation Trial, *Chromatographia, 6*(1973)483–488.

79. Gough, T.A. and Jones, C.E.R., Precision of Py–GC of Polymers. Part IV. Assessment of the Results of the Fourth Correlation Trial Organised by the Pyrolysis Sub-Group of the Chromatography Discussion Group (London), *Chromatographia, 8*(1975)696–698.

80. Stewart, Jr., W.D., Py–GC Techniques for the Analysis of Automobile Finishes, *J. Ass. Offic. Analyt. Chem., 59*(1976)35–41.

81. Walker, J.Q., Fourth Correlation Trial of the American Society for Testing and Materials, personal communication, 1978, from McDonnell-Douglas, St. Louis, Missouri.

82. Levy, E.J., Concept of a Model Molecular Thermometer, *Conference on Analytical Chemistry and Spectroscopy,* Pittsburgh, Pennsylvania, 1977.

83. Vorhees, K.J., Hileman, F.D. and Einhorn, I.N., Generation of Retention Index Standards by Pyrolysis of Hydrocarbons, *Analyt. Chem., 47*(1975)2385–2389.

84. Medley, E.E., Simmonds, P.G. and Manatt, S.L., A Py–GC–MS Study of the Actinomycete Streptomyces Longisporoflavus, *Biomed. Mass Spectrom., 2*(1975)261–265.

85. Bracewell, J.M. and Robertson, G.W., Pyrolysis Studies on Humus in Freely Drained Scottish Soils, in C.E.R. Jones and C.A. Cramers (Eds.), *Analytical Pyrolysis,* Elsevier, Amsterdam, 1977, pp. 167–178.

86. Needleman, M. and Stuchberry, P., The Identification of Microorganisms by Py–GC, in C.E.R. Jones and C.A. Cramers (Eds.), *Analytical Pyrolysis*, Elsevier, Amsterdam, 1977, pp. 77–88.

87. Nelson, D.F. and Kirk, P.L., Identification of the Pyrolysates of Substituted Barbituric Acids by GC, *Analyt. Chem., 36*(1964) 875–878.

88. Irwin, W.J. and Slack, J.A., Identification of Ibuprofen and Analogues by Py–GC–MS, *Biomed. Mass Spectrom., 5*(1978)654–657.

89. Irwin, W.J. and Slack, J.A., Py–GC–MS Study of Medicinal Sulphonamides, *J. Chromatogr., 139*(1977)364–369.

90. Szilagyi, P.I.A., Schmidt, D.E. and Green, J.P., Microanalytical Determination of Acetylcholine, other Choline Esters and Choline by Py–GC, *Analyt. Chem., 40*(1968)2009–2013.

91. Schmidt, D.E., Szilagyi, P.I.A. and Green, J.P., Identification of Submicrogram Quantities of Onium Compounds by Py–GC, *J. Chromatogr. Sci., 7*(1969)248–249.

92. Polak, R.L. and Molenaar, P.C., Pitfalls in Determination of
     Acetylcholine from Brain by Py-GC-MS, J. Neurochem., 23(1974)
     1295-1297.

93. Szilagyi, P.I.A., Green, J.P., Brown, O.M. and Margolis, S.,
     The Measurement of Nanogram Amounts of Acetylcholine in Tis-
     sues by Py-GC, J. Neurochem., 19(1972)2555-2566.

94. Stavinoha, W.B. and Weintraub, S.T., Estimation of Choline and
     Acetylcholine in Tissue by Py-GC, Analyt. Chem., 46(1974)757-
     760.

95. Miledi, R., Molenaar, P.C. and Polak, R.L., An Analysis of
     Acetylcholine in Frog Muscle by Mass Fragmentography, Proc.
     Royal Soc., B197(1977)285-297.

96. Szilagyi, P.I.A. and Green, J.P., The Quantitative Analysis of
     Choline Esters in Biological Samples by Py-GC-MS, in C.E.R.
     Jones and C.A. Cramers (Eds.), Analytical Pyrolysis, Elsevier,
     Amsterdam, 1977, p. 417.

97. Fidone, S.J., Weintraub, S.T. and Stavinoha, W.B., Acetylcho-
     line Content of Normal and Denervated Cat Carotoid Bodies, J.
     Neurochem., 26(1976)1047-1049.

98. Bianchi, W., Boniforti, L. and diDomenco, A., Analysis of Some
     Quaternary Ammonium Compounds by GC-MS, in A. Frigerio and N.
     Castagnolli (Eds.), Mass Spectrometry in Biochemistry and
     Medicine, Raven Press, New York, 1974, pp. 183-196.

99. Cannard, A.J. and Criddle, W.J., A Rapid Method for the Simul-
     taneous Determination of Paraquat and Diquat in Pond and River
     Waters by Pyrolysis and GC, Analyst, 100(1975)848-853.

100. Marteus, M.A. and Heyndrickx, A., Determination of Paraquat in
     Urine by Py-GC, J. Pharm. Belg., 29(1974)449-454.

101. Choi, P., Criddle, W.J. and Thomas, J., Determination of Cet-
     rimide in Typical Pharmaceutical Preparations using Py-GC,
     Analyst, 104(1979)451-455.

102. Ericsson, I., Py-GC--A Potential Technique for Characterisa-
     tion of Complex Organic Material, Acta Path. Microbiol. Scand.
     Sect. B. (Suppl.), 259(1977)37-42.

103. Roy, T.A. and Szinai, S.S., Py-GLC Identification of Food and
     Drug Ingredients. II. Qualitative and Quantitative Analysis
     of Penicillins and Cephalosporins, J. Chromatogr. Sci., 14
     (1976)580-584.

104. Stanford, F.G., An Improved Method of Py-GC, Analyst, 90(1965)
     266-269.

105. Radecka, C. and Nigam, I.C., Reaction GC. III. Recognition of
     Tropane Structure in Alkaloids, J. Pharm. Sci., 56(1967)1608-
     1611.

106. Liddicoet, T.H. and Smithson, L.H., Analysis of Surfactants
     using Py-GC, J. Amer. Oil Chem. Soc., 42(1965)1097-1102.

107. Nakagawa, T., Miyajima, K. and Uno, T., Py-GC of Long Chain Fatty Acid Salts, *J. Chromatogr. Sci.*, *8*(1970)261-265.

108. Hirayama, S. and Imai, C., Rapid Determination of Residual Hexane in Oils by GC Using Pyrolyser, *J. Amer. Oil Chem. Soc.*, *54*(1977)190-192.

109. Thomas, R.J., Hagstrom, R.A. and Kuchar, E.J., Rapid Pyrolytic Method to Determine Total Mercury in Fish, *Analyt. Chem.*, *44* (1972)512-515.

110. Lysyj, I. and Nelson, K.H., Py-GC Determination of Organics in Aqueous Solution, *Analyt. Chem.*, *40*(1968)1365-1367.

111. Lysyj, I., Nelson, K.H. and Webb, S.R., Determination of Multi-component Organic Compositions in Aqueous Media, *Water Res.*, *4* (1970)157-163.

112. Lysyj, I., Newton, P.R. and Taylor, W.J., Instrumental-Computer System for Analysis of Multicomponent Organic Mixtures, *Analyt. Chem.*, *43*(1971)1277-1281.

113. Lysyj, I., A Pyrographic Instrument for Analysis of Water Pollutants, *Int. Lab.*, July-Aug. (1971)33-35.

114. Ignasiak, B.S., Nandi, B.N. and Montgomery, D.S., Direct Determination of Oxygen in Coal, *Analyt. Chem.*, *41*(1969)1676-1678.

115. Pella, E. and Andreoni, R., Developments in the Determination of Organic Oxygen by Py-GC, *Mikrochim. Acta*, *II*(1976)175-184.

116. Tsuji, K., Fujinaga, K. and Hara, T., Simultaneous Determination of the Atomic Ratio between C, H, O, and N by the Pyrolytic Sulphurisation Method, *Bull. Chem. Soc. Jap.*, *50*(1977) 2292-2298.

117. Hara, T., Fujinaga, K. and Tsuji, K., Pyrolytic Sulphurisation GC. II. Simple Determination of the Correction Factors Using 5 Standard Compounds, *Bull. Chem. Soc. Jap.*, *51*(1978)1110-1113, 2951-2956, 3079-3080.

118. Blau, K. and King, G.S. (Eds.), *Handbook of Derivatives for Chromatography*, Heyden, London, 1977.

119. van den Heuvel, W.J.A. and Zaccei, A.G., GLC in Drug Analysis, *Adv. Chromatogr.*, *14*(1976)199-263.

120. Kralovsky, J. and Matousek, P., GC of Aromatic Acids after Pyrolysis of their Trimethylphenylammonium Salts, *J. Chromatogr.*, *147*(1978)404-407.

121. Wickramasinghe, J.A.F. and Shaw, S.R., GC Behaviour of Buformin HCl, Phenformin HCl and Phenylbiguanide, *J. Chromatogr.*, *71*(1972)265-273.

122. Levitt, M.J., Josimovich, J.B. and Broskin, K.D., Analysis of Prostaglandins by ECD-GC. Thermal Decomposition of Heptafluorobutyrate of Methyl Esters of $F_{1\alpha}$ and $F_{2\beta}$, *Prostaglandins*, *1* (1972)121-131.

123. Lehrle, R.L. and Robb, J.C., Direct Examination of the Degrad-
     ation of High Polymers by GC, *Nature, 183*(1959)1671.

124. Evans, D.L., Weaver, J.L., Mukherji, A.K. and Beatty, C.L.,
     Compositional Determination of Styrene-Methacrylate Copolymers
     by Py-GC, 'H NMR Spectroscopy and Carbon Analysis, *Analyt.
     Chem., 50*(1978)857-860.

125. Tanaka, M., Nishimura, S. and Shono, T., Py-GC of Vinyl
     Chloride-Methyl Methacrylate and Vinyl Chloride-Acrylonitrile
     Co-polymers, *Analyt. Chim. Acta, 74*(1975)119-124.

126. Dimbat, M., Py-Hydrogenation-Capillary Chromatography. A
     Method for the Determination of the Isotacticity and Isotactic
     and Syndiotactic Block Length of Polypropylene, in R. Stock
     (Ed.), *Gas Chromatography 1970*, Institute of Petroleum,
     London, 1971, pp. 237-246.

127. Krishen, A., Quantitative Determination of Natural Rubber,
     Styrene-Butadiene Rubber and Ethylene-Propylene Terpolymer
     in Compounded Cured Stocks by Py-GC, *Analyt. Chem., 44*(1972)
     494-497.

128. Coulter, G.L. and Thompson, W.C., Tyre Polymer Analysis by Py-
     GC, *Column, 3*(1970)6-8.

129. Haken, J.K. and Ho, D.K.M., Quantitative Pyrolysis Studies of
     Styrene, Acrylate Ester Systems and their α-Methyl-substituted
     Homologues, *J. Chromatogr., 126*(1976)239-247.

130. Shibasaki, Y., Boundary Effect on the Thermal Degradation of
     Copolymers, *J. Polym. Sci., Part A-1, 5*(1967)21-34.

131. Tsuge, S., Hiramitsu, S., Horibe, T., Yamaoka, M. and Takeuchi,
     T., Characterisation of Sequence Distribution in Methyl
     Acrylate-Styrene Copolymers to High Conversion by Py-GC,
     Macromolecules, 8(1975)721-725.

132. Okumoto, T., Tsuge, S., Yamamoto, Y. and Takeuchi, T., Py-GC
     Evaluation on Sequence Distributions of Dyads in Vinyl-type
     Copolymers: Acrylonitrile-*m*-chlorostyrene Copolymers, *Macro-
     molecules, 7*(1974)376-380.

133. Tsuge, S., Kobayashi, T., Nagaya, T. and Takeuchi, T., Py-GC
     Determination of Run Numbers in Methyl Methacrylate-Styrene
     Co-polymers Using the Boundary Effect, *J. Anal. Appl. Pyrol.,
     1*(1979)133-141.

134. Kalal, J., Zachoval, J., Kubat, J. and Svec, F., Application
     of Py-GC in the Analysis of Diad Sequence Distribution in
     Styrene-Glycidyl Methacrylate Co-polymers, *J. Anal. Appl.
     Pyrol., 1*(1979)143-157.

135. Shimono, T., Tanaka, M. and Shono, T., Py-GC of Butadiene Co-
     polymers, *Analyt. Chim. Acta, 96*(1978)359-365.

136. Harwood, H.J., The Characterisation of Sequence Distribution
     in Copolymers, *J. Polym. Sci., Part B, 2*(1964)601-607.

137. Senoo, H., Tsuge, S. and Takeuchi, T., Estimation of Chemical Inversions of Monomer Placement in Polypropylene by Py-GC, *Makromol. Chem.*, *161*(1972)185-193.

138. Okumoto, T., Ito, H., Tsuge, S. and Takeuchi, T., On the Distribution of Chlorine Atoms in Chlorinated PVC Prepared by Various Methods, *Makromol. Chem.*, *151*(1972)285-288.

139. Blackwell, J.T., Quantitative Determination of the Monomer Composition in Hexafluoropropylene-Vinylidene Fluoride by Py-GC, *Analyt. Chem.*, *48*(1976)1883-1885.

140. Lehrle, R.S. and Robb, J.C., The Quantitative Study of Polymer Degradation by GC, *J. Gas Chromatogr.*, *5*(1967)89-95.

141. Ferlauto, E.C., Lindemann, M.K., Lucchesi, C.A. and Gaskill, D.R., Pyrolysis of Poly(methyl Methacrylate co Ethyl Acrylate), *J. Appl. Polym. Sci.*, *15*(1971)445-453.

142. Haken, J.K. and McKay, T.R., Quantitative Py-GC of Some Acrylic Copolymers and Homopolymers, *Analyt. Chem.*, *45*(1973)1251-1257.

143. Sugimura, Y., Tsuge, S. and Takeuchi, T., Characterisation of Ethylene-Methyl Methacrylate Copolymers by Conventional and Stepwise Py-GC, *Analyt. Chem.*, *50*(1978)1173-1176.

144. Shimono, T., Tanaka, M. and Shono, T., Py-GC of Methyl Methacrylate-Styrene and Methyl Methacrylate-α-Methylstyrene Co-polymers, *J. Anal. Appl. Pyrol.*, *1*(1979)77-84.

145. Braun, D. and Disselhoff, R., Analyse von Acrylonitril: Styrol-Copolymeren mit Hilfe Py-GC, *Angew. Makromol. Chem.*, *23* (1972)103-115.

146. Eustache, H., Robin, N., Daniel, J.C. and Carrega, M., Determination of Multipolymer Structure Using Py-GC, *Eur. Polym. J.*, *14*(1978)239-243.

147. Krishen, A. and Tucker, R.G., Quantitative Determination of the Polymeric Constituents in Compounded Cured Stocks by Curie-point Py-GC, *Analyt. Chem.*, *46*(1974)29-33.

148. Alexeeva, K.V., Khramova, L.P. and Solomatina, L.S., Determination of Polymers by Py-GC, *J. Chromatogr.*, *77*(1973)61-67.

149. Zulaica, J. and Guiochon, G., Fast Qualitative and Quantitative Microanalysis of Plasticisers in Plastics by GLC, *Analyt. Chem.*, *35*(1963)1724-1728.

150. Senoo, H., Tsuge, S. and Takeuchi, T., Py-GC Analysis of 6-66-Nylon Copolymers, *J. Chromatogr. Sci.*, *9*(1971)315-318.

151. Martinex, J. and Guiochon, G., Identification of Phenol-Formaldehyde Polycondensates by Py-GC, *J. Gas Chromatogr.*, *5*(1967) 146-150.

152. O'Neil, D.J., Determination of Polymer Content by Py-GC, *J. Compos. Mater.*, *2*(1969)502-505.

153. Burg, K.H., Fischer, E. and Niessermel, K., Bausteinanalyse bei Homo-und Copolymerisation des Trioxans mit Hilfe der Katalytischen Py-GC, *Makromol. Chem.*, *103*(1967)268-278.

154. Mokeeva, R.N. and Tsarfin, T.A., GC Determination of Oxyethylene and Oxypropylene Groups in an Ethylene Oxide-Propylene Oxide Copolymer, *Plast. Massy*, *3*(1970)52-53.

155. Takeuchi, T., Tsuge, S. and Okumoto, T., Identification and Analysis of Polyurethane Foams by Py-GC, *J. Gas Chromatogr.*, *6*(1968)542-547.

156. Fischer, W. and Meuser, H., Determination of Plasticisers in Vulcanisates by Py-GC, *Gummi Asbest. Kunstst.*, *20*(1967)17-20.

157. Tesarik, K. and Borkovcova, I., Quantitative Determination of Organic Materials Present in Chromatographic Packings or Columns by the Thermal Destruction Method, *Chromatographia, 8* (1975)286-289.

158. Tsuge, S., Okumoto, T., Sugimura, Y. and Takeuchi, T., Py-GC Investigation of Fractionated Polycarbonates, *J. Chromatogr. Sci.*, *7*(1969)253-256.

159. Tai, H., Powers, R.M. and Protzman, T.F., Determination of Hydroxyethyl Group in Hydroxyethyl Starch by Py-GC Technique, *Analyt. Chem.*, *36*(1964)108-110.

160. King, W.D. and Stanonis, D.J., Determination of the Substituent Content of Certain Cellulose Esters and Ethers by Py-GC, *Tappi*, *52*(1969)465-467.

161. Neumann, E.W. and Nadeau, H.G., Analysis of Polyether and Polyolefin Polymers by GC Determination of the Volatile Products Resulting from Controlled Pyrolysis, *Analyt. Chem.*, *35* (1963)1454-1457.

162. Groten, B., Application of Py-GC to Polymer Characterisation, *Analyt. Chem.*, *36*(1964)1206-1212.

163. Levy, R.L., Gesser, H., Halevi, E.A. and Saidman, S., Py-GC of Porphyrins, *J. Gas Chromatogr.*, *2*(1964)254-255.

164. Danielson, N.D. and Rogers, L.B., Determination of Tryptophan in Proteins by Py-GC, *Analyt. Chem.*, *50*(1978)1680-1683.

165. Hanson, R.L., Brookins, D.G. and Vanderborgh, N.E., Stoichiometric Analysis of Oil Shales by Laser Py-GC, *Analyt. Chem.*, *48*(1976)2210-2214.

166. Hanson, R.L., Vanderborgh, N.E. and Brookins, D.G., Characterisation of Oil Shales by Laser Py-GC, *Analyt. Chem.*, *47* (1975)335-338.

167. Coulter, G.L. and Thompson, W.C., Automatic Analysis of Tyre Rubber Blends by Computer-linked Py-GC, in C.E.R. Jones and C.A. Cramers (Eds.), *Analytical Pyrolysis*, Elsevier, Amsterdam, 1977, pp. 1-15.

168. Jenkins, A. and Hunt, R.J., An Automatic Injection System for Solid Samples, *Column, 2*(1968)2-5.

169. Rushneck, D.R., Diaz, A.V., Howarth, D.W., Rampacek, J., Olsen, K.W., Dencker, W.D., Smith, P., McDavid, L., Tomassian, A., Harris, M., Bulota, K., Biemann, K., Lafleur, A.L., Biller, J.E. and Owen, T., Viking Gas-Chromatograph-Mass Spectrometer, *Rev. Sci. Instrum., 49*(1978)817-834.

170. Dencker, W.D., Rushneck, D.R. and Shoemaker, G.R., Electro-chemical Cell as a GC-MS Interface, *Analyt. Chem., 44*(1972) 1753-1758.

Chapter 4

Pyrolysis Mass Spectrometry

Pyrolysis gas chromatography has undeniable attributes in its ability
to resolve many components and to yield precise, quantitative assessments. Combination with mass spectrometry (Py-GC-MS) enables fragment identification to be achieved and makes available a highly
specific detector to allow quantification at very low levels. Many
important applications of the technique have appeared, and Py-GC is
now an essential analytical tool in many fields. Nevertheless, pyrolysis gas chromatography has certain disadvantages. Perhaps the
most compelling of these is the time required for an analysis. This
is particularly long when slowly eluting complex products--as with
pyrograms from microorganisms, which may contain over 200 components--
are encountered. The use of temperature programming exacerbates the
problem, in that a further period to obtain thermal equilibration on
cooling must also be allowed. Although automated systems give an increased through-put of work, the analysis load per instrument is
strictly finite. The taxonomic applications of Py-GC may also be
hindered, due to the difficulties associated with the detection of
high-molecular weight, polar, or unstable products. Such compounds
may well be expected to be of value in the characterization of the
organism, but generally they will not be transmitted efficiently by
the column. Moreover, the variable sampling and resolution of such
components, depending upon GC conditions and transfer, is a major
factor when low reproducibility is observed. Variations in retention
time and resolution, which may be caused by small changes in

temperature, carrier gas flow, or column degeneration, add to these
problems and necessitate much care when comparisons with library data
are undertaken. The lack of a fixed base line due to column bleed
and overlapping peaks adds to the difficulties and makes quantification, particularly by automatic data-logging systems, difficult to
control [1].

Reports have appeared on the direct transmission of pyrolysis
products to a GC detector, without prior separation of the components [2-4]. The use of a mass spectrometer in these circumstances
offers considerable advantages [5]:

1. The mass pyrograms so obtained are characteristic and are
   suitable for fingerprint identification purposes.

2. Direct chemical information (from single peaks or from ion
   series) is available.

3. Rapid analysis is possible with a through-put of up to one
   sample per minute.

4. Components are separated according to mass on a stable,
   linear, and reproducible mass scale.

5. The mass scale, together with a uniform base line, simplifies data collection and coding and facilitates computer
   manipulation.

6. Transfer efficiency is very high, so that considerable increases in sensitivity may be available.

7. A more faithful analysis of polar or less-stable fragments
   may be expected.

8. The system may be readily automated.

9. The rapid through-put enables reference standards to be
   incorporated during each run rather than place total re-
   liance on a data bank.

Until recently, the success of Py-GC eclipsed the use of mass
spectrometry as a detection system, despite the fact that pyrolysis
mass spectrometry was the first of the integrated techniques of analytical pyrolysis to be described. Progress in the design of mass
spectrometers and the availability of online computer systems initiated a reappraisal of the role of Py-MS, and the work of Meuzelaar
and his group has largely been responsible for the reemergence of
this technique as a viable taxonomic tool [6].

4.1. TECHNIQUES

The usual sequence of events in pyrolysis-mass spectrometry is shown
in Fig. 4.1. Important features to note are that

> A single pyrolysis product may give rise to several peaks in
> the mass pyrogram due to fragmentation within the ion
> source.
>
> Each fragment ion may be produced by more than one pyrolysis
> product.
>
> Isomeric products will give a single peak in the mass pyrogram,
> although they may be well separated in a Py-GC analysis.

4.1.1. Pyrolysis

In contrast to Py-GC, pyrolysis in Py-MS applications is undertaken
in vacuo and many of the reported applications have used Curie-point
systems.

Curie-Point Systems

Although the original work essentially used custom-built apparatus,
pyrolyzers suitable for Py-MS are now available from several manufacturers [7-11]. The requirements for effective pyrolysis have
already been discussed in Chap. 2. It should be noted, however,
that in a vacuum, cooling processes are considerably slower, so that
temperature profiles measured under Py-GC conditions may not correspond to the Py-MS case. For maximum transfer efficiency, pyrolysis
should be as near to the ion source region of the mass spectrometer
as possible. However, pyrolysis directly within the ion source is
generally not recommended for continuous fingerprint use, due to the
rapid contamination which may result. Early systems [6] and a modern
commercial development (Fig. 4.2) [7] thus use an intermediate buffer
volume between the pyrolyzer and the ion source, to act as an expansion chamber. The schematic layout is depicted in Fig. 4.3. The
walls of the expansion chamber must be inert (gold coated) to prevent catalytic decomposition of pyrolysis products, and should be
heated (150 to 200°C) to reduce condensation of the pyrolysate. In
practice, most of the condensable material is probably lost to the

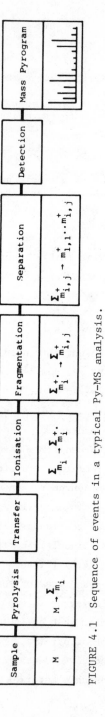

FIGURE 4.1 Sequence of events in a typical Py-MS analysis.

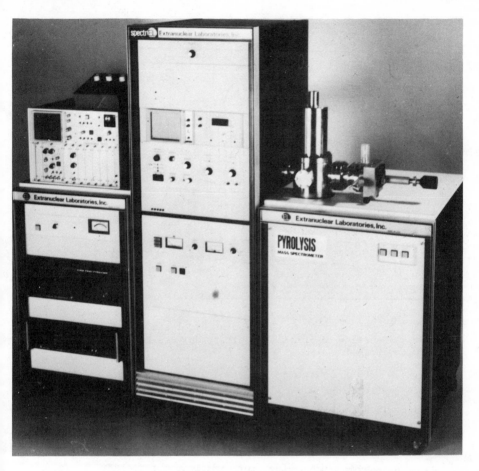

FIGURE 4.2 Curie-point pyrolysis mass spectrometry system. (Courtesy of Extranuclear Export Corporation, Pittsburgh, Pa.)

cool pyrolysis reaction tube [12], thus minimizing contamination. The expansion chamber serves a further purpose. The entry of the pyrolysis products into the ion source is extended in time to allow successive mass scans of the pyrolysate to be undertaken. The experimental arrangement is a compromise between direct pyrolysis with the possibilities of source contamination and hence low reproducibility, or the loss of molecules with low-volatility which may have important structural significance.

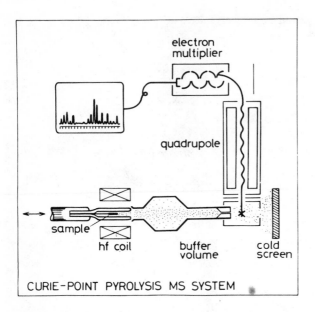

FIGURE 4.3 Arrangement of Curie-point pyrolysis mass spectrometry
system. (Courtesy of H.L.C. Meuzelaar, University of Utah, Salt
Lake City, Utah.)

In applications where a more efficient sampling is essential,
provision is also made to heat the reaction tube and for the removal
of the expansion chamber. The modular design of the instrument enables pyrolysis to be effected directly in front of the ion source.
Such an arrangement is useful, for example, in the study of time-
resolved pyrolysis phenomena which, to date, have essentially been
monitored by direct insertion-probe techniques. Under normal conditions of use the heating is controlled by a 1.5-kW, 1.1-MHz induction coil. This allows practical temperature rise times of less than
100 µs to be achieved. The TRT may be controlled by means of a
second induction coil in series with the first. Power attenuation
is obtained by changing the inductance of the second coil by insert-
ing an iron rod into the core. Calibration of the rod position versus the filament heating profile enables a wide range of TRTs to be
selected through control of the field strength of the pyrolysis coil
[13]. Temperature rise-times greater than 1.5 s are available [14],

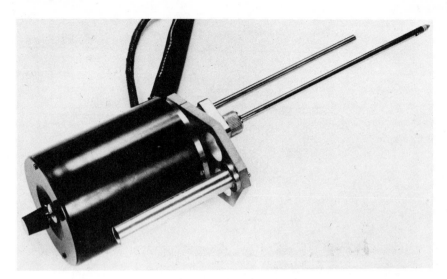

FIGURE 4.4 Curie-point pyrolysis probe. (From Ref. 8, courtesy of
VG-Micromass Company, Altrincham, Cheshire, U.K.)

and it may well be that this control will greatly facilitate time-
resolved studies. Heating rates greater than those available by
probe heating may be advantageous in that the production of nonvolatile carbonaceous degradation products (char) increases with TRT
[15]. The completed system has a demountable pyrolysis probe which
can be adapted to automated use, or which can be interchanged to
allow laser Py-MS or direct insertion-probe techniques to be undertaken [7].

The replacement of the direct insertion probe of a mass spectrometer by a probe suitable for pyrolysis is a second approach to
Py-MS. In one arrangement (Fig. 4.4) the probe consists of a por-
table induction coil (500 kHz) which surrounds an expansion chamber
(1 ml) containing the coated ferromagnetic wire [8]. The expansion
chamber is connected to the mass spectrometer ion source by a glass-
lined inlet tube (0.5-mm bore). The temperature of the inlet tube
and expansion chamber may be controlled and is variable up to 200°C.
The pressure after pyrolysis decays at 3% s^{-1} allowing repetitive
scans to be taken. A direct insertion probe based on the Pyroprobe

a)

PYROLYSIS CHAMBER   VACUUM LOCK

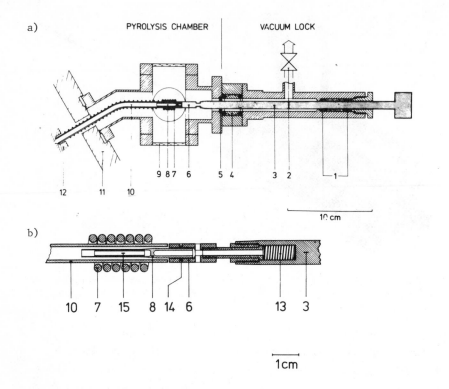

FIGURE 4.5 High vacuum Curie-point Py-MS probe: 1, Seals; 2, high-vacuum valve; 3, insertion rod; 4, ball valve; 5, Cu bearing; 6, quartz tube carrier; 7, induction coil; 8, quartz tube; 9, viewing port; 10, heated inlet tube; 11, ion-source housing; 12, ion source; 13, spring;, 14, clamp ring; 15, ferromagnetic tube. (From Ref. 17.)

heated filament is also available [16] and other systems, too, may be interfaced to a range of different mass spectrometers [9,10].

    A Curie-point Py-MS system which combines the advantages of small sample size and high-vacuum pyrolysis has also been designed [17]. Pyrolysis is undertaken on submicrogram amounts of sample deposited onto the inside surface of a ferromagnetic tube. This is held within a quartz tube attached to an insertion probe (Fig. 4.5). The sample is inserted into the center of the induction coil, flanged directly onto the MS inlet, and is tightly butted to the inlet line. This design was introduced in order to suppress intermolecular reactions. On pyrolysis the vapors pass rapidly into the source, thus minimizing the time between the formation of fragments and their analysis. In conjunction with small sample sizes, this configuration achieves very low pressure pyrolysis conditions (Knudsen reactor [18]), in which wall collisions rather than intermolecular collisions predominate. The success of this approach is illustrated with a comparison of Py-MS studies of 4-phenylbutyric acid. Figure 4.6 records the mass pyrograms of this compound, obtained using direct high vacuum pyrolysis (b) and pyrolysis into an expansion chamber (c) [19]. The latter mass pyrogram shows the presence of a fragment (m/z 182) due to the formation of dibenzyl, a dimeric product. This reaction is not observed to any significant degree in the direct analysis. This technique, however, did not involve the determination of the mass pyrogram by repetitive scans. Rather, a delay of 0.5 s was allowed to maximize entry of the fragments into the source (measured for m/z 146) before a spectrum was recorded.

Probe Analysis

The direct insertion probe of a mass spectrometer is perhaps the most readily available pyrolysis probe for Py-MS studies. These probes are fitted with heating units capable of controlling the sample temperature up to about 500°C. Calibration and the addition of a ramp control where necessary to enable precise programming of the temperature increase allows pyrolysis to be achieved under isothermal or temperature-programmed modes of operation. These techniques have, in the main, been applied to the study of synthetic polymers, particularly condensates, and are useful to achieve structural (isothermal) or mechanistic (temperature-programmed) information [20-24]. At an elevated temperature the macromolecule undergoes cleavage to produce a range of volatile fragments which characterize the original polymer. The sensitivity of the mass spectrometer is such that a decomposition rate of about 1% min^{-1} may be readily followed. This allows temperatures in the range of 250 to

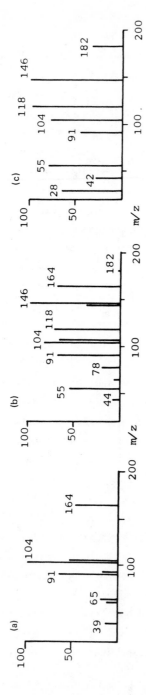

FIGURE 4.6 4-Phenylbutyric acid: mass spectrum and mass pyrograms [a, mass spectrum (70 eV); b, very low pressure pyrolysis mass pyrogram (15 eV) (from Ref. 17); c, conventional Curie-point mass pyrogram (13 eV)]. (From Ref. 19, by permission of Heyden & Son Ltd.)

350°C to be used for many polymers, although higher temperatures may be required for thermally resistant systems.

The volatile products, consisting essentially of monomeric and oligomeric subunits, undergo further fragmentation on ionization to yield the mass pyrogram. Typical is the case of polyethyleneterephthalate. Pyrolysis of this polymer essentially follows a Chugaev-type elimination to yield vinyl ester and carboxylic acid end groups (Fig. 4.7). An alternative pyrolysis pathway generates ethylene glycol endgroups via hydrogen migration. Electron-impact fragmentation proceeds through sequential loss of $O-CH=CH_2$ and CO characteristic of the ester function. The simplified mass pyrogram is depicted in Fig. 4.8. Mechanistic information may be obtained by monitoring the mass spectral changes as the probe is heated (typically 50°C min^{-1}). The data may be displayed as a mass fragmentogram (Fig. 4.9), which shows the production of various ions as a function of temperature [25]. The maxima observed at 240 to 260°C are due to the presence of cyclic oligomers which evaporate before thermal fragmentation (>300°C) begins.

Probe pyrolysis has also been used for the identification of bacteria [26-28]. Isothermal analysis (350°C) revealed the presence of large polar fragments (probably ubiquinones and phosphodiglyceride residues) which enabled differentiation of the organisms to be achieved [26]. Lipids were also the basis of characteristic patterns from the linear programmed thermal degradation mass spectrometry (LPTDMS) of microorganisms [27,28]. Here, the samples were heated to 400°C at 20°C min^{-1} and mass spectra were continuously monitored. Mass fragmentograms could thus be determined to generate a characteristic profile for each organism, enabling identification to be achieved. Although this procedure is time consuming in relation to a Curie-point Py-MS analysis, it does have the advantage that multidimensional data is available and useful if organisms have very similar mass pyrograms. It may well be that time-resolved Curie-point or heated-filament pyrolysis (with longer temperature rise times) may generate such information within an acceptable analysis time for batch work [7,16].

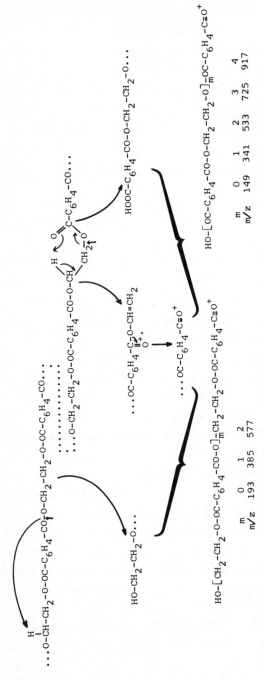

FIGURE 4.7 Probe pyrolysis-Py-EIMS fragmentation of polyethyleneterephthalate. (From Ref. 25.)

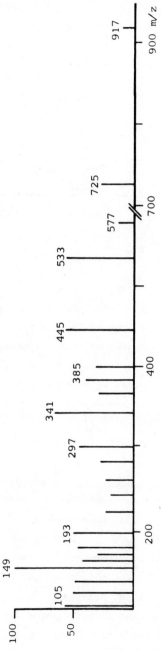

FIGURE 4.8 Probe pyrolysis: simplified mass pyrogram of polyethyleneterephthalate. (From Ref. 25.)

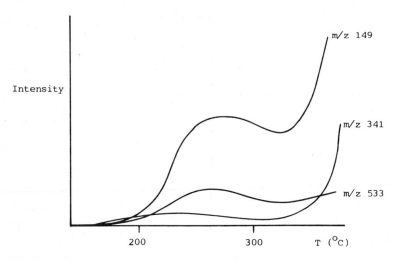

FIGURE 4.9   Probe pyrolysis mass fragmentogram of polyethyleneterephthalate.   (From Ref. 25.)

Probe pyrolysis may also occur during attempted mass spectrometry of thermolabile or involatile samples. Biopolymers are a useful example here. The mass spectrum of DNA has been described [29], but many fragments resulted from the thermal extrusion of phosphodiester groups and no phosphorus-containing ions were detected. This clear pyrolysis has been recognized as such by later workers [30-32], and pyrolytic conversions before mass spectrometry are now providing useful information on RNA sequences [33], or the change in degradation profiles of normal and flame-retardant cotton [34].

Laser Pyrolysis

The use of laser pyrolysis enables rapid, direct heating of the sample to be achieved. The thermal reservoir caused by the slower cooling of filament systems in vacuo is, however, absent. The laser may be interfaced to the mass spectrometer as shown in Fig. 4.10 [35,36]. In this configuration the cold screen (Fig. 4.3) contains an aperture through which the laser beam is focused. The sample is held on a probe, which may be used with the expansion chamber in position, or else within the ion source of the mass spectrometer.

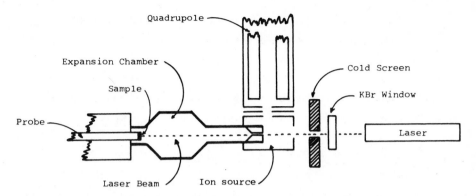

FIGURE 4.10   Laser Py-MS.   (From Ref. 36, by permission of Heyden
& Son, Ltd.)

In practice, the mass pyrograms from Curie-point and laser pyrolysis
with the expansion chamber in position were very similar. Larger
fragments suitable for structure elucidation were obtained only when
pyrolysis was undertaken directly into the ion source.

   Laser microprobe mass analysis (LAMMA) has also been described.
The analysis of milligram amounts of sample may be achieved in the
normal way with ionization and mass analysis following pyrolysis
[37-39]. The pyrolysis products are modulated by a mechanical chop-
per before ionization. This interuption enables background effect
to be eliminated. The pressure of the pyrolysis chamber is also
monitored continuously. Differences between this total pressure and
the abundance profile of selected ions (mass fragmentogram) enable
the origin of peaks in the mass pyrogram to be determined. This
allows distinction between the evolution of absorbed gases, pyrolysis products and fragment ions to be made, and enables kinetic data
via time-resolved mass fragmentograms to be obtained. The key is
the use of phase spectrometry. This uses the phase lags (ϕ) which
develop between the modulated ion signals, due to the different
velocities of species within the molecular beam. The phase lag
depends upon the flight time within the molecular beam, and has components before (ϕ_n) and after (ϕ_i) ionization:

$$\phi = \phi_n + \phi_i$$

If m_n and m_i are the masses of the neutral and ionized species at constant source temperature, this can be rewritten as

$$\phi = Am_n^{\frac{1}{2}} + Bm_i^{\frac{1}{2}}$$

where A and B are constants depending upon instrumental parameters. If the ionized species is a molecular ion $m_i = m_n$ and

$$\phi = m_i^{\frac{1}{2}}(A + B)$$

thus a plot of ϕ versus $m_i^{\frac{1}{2}}$ generates a linear plot of slope A + B. When daughter ions are detected, the measured phase (ϕ) for any particular m/z value is larger ($m_n > m_i$), so that fragment ions may be detected as deviants from the plot. Moreover, if different fragment ions result from the same unionized molecule, then

$$\phi = A' + Bm_i^{\frac{1}{2}}$$

where $A' = Am_n^{\frac{1}{2}}$ = constant

Such ions will lie on a linear plot exhibiting a lesser slope (B). A typical plot for the degradation of PVC is depicted in Fig. 4.11. The technique is also useful for the detection of contaminants in polymers [38].

An alternative approach is the detection of very small (pg) amounts of material by direct ionization due to laser irradiation [40,41]. This process, however, differs considerably from the currently accepted modes of Py-MS. With laser-desorption mass spectrometry, a short (0.15 µs) laser burst ionizes and volatilizes polar, nonvolatile macromolecules. This technique does not induce pyrolysis, but rather causes mild ionization which leads to the detection of large intact species. Thus digitonin (M 1228), adenyl-(3'-5')-cytidine (M 572) and sucrose (M 342) were readily detected--usually as a sodium or potassium complex [42].

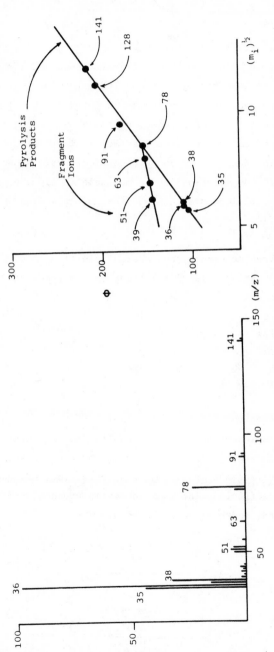

FIGURE 4.11 Laser Py-MS of PVC. The use of phase measurements (ϕ) to distinguish pyrolysis products and fragment ions. (From Ref. 37.)

4.1.2. Ionization

Electron-Impact Ionization

Ionization by electron impact (EI) is the most common mode of operation. The molecular beam from the pyrolysate is passed through a high-energy beam of electrons generated within the ion source. For normal mass spectra an ionization energy of 70 eV (1 eV = 1.602 × 10^{-19} J; 96.5 kJ mol^{-1}) is often used, although 20 eV is common in GC-MS work so that residual carrier gas (He, ionization potential 24.6 eV) is not ionized. Under these conditions, collisions result in the expulsion of an electron from a molecule (M) in the beam to form an electron-deficient radical ion, the molecular ion ($M^{+\cdot}$):

$$M \xrightarrow{\quad e \quad} M^{+\cdot} + 2e$$

(Under milder operating conditions, thermal electron gain, with the production of a negative molecular ion may be achieved, and negative-ion mass spectrometry results [43]).

The ionization energy available is in excess of the ionization potential of organic molecules:

$$R-\ddot{N}H_2 \longrightarrow R-\overset{+\cdot}{N}H_2 \qquad (9.5 \text{ eV})$$

$$R-\ddot{O}H \longrightarrow R-\overset{+\cdot}{O}H \qquad (10.7 \text{ eV})$$

so that on formation the molecular ions may have high internal energies. Dissipation of this energy results in fragmentation of the molecule, with the ejection of a radical or neutral species to yield daughter or fragment ions. These species, too, may undergo fragmentation if sufficient internal energy is present, and each ion may display several modes of decomposition:

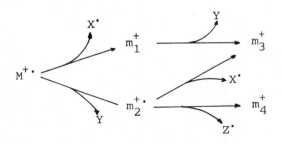

Separation based upon mass (m/z) and measurement of the abundance of each ion yields the mass spectrum, normalized to the most
intense fragment (base peak). The mass of the molecular ion reveals
the molecular weight of the compound, unless the ion is too unstable
for detection. As a test, the molecular ion should be of even mass
(unless an odd number of N atoms are present), and decomposition
pathways should be feasible knowing the elemental composition of the
molecule. The loss of a neutral fragment, however, may generate an
apparently acceptable molecular ion (e.g., alcohol $\longrightarrow$ olefin). To
detect this, other ionization conditions may be necessary. Each ion
is, however, accompanied by natural abundance isotope contributions.
These are highly distinctive for chlorine ($^{35}Cl:^{37}Cl$, 3:1) and bro-
mine ($^{79}Br:^{81}Br$, 1:1), with pairs of peaks two mass units apart, indicating the halogen containing fragments. Sulfur is also characteristic ($^{32}S:^{33}S:^{34}S$, 100:0.8:4.44). The most common isotope contribution is due to ^{13}C ($^{12}C:^{13}C$, 100:1.12). This far exceeds isotopic contributions by other common elements so that, as a rule of
thumb, M +1 intensities are 1.12n.I, where n is the number of carbon
atoms in the molecule and I is the relative intensity (%) of the
peaks. Exact calculations based upon natural abundancies allow the
intensities of M +1 and M +2 ions (due to the incorporation of two
isotopes in the molecule) to be tabulated for any combination of
elements. Measurement of these peaks from the mass spectrum thus
enables an estimation of the molecular formula of the molecule to be
made. Typical data for m/z 100, methyl methacrylate ($C_5H_8O_2$) are
displayed in Table 4.1 [44]. This procedure requires high-precision
measurements for definitive assignments, and therefore is prone to
error, but elimination of unlikely structures is possible on this
basis, and also double-charged ions (m/2z) may be revealed in this
way.

The fragmentation pattern may be used to elucidate the structure of the original molecule [45,46]. It is possible to account
for the appearance of many of the fragment ions by assuming that:

1. Ionization preferably involves the most weakly held elec-
    trons (n $\longrightarrow$ π $\longrightarrow$ σ).

TABLE 4.1 Molecular Formula Determination from Isotope Contributions
(Intensities Normalized to M = 100%)

FORMULA	$C_4H_4O_3$	$C_5H_8O_2$	$C_6H_{12}O$	C_7H_{16}	C_8H_4
M+1	4.504	5.609	6.715	7.820	8.709
M+2	0.680	0.529	3.909	0.2649	0.333
(M+1)/(M+2)	6.63	10.60	17.18	29.52	26.19

Source: Ref. 44.

2. Fragmentation initiated by the electron-deficient center
    involves the production of a charge-stabilized ion or the
    ejection of a stable neutral molecule.

Typical processes are

β-FISSION

$CH_3-CH_2-\overset{\cdot}{\overset{+}{O}}H$ $\xrightarrow{-CH_3^{\cdot}}$ $CH_2=\overset{+}{O}H$ (m/z 31)

α-FISSION

$CH_3-CH_2-\overset{+\cdot}{Br}$ $\xrightarrow{-Br^{\cdot}}$ $CH_3-CH_2^+$ (m/z 29)

REARRANGEMENT

$\xrightarrow{-CH_2=CH_2}$

$\rightleftharpoons$ $\overset{CH_3}{\underset{CH_3}{>}}C=O^{+\cdot}$ (m/z 58)

Within any one fragmentation sequence it is usual to find only
one radical loss (change of parity, even ⟶ odd if no N atoms are
involved), with all other losses being due to neutral fragments.
This is due to the usually greater stability of even-electron species,
although many exceptions are now recognized [47]. Thus, rearrangements frequently follow a fission process:

$$\underset{\substack{\text{M}^{+\cdot}\ 87}}{\overset{}{\underset{}{\begin{array}{c} \text{HN}^{+\cdot}\text{-CH}_2\text{-CH}_3 \\ | \\ \text{CH}_2\text{-CH}_2\text{-CH}_3 \end{array}}}} \quad \xrightarrow{\ -\text{C}_2\text{H}_5^{\boldsymbol{\cdot}}\ } \quad \underset{\substack{\text{m/z}\ 58}}{\overset{}{\begin{array}{c} \text{H}\!-\!\!-\text{CH}_2 \\ \text{HN}^{+}\!\!-\!\!\text{CH}_2 \\ \| \\ \text{CH}_2 \end{array}}} \quad \xrightarrow{\ -\text{CH}_2\text{=CH}_2\ } \quad \underset{\substack{\text{m/z}\ 30}}{\overset{}{\begin{array}{c} \text{H} \\ | \\ \text{HN}^{+} \\ \| \\ \text{CH}_2 \end{array}}}$$

These processes are illustrated for the fragmentation of methyl butyrate in Fig. 4.12. In many molecules, quite complex sequences of events occur before final fragmentation, e.g., phenols (-CO), anilines (-HCN), stilbene (Ph-CH=CH-Ph) (-Me!), and cyclization and scrambling reactions abound. Nevertheless, considerable structural information may be obtained by an empirical interpretation of the spectrum.

With Py-MS, such fragmentations complicate the mass pyrogram, the appearance of which depends both on pyrolysis and electron-impact fission processes. The electron-impact-induced fragmentation may be minimized by a reduction of the ionization energy to a level approaching the ionization potential of the molecules in the pyrolysate. Electron energies in the 10 to 15 eV range are sufficient to initiate ionization, without substantially increasing the internal energy of the molecule. Thus fragmentation is minimized. The simplification in the appearance of the pyrogram which results from low-voltage electron-impact ionization is illustrated in Fig. 4.13. The main product is the monomer (methyl methacrylate, m/z 100) and the complex 70-eV spectrum indicates the presence of much electron-impact-reduced fragmentation. At 15 eV this is significantly reduced, with typical ester cleavage (M-OCH$_3$, m/z 69) being the predominant fission. The lowest voltage is below the appearance potential of most fragments, and the molecular ion is the major contributor to the mass pyrogram; m/z 56 is probably due to methylketene, another molecular ion. For bacterial fingerprint purposes in particular, where many pyrolysis products are obtained, ionizing voltages in the 13- to 15-eV range are common. It should be noted, however, that in some cases the ionization voltage may be below the appearance potential of molecular ions. Thus, for example, benzene

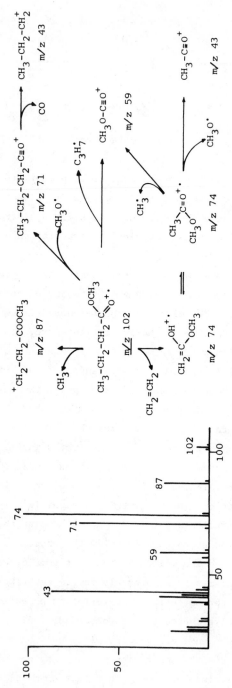

FIGURE 4.12 Mass spectral fragmentation of methyl butyrate (70 eV).

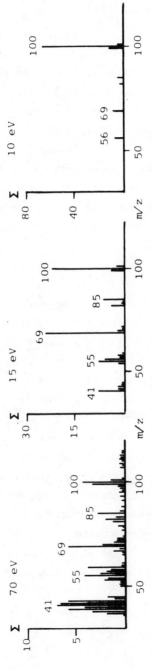

FIGURE 4.13 Mass pyrograms of polymethyl methacrylate: effect of ionization voltage. (Adapted from Ref. 36, by permission of Heyden & Son Ltd.)

(m/z 78) is observed in a polystyrene mass pyrogram at 15 eV but is
not detected at 10 eV [35]. At these low voltages, small errors may
have a dramatic influence upon the appearance of the mass pyrogram
and therefore every effort should be made to standardize this variable precisely. In other work using synthetic polymers, high voltage ionization (70 eV) has been standard [48,49], and the rather
more complex mass pyrograms were useful for characterization purposes. From the point of view of using mass pyrograms to provide
chemical information for structure elucidation, it would appear that
the suppression of fragmentation is advantageous. Comparisons of
low-voltage and high-voltage mass pyrograms may also be useful in
data analysis to distinguish pyrolysis products from fragment ions
[50].

Chemical Ionization

With typical source pressures of 10^{-6} torr (1.333×10^{-4} Nm^{-2}) mass
spectral fragmentation is essentially unimolecular. If the source
pressure is increased, the mean free path (200 mm at 10^{-6} torr) is
substantially reduced (50 μm at 1 torr), and ion-molecule reactions
result (Fig. 4.14). This results in the generation of a series of
Bronsted and Lewis acids, which are capable of reaction with
electron-donor molecules. These processes are the basis of chemical
ionization (CIMS) [51,52]. In this mode, the sample to be ionized
is introduced into the ion source in the presence of a large amount
of a reagent gas (1:1000). The source pressure is such that ion-
molecule reactions within the reagent gas generate the acids, which
eventually may collide with the sample. The reagent gas ions lose
energy during the various collision processes, so that sample ionization is essentially a low-energy chemical process. The ionized
species thus formed have a low internal energy, and fragmentation
is, therefore, much reduced in comparison with electron-impact
processes. It must be noted, however, that the ions are not true
molecular ions but rather they are even-electron species with a
higher mass. These are termed *quasimolecular* ions (M + 1).

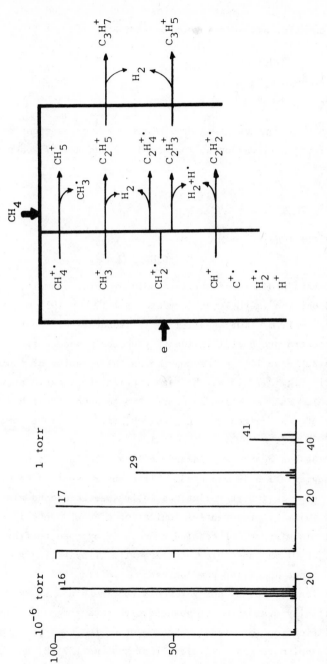

FIGURE 4.14 Ion-molecule reactions of methane: dependence upon source pressure.

Moreover, higher molecular weight species may also be produced. Methane, for example, also displays ions at M + 29 and M + 41:

$$M + \begin{cases} CH_5^+ & \rightarrow \overset{+}{M}\text{-H} + CH_4 \\ C_2H_5^+ & \rightarrow \overset{+}{M}\text{-}C_2H_5 \\ C_3H_5^+ & \rightarrow \overset{+}{M}\text{-}C_3H_5 \end{cases}$$

In contrast, isobutane gives M + 1 ions via hydride transfer from the tertiary butyl carbonium ion and M - 1 and M + 57 peaks may also be seen:

$$(CH_3)_3CH^{+\cdot} \rightarrow (CH_3)_3C^+ + M \begin{cases} \rightarrow \overset{+}{M}\text{-H} + \overset{CH_3}{\underset{CH_3}{>}}C{=}CH_2 \\ \rightarrow (M - 1)^+ + C_4H_{10} \\ \rightarrow M^+\text{---}C_4H_9 \end{cases}$$

The pyrolysis products of acrylate polymers, for example, effectively showed one ion (methyl methacrylate; M + 1, m/z 101) or two ion (styrene; M + 1, m/z 105; $C_7H_7^+$; m/z 91) isobutane CI spectra [53]. Such spectra are useful in identifying the products from complex polymers but give little information on the identity of isomeric components (cf. Fig. 3.7, in which fragmentation of isomeric acrylates is distinctive). Fragmentation is controlled to some extent by the acid strength of the ionized reagent gas. H_3^+ (from H_2) is a powerful donor and CH_5^+ is able to promote the ionization of most organic molecules, including hydrocarbons. Isobutane is a mild ionization medium and little fragmentation is expected. Different reagent gases may, therefore, be used to impart variations in fragmentation and also to enhance selectivity by preferential ionization of components in a complex pyrolysis mixture. Excessive amounts of byproducts from pyrolysis (e.g., H_2O) may possible interfere with CI by acting as a reagent gas and hence modifying the ionization profile [54].

Chemical ionization has not yet won a routine place in Py-MS, but the sensitivity available for quantitative mass fragmentography and the simplification, or signal enhancement of the mass pyrograms from synthetic polymers [55], in biological systems [27,28] and in forensic work [53] promise rapid consolidation, particularly in structure elucidation.

Field Ionization

Field ionization (FIMS) is a yet more specialized soft ionization technique [56]. In a typical system, a tungsten wire (10 μm) is activated by the high-field polymerization or pyrolysis of benzonitrile, which results in the deposition of pyrocarbon needles. This activated emitter (anode) is placed a short distance in front of the cathode which also contains an exit slit. A voltage of 8 to 10 kV is applied across the electrodes. The needlelike deposit enhances electric field gradients and potentials of up to 10^8 V cm^{-1} are created. When sample molecules enter this nonhomogeneous field a strong interaction with outer electrons is experienced. As the molecules near the anode, the potential wall around the molecule becomes weaker and eventually an electron is rapidly lost (10^{-12} s) from the molecule, a process known as *quantum mechanical tunneling*. The positive radical thus formed is rapidly repelled from the anode and into the analytical segment. Very little excess internal energy is gained by the molecule, so that fragmentation processes are minimized. Thus molecular ions generally characterize FI mass pyrograms, although surface interactions may result in the formation of quasimolecular ions (M + 1). Figure 4.15 compares the EI, CI [55], and FI [57] mass pyrograms from polystyrene. The degradation essentially gives monomer ($M^{+ \cdot}104$), dimer ($M^{+ \cdot}208$) and trimer ($M^{+ \cdot}312$). These peaks characterize the FI pyrogram, but the oligomers are quite weak in the EI spectrum, which shows significant fragmentation. The CI mass pyrogram (with isobutane) is complicated by butylation of the pyrolysis products. This is, therefore, characterized by M + 1 and M + 57 pairs.

Field ionization following pyrolysis directly within the mass spectrometer ion source has been described [58] and applications of Py-FIMS include studies of the pyrolysis products from bacteria [59], biological molecules such as nucleic acids [60] and polysaccharides

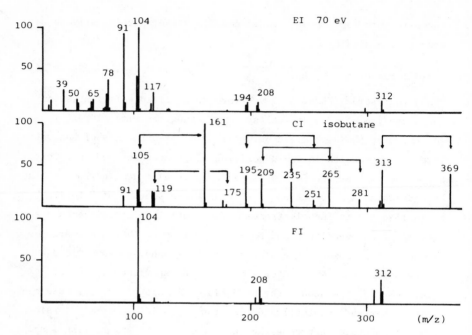

FIGURE 4.15 Comparison of mass pyrograms from polystyrene. (From Refs. 55 and 57.)

[58], and synthetic polymers [61]. The technique is highly specialized, and although for nonpolar compounds it has quantitative super-
iority over low-voltage EI [62], it is unlikely to find routine application in finger printing unknown samples. It does, however,
offer an elegant means to detect molecular ions from labile or polar
materials, compounds which are particularly important in the pyrolysis of biological samples. The principal use of Py-FIMS to date has
been to identify pyrolysis products through accurate mass determination, in conjunction with high-resolution mass spectrometry. In
those cases reported, surprisingly close relationships between Py-
FIMS products and those from Py-GC [59] and Py-FIMS [60] were
obtained.

Field Desorption

The previously described modes of Py-MS have involved initial pyrolysis, migration of the vapors into the ionization region of the source

and then ionization, so that mass analysis may produce the mass pyrogram. Under these conditions, losses due to adsorption or decomposition of large, polar, involatile, or unstable fragments (initiated by
thermal or ionization effects) may occur [35]. Such losses would be
expected to involve labile fragments which may hold much structural
information. These disadvantages may be overcome by the use of py-
rolysis-field-desorption mass spectrometry (Py-FDMS) in which pyrolysis, volatilization and ionization are accomplished almost simultaneously on an activated emitter [56,63].

The mode of ionization is similar to FI, but the emitter is
directly coated with a small amount of the sample (1 to 10 ng, either
as a solution or as a suspension), rather than being introduced as a
vapor. Thermal effects are thus significantly reduced. A combination of gentle heat to the filament, the high vacuum in the source
and quantum tunneling leads to pyrolysis, desorption and ionization.
The pyrolysis products so formed remain in the pyrolysis zone for
very short periods of time (10^{-11} s), they have a minimum transfer
distance before ionization and the excess energy gained by the ions
is very low (2 eV). This combined pyrolysis-ionization technique
avoids secondary pyrolysis, and transfer losses of polar products
and minimizes fragmentation of molecular ions. Py-FDMS pyrograms
are, therefore, characterized by the presence of large informative
fragments, composed mainly of the molecular ions (or M + 1 ions, due
to surface effects) of the primary pyrolysis products. It should
also be noted that the pyrolysis of large molecules is not the normal mode used in FDMS analysis. Direct desorption is frequently
achieved and leads to the production of abundant molecular ions.
Complex species, including oligopeptides such as the hormone brady-
kinin (M + 1, m/z 1060) may, thus, be readily analyzed.

Py-FDMS is a recently developed technique, and as few mass
spectrometers are yet equipped with field ionization sources, applications to date are restricted. These include studies on nucleic
acids [64], polysaccharides [65,66], blood [56], and synthetic polymers [56]. The potential of the method, however, has been amply
established. In particular, the detection of dimeric units from

glycogen and the appearance of dinucleotide residues in the mass
pyrogram of DNA indicates the potential for sequence analysis.
Figure 4.16 compares the Curie-point Py-EIMS of glycogen [14] with
the corresponding Py-FDMS data, which shows that the former reveals
significant degradation of the monomer (glucose), whereas substantial
fragments remain in the Py-FDMS pyrogram.

Py-FDMS is perhaps the most sophisticated of the pyrolysis
methods available to date. The production of FDMS mass pyrograms
requires emitter activation, a specially modified source and emitter
tuning before analysis. Moreover, the low ion currents, due to small
sample loads, demand sensitive detection and the summation of response offered by a photoplate is currently the preferred method.
The technique is not, therefore, for general routine use. However,
in combination with high-resolution mass spectrometry, the composition of large fragments may be determined, and it is probable that
future work will be concerned with the detailed structural and mechanistic implications revealed by pyrolysis processes.

Progress in soft ionization techniques is rapid, and laser-
desorption [42] and desorption-chemical ionization processes [67]
have also been described, but not in combination with pyrolysis.

4.1.3. Separation

On ionization, the pyrolysis fragments are propelled from the ion
source and into the analytical segment by an applied potential. Mass
analysis of the resulting ion beam has been undertaken in several
ways. These are illustrated in Fig. 4.17. The analytical system
used has clearly been influenced by instrumental availability, but
each device has different characteristics. Thus, such factors as

 Purpose of analysis: fingerprint or mechanistic work
 Mass range: for the detection of high-molecular-weight
  fragments
 Resolution: to determine precise molecular formulae
 Scan speed: to allow spectral summation to give an average
  mass pyrogram

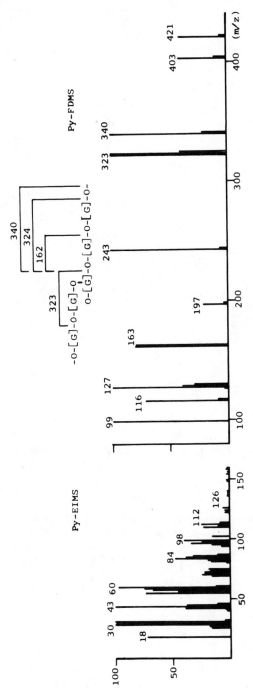

FIGURE 4.16 Py-EIMS and Py-FDMS mass pyrograms from glycogen. [G] are glucose residues. (Adapted from Refs. 14 and 65.)

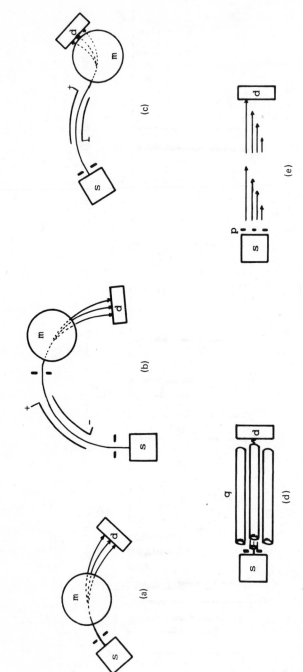

FIGURE 4.17 Types of mass spectrometer used in Py-MS: (a) single-focusing magnetic sector; (b) double-focusing (Nier-Johnson); (c) double-focusing (Mattauch-Herzog); (d) Quadrupole; (e) time of flight; s, source; d, detector; e, electric sector; m, magnetic sector; p, pulsed grid; q, quadrupole.

are considered, together with pyrolysis and detection variables.
For fingerprint work, both quadrupole [14] and magnetic sector [48]
instruments have been used. These latter devices have also been essential in the more specialized applications requiring high-resolution instruments for accurate mass measurements [58].

Magnetic Sector Mass Analysis

The simplest arrangement in this mode of separation is the single-
focusing mass spectrometer (Fig. 4.17a). Here, the ions are accelerated from the ion source under a potential of usually 2 to 4 kV--
the accelerating voltage (V). The ion beam is collimated and passes
into the flight tube, which is held within the field (H) of an external electromagnet. Ions in this field, which is perpendicular to the
flight path, follow a curved trajectory (of radius r), depending upon
their mass-to-charge ration (m/z) such that

$$\frac{m}{z} = \frac{H^2 r^2}{2V}$$

Thus, if ions are subjected to the same accelerating voltage and
field strength, singly charged ions, which represent the majority of
species present, are deflected along different paths (r), depending
upon mass. Double-charged ions appear at half the nominal mass.
Such ions may be detected simultaneously on a photographic plate,
but the more usual system is to focus each component of the ion beam
separately onto an electron multiplier. The flight path (r) is thus
held constant, and focusing is achieved by either accelerating voltage of magnet current scans. A decreased sensitivity accompanies a
fall in accelerating voltage so that, in practice, although a voltage
scan may be undertaken rapidly, magnetic scanning is more usual. For
the rapid scans required for Py-MS pyrogram accumulation, the recycling time of the magnet may be a problem. Modern spectrometers designed for GC-MS work, which also requires fast, repetitive scanning,
have improved performance by means of laminated magnets which are
driven, rather than being allowed to relax, to the initial field
strength. The output from the multiplier drives a series of mirror

galvanometers (each with a different attenuation) and the spectrum
is displayed on UV-sensitive paper. Other peripherals which may be
used are a chart recorder for very slow scans over a small mass
range, an oscilloscope for tuning purposes, and a computer for on-
line data storage and manipulation. Only the last of these allows a
time-averaged mass pyrogram to be obtained.

The usable mass range depends upon the magnetic field strength
and, due to the nonlinear relationship between m/z and H, the separation between adjacent masses exponentially decreases as the mass
range increases. The performance is usually quantified in terms of
resolving power (Rp). If two peaks (m_1, m_2) are partially resolved
by the mass spectrometer so that the height of the overlap valley is
10% of the intensity of m_1, then

$$Rp = \frac{m_1}{m_2 - m_1}$$

Alternatively, resolution may be estimated from

$$Rp = \frac{m}{W_m^{\frac{1}{2}}}$$

where $W_m^{\frac{1}{2}}$ is the width of peak m (in amu) at half-height.

In most cases, single-focusing instruments have resolving powers
up to 1000. This enables the transmission and separation of integral
masses up to about m/z 500 to be accomplished routinely. This is
adequate for fingerprint applications and many mechanistic studies.
The disadvantage of these instruments is that the resolving power is
sufficient to resolve integral masses only. Such spectrometers are
termed *low-resolution instruments*. More information is available
with a higher level of resolving power. This arises because atomic
weights, apart from the standard ^{12}C = 12.0000, have nonintegral
masses (Table 4.2). Different combinations of these elements therefore, produce structures with unique molecular weights, but which
have identical integral masses. Such data is available in tabular
form [68,69]. The molecular formula of an ion may thus be revealed
if the accurate, rather than the integral, mass is measured. This

TABLE 4.2 Precise Atomic Weights

Isotope	^1H	^{12}C	^{14}N	^{16}O
At wt	1.0078246	12.0000000	14.0030732	15.9949141

is shown in Table 4.3 for the pyrolysis products of DNA [70]. The base peak in the mass pyrogram is found at m/z 82 (Fig. 4.18.) The possible formulae for this species are displayed and the determination of the precise mass of the fragment enables its constitution to be revealed as C_5H_6O (2-methylfuran).

To perform these measurements, a high-resolution double-focusing mass spectrometer is employed. A typical system (Nier-Johnson configuration) is shown in Fig. 4.17b, in which an electric field (E) precedes the magnetic sector. The electric sector deflects ions (of velocity v) according to their kinetic energies (nominally zV) and achieves directional focusing of the beam. The path (of radius r) through this segment is

$$r = \frac{mv^2}{zE}$$

This arrangement serves to narrow the kinetic energy spread within the ion beam, which is responsible for peak broadening and low resolution. A manual comparison involves determining the ratio of the accelerating voltages necessary to transmit on the same trajectory through a fixed magnetic field, an ion of known precise mass, and the unknown ion. Resolving powers of 50,000 are available, and relative masses, using calibrated standards, may be determined with a precision of 2 to 5 ppm. This type of mass spectrometer uses magnetic scanning to produce a mass spectrum. An alternative arrangement is the Mattauch-Herzog geometry (Fig. 4.17c), in which a fixed magnetic field is employed. This system allows simultaneous detection of ions by means of photographic detection [58]. The masses of ions on the plate (m_1, m_2, m_3) are related by

$$m_2 = \left[\frac{d_{1,2}}{d_{1,3}} (m_3^{\frac{1}{2}} - m_1^{\frac{1}{2}}) + m_1^{\frac{1}{2}} \right]^2$$

TABLE 4.3 Possible Formulas and Masses for DNA Pyrolysis Product (m/z 82)

Formula	$C_2H_2N_4$	$C_3H_2N_2O$	$C_3H_4N_3$	$C_4H_2O_2$	C_4H_4NO	$C_4H_6N_2$	C_5H_6O	C_5H_8N	C_6H_{10}
Mass	82.0542	82.0429	82.0668	82.0317	82.0555	82.07973	82.0681	82.0919	82.1045

Source: Ref. 70.

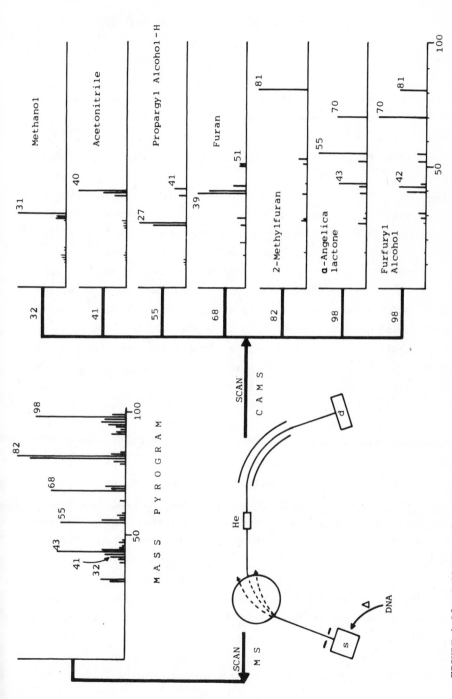

FIGURE 4.18 Collisional activation mass spectrometry: pyrolysis of DNA. (From Ref. 70, by permission of Heyden & Son Ltd.)

where $d_{1,2}$ and $d_{1,3}$ are the separations between ions m_1, m_2 and m_1, m_3. The use of high-resolution measurements generally requires that the pyrolysis vapors are held in the source for an extended period. This may mean that differences in product distribution may be observed. In practice, surprisingly good correlations are obtained [59,60,70], although propargyl alcohol has been identified under specific conditions only [70] in the pyrolysis of DNA.

High-resolution mass spectrometry enables the molecular formulae of pyrolysis products to be established. This is clearly of great importance in the establishment of pyrolytic fragmentation pathways and in the identification of specific structure-indicating fragments. This information alone, however, is generally insufficient to assign molecular structures. The fact that mass pyrograms are composed of many components means that individual mass spectral fragmentation pathways are difficult to elucidate, and hence structural evidence is sparse. Collisional activation mass spectrometry (CAMS), also known as direct analysis of daughter ions (DADI) or mass selection-ion kinetic energy analysis (MIKE), is a technique which is useful in the identification of mass spectral fragments and in mixture analysis and applications hold much promise in this area [71-74].

In this technique, a reversed-geometry double-focusing mass spectrometer (magnetic sector preceding electric sector) is used (Fig. 4.18). The pyrolysis products are leaked into the ion-source and ionized to yield the components of the mass pyrogram. The magnetic sector is adjusted so that ions of a specific mass are transmitted and these are led into the collision chamber. Here, the ions collide with helium, and kinetic energy is converted into internal energy, inducing fragmentation. The fragments may be mass analyzed by scanning the electric sector, to give a spectrum which is characteristic of the ion under study. This process may be repeated for each ion of interest in the mass pyrogram so that identification may be realized. In the mass pyrogram of DNA, comparison with authentic CAMS spectra enabled the identity of seven major components to be established, including an M-H ion from propargyl alcohol [70].

At present, this technique is highly specialized, although commercial instruments are available [8,74], and certain disadvantages are apparent with the present state of the art. In particular, separation is based upon kinetic energy differences. Thus, ions of the same mass may have different translational energies, depending upon their collision and fragmentation histories. Such energy spreads decrease resolution and mean that CAMS on ions of high mass (~200) is difficult. Moreover, isomeric and isobaric ions are transmitted into the collision chamber together, so that complex fragmentation patterns may result. The component at m/z 98 in the DNA pyrogram, for example, was found to be composed of α-angelicalactone and furfuryl alcohol. To improve the resolution capabilities, McLafferty has outlined proposals for a tandem double-focusing CAMS system [75]. High-resolution data are available with three-sector instruments [76] and commercial double-focusing instruments already have provision for extra electric, magnetic, or quadrupole sectors [8].

One more feature of mass pyrograms may be evident: the appearance of metastable ions. These are broad, low intensity peaks usually found at nonintegral masses and are indicative of mass spectral fragmentation processes. Such ions are produced by decomposition in a field-free region of the flight path, rather than within the ion source before acceleration, as with normal ions. Daughter ions (m_2^+) may thus have two modes of formation. Those produced in the ion-source gain kinetic energy zV, whereas those produced after acceleration (from ion m_1^+ which also has kinetic energy, zV) have less translational energy, as some is lost to the other fragment involved. These ions are deflected more by the field than those of equivalent mass, but with normal energy, and appear at a lower m/z value. Such species are the metastable ions (m^*) and their nominal mass position is given by

$$m^* = \frac{(m_2^+)^2}{(m_1^+)}$$

The appearance of such ions is useful in confirming the assignment of a single-step fragmentation process $(m_1^+ \rightarrow m_2^+)$. Techniques

are available for defocusing the electric sector of a double-focusing
instrument to reveal metastable ions which are normally not transmitted to the detector [77]. These techniques, however, have not
been applied to date in Py-MS studies.

Quadrupole Mass Analysis

The quadrupole mass analyzer, which separates ions by means of electric fields only, is shown in Fig. 4.17d [78]. The mass filter involves four parallel electrodes, held in a square array in which opposite electrodes are interconnected. Each pair of rods is subjected
to a combination of direct current voltage (Vdc) of increasing ampli-
tude and a radiofrequency voltage (Vrf). One pair of rods receives
Vrf and a positive Vdc (i.e., Vrf + Vdc) while the other pair are
subjected to a negative dc voltage, with Vrf 180° out of phase [i.e.,
-(Vrf + Vdc)]. These voltages give rise to an electrostatic field
which controls the motion of ions within the mass filter. This motion depends upon m/z, Vdc, Vrf, the rf frequency, the diagonal separation of the rods and the position of the ion within the field.
Under fixed operating conditions, most ions are collected and dis-
charged onto the electrodes as they pass through the mass filter
(unstable). For one particular combination, ions of a specific m/z
are subjected to bounded oscillations (stable) and pass through the
quadrupole to the detector. The residence time of ions within the
flight tube must be sufficiently long for unwanted masses to be fil-
tered out. The accelerating voltage which enables this to be
achieved has a maximum value and, in contrast to magnetic sector instruments, low voltages (5 to 15 V) are employed.

The mass spectrum is obtained by variation in the electrode
voltage. This is directly related to the mass transmitted by the
instrument, so that mass spectra from quadrupole instruments have a
linear, rather than exponential, presentation. This greatly facilitates counting and presentation of mass pyrograms. The resolution
of the instrument is controlled by the ratio Vdc/Vrf. As peak profile and separation are independent of mass, quadrupole mass spectrometers exhibit variable resolution ($Rp = m/Wm^{\frac{1}{2}}$) which increases

with mass. Resolution varies inversely with sensitivity so that, in comparison with a fixed resolution instrument, quadrupoles display enhanced sensitivity at lower mass values. Sensitivity is also increased because performance is less dependent upon the entrance angle and the energy distribution of the ions. Thus, finely collimated molecular beams are not required and, in comparison with sector instruments, more ions are transmitted for detection. The continuous focusing of the quadrupole and the low accelerating voltage requirements allow these instruments to operate at higher source pressures (10^{-4} torr) if necessary. Quadrupoles have a reputation for mass discrimination and poor performance at higher masses. In practice, for low resolution work, quadrupole performance is excellent. Most applications of Py-MS fingerprinting have used quadrupole mass analyzers, and a mass range of 16 to 200 is typical. Available instruments are also capable of unit resolution to masses of over 1000 Daltons, but with reduced sensitivity. One further advantage is apparent with the quadrupole mass analyzer for Py-MS work. Fast, repetitive scans for the accumulation and averaging of successive mass pyrograms may readily be undertaken. Moreover, there is no reset time which may be necessary with other systems. This enables spectral accumulation at a rate of 5 to 10 spectra per second. Such a rate demands computer assistance and here, too, quadrupoles are efficient and are readily interfaced to an online system.

Time-of-Flight Mass Analysis

When ions are accelerated across a potential difference, they acquire the same kinetic energy ($zV = \frac{1}{2}mv^2$). The ions travel at different velocities through the spectrometer (Fig. 4.17e); in principle this difference is used to separate the ions. Pulses of ions are analyzed at intervals of 30 to 50 µs via gated collectors, so that ions arriving at various time intervals may be resolved. The mass of a particular ion is determined from the time required to pass through the flight tube. Although such systems may be used for GC-MS applications, mass range is narrow, resolution is frequently low and they are more usually used for short duration phenomena. Their advantages have been exposed with pulsed laser Py-MS studies [79] and LAMMA [40], but it would appear unlikely that these instruments will see universal adoption.

4.1.4. Detection

Electron-multiplier detectors are perhaps the commonest detectors used in traditional mass spectrometers and an analogue display of ion-current is presented. One disadvantage of these detectors is that a large fraction of the ions is lost during scanning. In particular, the low bandwidths used at high amplification limit scan speed or resolution. When rapid phenomena are to be recorded, or when minute amounts of sample are available, this can be unacceptable. Photographic recording enables the simultaneous detection of all ions with a wide range of intensities (10^8), and produces an integrated mass spectrum. Although this has been used to effect, particularly in high-resolution studies, it is considerably less sensitive than electrical detectors, and the care required in use (degassing, development, measurement) has limited photographic recording to specialized applications.

Ion-counting has been proposed as an alternative [11,71,80]. Here, a high-speed pulse amplifier system (Channeltron electron multiplier array, CEMA) enables rapid ion-counting (2×10^7 s^{-1}) to be achieved. This acquisition rate allows a minimum intensity peak (100 to 200 ions collected) to be scanned in 5 to 10 μs, and means that over the duration of the pyrolysis event spectra may be accumulated at the scanning rate of the mass spectrometer. With this type of detector, the rapid scanning capabilities of quadrupole instruments enable maximization of analytical information, including time-resolved variations in single-ion intensities.

Ion-cyclotron resonance MS, in which ion flight-time is extended to 5 to 10 ms via a cycloidal trajectory imposed by crossed magnetic and electric fields, is useful for the study of ion-molecule interactions. In addition, almost all ions arrive at the detector, giving improved sensitivity. This approach, or indeed many of the

variations available today [51], such as secondary ion MS [81], have
not as yet found applications in analytical pyrolysis.

4.2. AUTOMATION

An automated system for Py-MS analyses has been described by Meuzelaar and co-workers [80]. This has found wide application [14], and
the layout is displayed in Fig. 4.19. The system is based on a quadrupole mass spectrometer, interfaced with a Curie-point pyrolyzer via
an expansion chamber. The coated wires, held in reaction tubes are
loaded into a rotary turntable, which is mounted within the vacuum
chamber. An electromagnetic arm transfers tubes from the turntable
to the pyrolysis coil in sequence. Pyrolysis is initiated and the
quadrupole is rapidly scanned to yield the mass pyrogram. Typically,

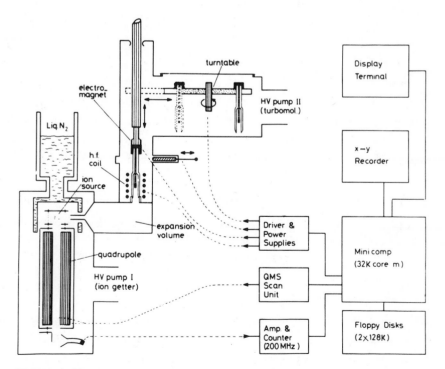

FIGURE 4.19 Automated Py-MS apparatus. (Reproduced with permission
from Ref. 14.)

the expansion chamber pressure is maintained for 5 to 10 s after pyrolysis, and during this time spectra are collected at a rate of 10
per second, with about 10 data points per mass unit. The rapid
counting ability of the CEMA detector, and the rapid scanning ability
of the quadrupole MS are essential for this performance. This may be
compared with a magnetic sector instrument, where the scan rate was
1 s per decade with a magnet reset time of 1 s [49].

The synchronization of the system is organized by means of an
online minicomputer which controls the turntable, sample changing
and pyrolysis events. This unit also activates the mass spectrometer
scan and receives the ion counts from the detector. The raw data
(perhaps 1.6 Mbits if handled point by point) is summed and stored
on disk. When all samples have been analyzed, peaks are identified
from the summed data points and a time-averaged mass pyrogram is
produced for each sample. Thirty samples per hour may be processed
in this way, and long term reproducibility has been found satisfactory [14].

Although Py-MS studies have been undertaken without computer
assistance [17], it would appear that, for fingerprint identification
of complex samples, time-averaged mass pyrograms are demanded. Although data systems are now available with many mass spectrometers,
these have generally been designed with GC-MS, rather than Py-MS, in
mind. Ideally, for Py-MS studies the data system must be capable of
dealing with the rapid accumulation of a large volume of data over a
short period of time. Thus, either a large core memory, fast dumping
to disk, or signal-averaging capabilities must be available [11].

Automated Py-MS was also proposed for the Viking expeditions
[82], but this was not part of the final analysis programme.

4.3. REPRODUCIBILITY

The modern phase of Py-MS, where combined pyrolysis, analysis and
data handling have been used to develop a powerful analytical technique, has only recently been entered. Despite this, many instruments and operational variables have already been described. Where

new information is required--via, for example, high-resolution studies and soft ionization techniques--such developments are welcome.
However, for routine analytical use it would appear appropriate that
techniques should be standardized. This, in part, would enable the
proliferation of independent systems, which has so confused Py-GC
literature, to be controlled. Grossly divergent systems, such as
Curie point versus heated filament, expansion chamber versus empty
GC column and separator, low-voltage versus high-voltage ionization,
quadrupole versus magnetic sector, fast scanning versus slow scanning,
m/z 30 to 120 versus m/z 35 to 250, are in use [14,49]. The principle
variable would appear to be the ionization voltage (Fig. 4.20). With
this parameter controlled, it would seem that considerable qualitative information on chemical composition is available through comparisons with library data [83]. For taxonomic applications which rely
upon small differences between mass pyrograms, considerable control
of the variables is required.

Although no interlaboratory trials of Py-MS reproducibility
have, as yet, been organized, it is encouraging to note that the
long-term performance characteristics of the technique are now under
scrutiny [13,14,48,49,54,58]. The principle factors have been identified and studied by Meuzelaar and his group, and these are listed
in Table 4.4 [13,14]. Biological materials, such as glycogen and
bovine serum albumin (BSA), were used as the test samples, and variations in mass pyrograms were evaluated on the basis of distance
measurements (Chap. 5). Results for glycogen are summarized in
Table 4.5, which records the percentage increase in variation between
mass pyrograms. These were determined using a Curie-point system
with quadrupole analysis. In summary, the recommendations for the
pyrolysis of biological materials are concerned with the following:

1. *Cleaning of the wires.* The standard method involved pro-
   longed heating of the wires in a wet hydrogen atmosphere,
   at a temperature just below the Curie point. Successive
   ultrasonic washings with CS_2 (2 min) and methanol (2 min)
   with drying at 150°C, inductive heating (4 × 1 s) in a
   vacuum or flaming in a bunsen, were other alternatives
   examined. Significant variations were observed between

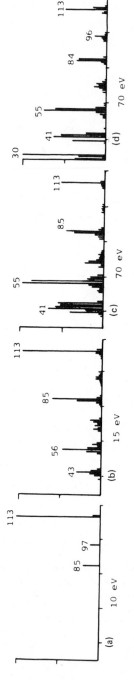

FIGURE 4.20 Comparison of mass pyrograms of Nylon-6: (a), (b), and (c) laser-Py-MS, with quadrupole analysis (from Ref. 36, by permission of Heyden & Son Ltd.); (d) heated filament Py-MS with magnetic sector analysis (from Ref. 48).

TABLE 4.4 Factors Governing Long-Term Reproducibility of Py-MS

Sample Preparation	Filament Heating	Product Transfer	Mass Analysis
Cleaning method	Equilibrium temperature	Inlet temperature	Ionisation
Solvent	Temperature rise time	Residence time	Extraction
Sample size	Total heating time	Surface activity	Separation

Source: Ref. 13, courtesy of H. L. C. Meuzelaar, University of Utah, Salt Lake City, Utah.

TABLE 4.5 Reproducibility of Py-MS Data for Glycogen. Variability Changes (%) Dependent upon Operating Conditions[a]

Variable	Condition	D(%)	Condition	D(%)	Condition	D(%)	Condition	D(%)
Wire cleaning	Reductive	0	Flame	373	Solvent	65	Inductive	0
Solvent	Methanol	0	Water	35	CS_2	27		0
Sample size (μg)	1	615	5	0	10	4	20	0
Equilibrium temp.(°C)	358	481	510	112	610	0	770	261
TRT (s)	0.1	0	0.5	-8	1.0	8	1.5	12
THT (s)	0.3	4	0.6	0	0.9	0	1.2	4
Inlet temp.(°C)	50	215	100	108	150	0	180	88
Long term (days)	0	0	12	92	26	81	34	127

[a]Reference conditions, 0; D, 2.6.

Source: Ref. 14.

mass pyrograms, and standardization is essential. The flaming method caused marked oxidation of the wire, with large differences between analyses being apparent. In contrast, the reductive method ensures the removal of oxide layers and eliminates tar formation.

2. *Solvent.* The appearance of mass pyrograms depended upon the nature of the solvent and ultrasonic suspension of the sample in methanol is recommended. Residual CS_2 was detected at m/z 76 and considerable variation was observed with BSA, e.g., H_2S at m/z 34 of low intensity, when water was used as the solvent. For biological samples whose pyrolysis varies with ionization, carefully standardized pH conditions are required for useful comparisons to be made and aqueous media are recommended.

3. *Sample size.* Background effects caused by residual contamination of the wire caused poor reproducibility at low (1 μg) loadings. Larger sample sizes overcame these problems. Five micrograms were recommended as optimum, but the latitude apparent in this parameter helps overcome the difficulty of precisely estimating the amount of a bacterial colony.

4. *Equilibrium temperature.* The effect of changes in this parameter depend upon the thermal stability of the sample. With glycogen, increases in T_{eq} result in more fragmentation and a decrease in the intensity of higher mass peaks. In contrast, BSA is of greater stability and exhibits maximum high mass contribution at 610°C. The recommended equilibrium temperature (500°C) is thus a compromise to maximize large fragments from both protein and carbohydrate components. The difference in temperature profile may be of potential use in emphasizing the contributions of these components in the mass pyrograms of microorganisms.

5. *Temperature rise time.* Although this parameter has been shown to be of considerable importance in Py-GC studies, the effect in these Py-MS studies was negligible. The effect of change in TRT from 0.1 to 1.5 s was not distinguishable from normal variations at constant TRT. This may be due to the greater volatility encouraged by vacuum pyrolysis, but is nevertheless a surprising result. It will be of interest to see if this behavior is characteristic of synthetic polymers which have precise structural definition and known kinetic and thermodynamic profiles.

6. *Total heating time.* This parameter, too, had little effect upon the reproducibility of glycogen or BSA mass pyrograms. In fact, at 610°C pyrolysis was completed within the shortest THT used (0.3 s).

7. *Inlet temperature.* The transfer of the pyrolysate into the ion source was effected through an expansion chamber held at 150°C. Substantial reduction of this temperature caused

products to condense onto the walls and to be excluded from the analysis. Higher temperatures increased the thermal energy of the products, resulting in enhanced electron-impact-induced fragmentation within the ion source.

The long-term reproducibility was also assessed by repeating analyses after 12, 26, and 34 days. Drift was clearly apparent, but a close similarity between mass pyrograms was evident. Contamination of the inlet and ion-source was felt to be the major cause of this variation, although up to 10,000 pyrolyses may be undertaken before this builds up to a troublesome level [13]. This was achieved by the use of cold reaction tubes to condense out very polar products, the expansion chamber to buffer the pyrolysate source pressure, the open design of the ion source to reduce surface interactions, and the cold screen. This latter device is extremely active in the removal of organic components which have escaped ionization—it is much more efficient than traditional diffusion pumps, which are limited by vapor pressure considerations—and in combination with the ion-getter pump, source pressures of 10^{-7} Pa (7.5×10^{-10} torr) are available. Mass spectral drift is another factor which may reduce long-term reproducibility. A change in ionizing voltage from 13.9 to 14.1 eV had a considerable effect (coefficient of variation of 4% for glycogen) upon the mass pyrogram, and although no MS tuning procedures were used, the speed of the automated instrument enables standards to be run regularly, so that changes may be readily observed.

An alternative approach to Py-MS reproducibility was taken by Hickman and Jane [49]. These workers compared mass pyrograms from three synthetic polymers, pyrolyzed with three different pyrolyzers. These were a low-power Curie-point system and the CDS Pyroprobe, both interfaced to the mass spectrometer through an empty GC column [4] (450×2 mm at 200°C with a flow rate of 5 ml min^{-1}), and the VG pyrolysis probe. Before analysis, the mass spectrometer was tuned to yield a standard toluene spectrum (Table 4.6).

The use of a simple vinyl toluene-styrene polymer enabled the origin of peaks in the mass pyrogram to be established. Styrene and

TABLE 4.6   Seventy-electron volt Toluene MS for Standardization of
Py-MS

m/z	92	91	65	63	39
I(%)	71-74	100	11-13	5-6	5

Source: Ref. 49.

vinylbenzene were the major products and were characterized by ions
at m/z 104, 103 and m/z 118, 117, respectively. The ion ratios 118/
104 enabled the product distribution to be assessed, while 117/118
estimated the MS reproducibility. In each case, the repeatability
of the mass spectrum was better than the component composition reproducibility. Pyrolysis into glass columns was more effective than
direct insertion, and a single scan estimate of the mass pyrogram
introduced at least twice as much variation as a time-averaged scan
(of 12 spectra). A significantly larger effect may be expected with
components which show a greater range of volatility. More complex
samples, such as alkyd paints, were more difficult to replicate,
with some ions showing a particularly wide distribution of intensities. Long-term reproducibility was monitored over a period of 6
weeks. This showed a small change in mass spectral operation (1 to
2%), with a somewhat greater change in the reproducibility of pyrolysis. This is illustrated for an acrylic paint in Table 4.7. Problems were encountered due to changeover between the pyrolysis systems, and because the mass spectrometer was also used for GC-MS work.
It was noted that a replacement source gave entirely different operating features.

    The long-term reproducibility of Py-MS, sufficient for compari-
sons with library data to be made, has yet to be established. Although further work is clearly necessary, sufficient information is
at hand to enable useful results to be obtained. Suggested operating
conditions are shown in Table 4.8 for a Curie-point system [14], although no definitive recommendations for sample transfer and mass
analysis are available. In design and use, it is imperative that

TABLE 4.7 Coefficient of Variation (%) for Repeat Pyrolysis of an Acrylic Paint

m/z	39		56		69		104		103/104	
Source	In-day	6-week	In-day	6-week	In-day	6-week	In-day	6-week	In-day	6-week
Curie-point	1.5	7.2	2.9	4.6	1.2	4.6	1.7	9.6	1.3	1.7
Pyroprobe	1.9	5.2	2.5	3.9	5.7	6.0	2.6	7.3	1.2	1.4
Insertion probe		4.4		6.8		6.3		9.7		2.1

Source: Ref. 49.

TABLE 4.8 Recommended Standardized Pyrolysis Conditions for Curie-point Py-MS

Parameter	Condition
Pyrolysis	Curie point
Wire cleaning	Reductive
Solvent	Methanol
Sample size	5 to 25 μg
Rf power	1.5 kW
T_{eq}	610°C
TRT	0.1 to 1 s
THT	0.3 to 1.2 s
Inlet temperature	150°C
Ionizing voltage	14 eV
Mass analysis	Quadrupole
Mass range	15 to 240
Maximum resolution	3000
Ion counting	20 MHz
Scan speed	2000 amu s^{-1}
Scans	250

Source: Ref. 14.

contamination from components of low volatility is minimized. The inclusion of standards (possible due to the short analysis time) enables compensation to be made. Furthermore, if a tuned system were developed--perhaps *via* MS standardization followed by Py-MS tuning with a sample such as KRATON 1107 [84]--relationships between different systems would be more readily revealed.

4.4. COMPARISON OF Py-MS AND Py-GC

The major features of Py-MS and Py-GC are listed in Table 4.9. The choice of technique clearly depends upon the facilities available and the nature of the study to be undertaken. The advantages of Py-MS are associated with the speed of analysis, the ease of automated

TABLE 4.9 Comparison of Py-MS and Py-GC

PARAMETER	Py-MS	Py-GC
Transfer efficiency	good	variable
Polar compounds	good	poor
Analysis time	short (1 min)	long (1 h)
Base-line stability	excellent	drifts
Background	low	column-bleed
Resolution	reproducible	variable
Retention scale	constant	variable
Mass scale	easy-essential	more difficult-optional
Computer handling	directly available	available with Py-GC-MS
Chemical information	yes	yes
Fingerprint pyrograms	easily incorporated	long analysis time
Reference samples	poorly studied	good
Quantitative analysis	yes	yes
Automation	high-requires MS-DS	low-requires GC
Cost	Py-FDMS	Py-GC-MS
Special techniques	CAMS	Mass fragmentography

data handling which results from a stable, linear mass scale, and
the chemical information which is available from the mass pyrogram.
Furthermore, any GC-MS system may be adapted to provide results, although it is essential to monitor performance by means of standard
samples. The strength of Py-GC lies with the widespread availability
of the instrumentation, its ability to separate isomers, and the excellent quantitative performance in mixture analysis. Py-GC-MS gives
structural information and extra sensitivity and specificity through
mass fragmentography.

The essential differences in the products detected appear to be
largely due to transfer losses between pyrolysis and analysis. Fig.
4.21 illustrates this for Py-GC-MS and various Py-MS techniques in
the analysis of nucleic acids [35]. The largest fragments detected
in each analysis are shown, and reveal that Py-GC-MS (N.B. furnace
pyrolysis) detects small, nonpolar degradation products [85]. Py-MS
into an expansion chamber provides low-pressure pyrolysis and polar
products are not removed by passage through a column. Nevertheless,
high molecular weight, or polar products, will condense out onto the
cold reaction tubes (a deliberate policy to reduce source contamination) so that small molecules, albeit more polar, are observed [60].
That this is due to transfer, rather than fragmentation induced by
electron impact, is shown by the fact that Py-FIMS has an almost
identical product distribution [60]. Probe pyrolysis, using 70 eV,
may detect larger fragments [86] and Py-FDMS, in which pyrolysis and
ionization are almost simultaneous, enables dinucleotide fragments
to be detected [64]. Such results herald the availability of sequencing information for biopolymers directly from Py-MS studies [33].

Direct comparisons of Py-MS and Py-GC using synthetic polymers
confirm the results [49]. An acrylic paint, which yielded butanol,
butyl methacrylate and styrene as major products in a Py-GC experiment, showed ions characteristic of these components in the mass
pyrogram (Fig. 4.22). An alkyd paint, however, was found to give
different profiles. Acrolein, methacrolein, and benzene were found
to be significant products in Py-GC, whereas the mass pyrogram

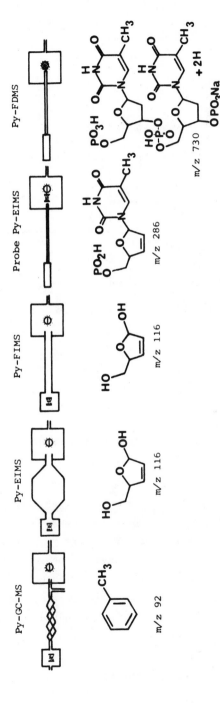

FIGURE 4.21 Large fragments detected in various modes of analytical pyrolysis of nucleic acids.

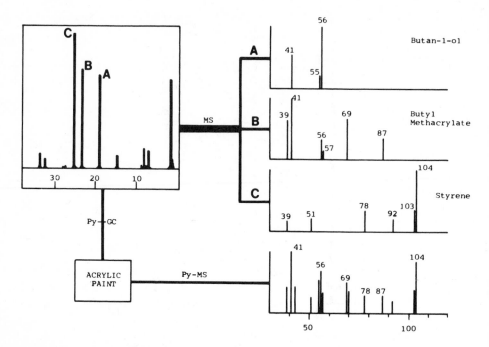

FIGURE 4.22 Pyrolysis of an acrylic paint: Py-MS versus Py-GC.
(From Ref. 49.)

(Py-MS) revealed the presence of phthalic anhydride and benzoic acid
as additional components.

The use of synthetic polymers with predictable pyrolysis profiles is an excellent way to compare Py-MS and Py-GC techniques.   In
view of reports which indicate that quantitative analysis by Py-MS
is achievable [25,86], a direct means is thus available to compare
the quantitative performance of the two methods.

Polyethylene and lipids behave differently.   Pyrolysis of these
compounds yields complex mixtures of hydrocarbons, olefins and dienes.
Py-GC analysis enables products with over 30 carbon atoms to be detected [87].   These products are highly susceptible to electron-
impact fragmentation, and mass spectra are characterized by homologous ion series with rapidly diminishing intensities.   Mass pyrograms are therefore distinguished by many fragment ions, with the
most intense fragment for polyethylene appearing at m/z 56 (15 ev)

[36], or m/z 57 (70 eV) [48]. Little evidence for species larger
than C_9 is available. Furthermore, the variation in the relative
intensity of the triplet peaks in Py-GC (above C_{21}) which enables
low-density and high-density polyethylene to be distinguished [88]
cannot be detected with conventional Py-MS.

In view of these considerations, it is perhaps surprising to
note the high degree of correspondence between the products identified in a Py-GC-MS study of *Bacillus subtilis* var. *niger* [89] and a
Py-FIMS analysis of *Pseudomonas putida* [59]. Of the 66 products
found with Py-GC-MS, 60 were included among the 113 molecular ions
detected by Py-MS.

The relationship between pyrolysis and mass spectrometry is
also of interest. Reference has already been made to the comparisons of Py-GC and EIMS in the identification of small molecules and
discussions comparing MS fragmentation with thermolysis and photolysis have appeared [90,91]. The comparison can only properly be made
with molecules which can be volatilized without decomposition into
the MS source, but which have a sufficiently low vapor pressure to
allow pyrolysis without evaporation. For example, in the study of
DNA pyrolysis it was noted that pyrolysis of the free bases mainly
resulted in volatilization. Further, the low-pressure pyrolysis of
4-phenylbutyric acid revealed the presence of the parent compound in
the mass pyrogram [17] (Fig. 4.6), whereas this component is not detected when an expansion chamber was used [19]. Caution must clearly
be exercised when detailed mechanistic interpretations are to be
made. Nevertheless, an approach to this problem has appeared. This
is based upon a detailed comparison of parent and deuterium-labeled
compounds which enables the origin of the various products to be exposed. Although used to compare MS fragmentation and pyrolysis of
4-phenylbutyric acid [19] or methionine [92], the technique is applicable to the study of the detailed mechanistic behavior of any
macromolecule which can be obtained with an isotopic label [24]. In
the case of 4-phenylbutyric acid, 2,2-, 3,3-, and 4,4-dideutero
derivatives were among compounds examined, and four major fragmentation pathways are evident:

1. Elimination of water:

$$C_6H_5-CH_2-CH_2-CH_2C=O \xrightarrow{\Delta} C_6H_5-CH_2-CH_2-CH=C=O + H_2O$$
                    H   OH

    M 164   M 146   M 18

The corresponding MS fragmentation involves loss of hydrogen from the benzylic position, but in a proportion which
indicates scrambling between the aromatic ring and the OH
prior to elimination. A 1-tetralone structure is not
indicated.

2. Elimination of formic acid:

$$C_6H_5-CH_2-CH-CH_2 \xrightarrow{\Delta} C_6H_5-CH_2-CH=CH_2 + HCOOH$$
              H   COOH

    M 164   M 118   M 46

This is a minor component of the mass spectrum and arises
via a CO loss from the ion at m/z 146.

3. Elimination of Acetic Acid Enol:

$$C_6H_5-CH \cdots CH_2-CH_2 \cdots C=O \xrightarrow{\Delta} C_6H_5-CH=CH_2 + CH_2=C{<}^{OH}_{OH}$$
         H   O
                  H

    M 164   M 104   M 60

In contrast, the stepwise loss of water and ketene characterized electron-impact-induced fragmentation.

4. Formation of benzyl radicals:

$$C_6H_5-CH_2-CH_2-CH=C=O \xrightarrow{\Delta} C_6H_5-CH_2^{\cdot} \xrightarrow[\times 2]{'H'} {}^{C_6H_5-CH_3 \quad M\ 92}_{C_6H_5-CH_2-CH_2-C_6H_5}$$

    M 146   M 91   M 182

The ion at m/z 91 in the mass spectrum is the tropylium
ion formed directly from the acid. In the mass pyrogram
this ion is produced by ionization of dibenzyl or toluene.

These studies reveal the subtle effects which accompany pyrolysis and mass spectral fragmentation processes. They show that, although fragments of the same mass are observed in both processes,
their formation pathway, and even their structure, may be very different. Such observations enhance the utility of analytical pyrolysis and indicate that, even for small molecules, the technique may
complement mass spectrometry rather than compete with it. At present, this is a largely neglected field.

REFERENCES

1. Meuzelaar, H.L.C., Kistemaker, P.G. and Tom, A., Rapid and
   Automated Identification of Microorganisms by Curie-point Py-
   rolysis Techniques. I. Differentiation of Bacterial Strains
   by Fully Automated Py-GC, in C.-G. Heden and T. Illeni (Eds.),
   New Approaches to the Identification of Microorganisms, John
   Wiley, New York, 1975, pp. 165-178.

2. Barker, C., Pyrolysis Techniques for Source-Rock Evaluation,
   Amer. Assoc. Petrol. Geol. Bull., 58(1974)2349-2361.

3. Nematollahi, J., Guess, W. and Autian, J., Pyrolytic Characterisation of Some Plastics by a Modified GC, *Microchem. J., 15*
   (1970)53-59.

4. Hughes, J.C., Wheals, B.B. and Whitehouse, M.J., Simple Technique for the Py-MS of Polymeric Materials, *Analyst, 102*(1977)
   143-144.

5. Kistemaker, P.G., Meuzelaar, H.L.C. and Posthumus, M.A., Rapid
   and Automated Identification of Microorganisms by Curie-point
   Pyrolysis Techniques. II. Fast Identification of Microbiological Samples by Curie-point Py-MS, in C.-G. Heden and T.
   Illeni (Eds.), *New Approaches to the Identification of Micro-
   organisms*, John Wiley, New York, 1975, pp. 179-191.

6. Meuzelaar, H.L.C. and Kistemaker, P.G., A Technique for Fast
   and Reproducible Fingerprinting of Bacteria by Py-MS, *Analyt.
   Chem., 45*(1973)587-590.

7. Extranuclear Laboratories, Inc., Pittsburgh, Pa.

8. VG Micromass (Analytical) Ltd., Tudor Road, Altrincham, Cheshire, U.K.

9. Fischer Labor-und Verfahrenstechnik, Meckenheim bei Bonn,
   Industriepark Kottenforst, BRD.

10. Japan Analytical Industry Co. Ltd., Tokyo, Japan.

11. Varian MAT Gmbh, Bremen, BRD.

12. Buhler, Ch. and Simon, W., Curie-point Py-GC, *J. Chromatogr. Sci.*, *8*(1970)323-329.

13. Meuzelaar, H.L.C., Py-MS; Prospects for Interlaboratory Standardisation, Proc. 26th. Annual Conference on MS and Allied Topics, St. Louis, May-June 1978, Plenary Lecture.

14. Windig, W., Kistemaker, P.G., Haverkamp, J. and Meuzelaar, H.L.C., The Effects of Sample Preparation, Pyrolysis and Pyrolysate Transfer Conditions of Py-MS, *J. Anal. Appl. Pyrol.*, *1*(1979)39-52.

15. Lincoln, K.A., Flash Pyrolysis of Solid Fuel Materials by Thermal Radiation, *Pyrodynamics*, *2*(1965)133-143.

16. Chemical Data Systems Inc., Oxford, Pa.

17. Schmid, P.P. and Simon, W., A Technique for Curie-point Py-MS with a Knudsen Reactor, *Analyt. Chim. Acta*, *89*(1977)1-8.

18. Golden, D.M., Spokes, G.N. and Benson, S.W., Very Low Pressure Pyrolysis--A Versatile Kinetic Tool, *Angew. Chem. Int. Ed.*, *12* (1973)534-546.

19. Posthumus, M.A., Nibbering, N.M.M. and Boerboom, A.J.H., A Comparative Study of the Pyrolytic and Electron Impact-Induced Fragmentations of 4-Phenylbutanoic Acid and some Analogues, *Org. Mass. Spectrom.*, *11*(1976)907-919.

20. Friedman, H.L., Griffith, G.A. and Goldstein, H.W., Thermal Analysis of Polymers by TOF-MS in R.F. Schwenker, Jr. and P.D. Garn (Eds.), Thermal Analysis, Academic, New York, Vol. I, pp. 405-416.

21. Luderwald, I., Recent Results and New Techniques in Py-EIMS of Synthetic Polymers, *Proc. Eur. Symp. Polym. Spectrosc.* (Weinheim, 1978), *5*(1979)217-255.

22. Ballistreri, A., Foti, S., Montaudo, G., Pappalardo, S., Scamporrino, E., Arnesano, A. and Calgari, S., Thermal Decomposition of Acrylonitrile Copolymers Investigated by Direct Pyrolysis in the MS, *Makromol. Chem.*, *180*(1979)2835-2842.

23. Mischer, G., Mass Spectrometry of High Polymers, *Adv. Mass Spectrom.*, *7B*(1978)1444-1451.

24. Luderwald, I., Przybylski, M. and Ringsdorf, H., Direct Pyrolysis of Polymers in the MS, *Adv. Mass Spectrom.*, *7B*(1978)1437-1443.

25. Buchhorn, G., Luderwald, I., Ringsdorf, H. and Willert, H.-G., MS Detection of Polymer Particles in Capsule Tissue Surrounding Joint Endoprothesis of Polyethyleneterephthalate, *Proc. Symp. Mass Spectrom.*, Tubingen, *1*(1977)319-329.

26. Anhalt, J.P. and Fenselau, C., Identification of Bacteria Using MS, *Analyt. Chem.*, *47*(1975)219-225.

27. Risby, T.H. and Yergey, A.L., Identification of Bacteria Using Linear-programmed Thermal Degradation MS. The Preliminary Investigation, *J. Phys. Chem., 80*(1976)2839-2845.

28. Risby, T.H. and Yergey, A.L., Linear Programmed Thermal Degradation MS, *Analyt. Chem., 50*(1978)327A-334A.

29. Charnock, G.A. and Loo, J.L., MS Studies of DNA, *Analyt. Biochem., 37*(1970)81-84.

30. Wiebers, J.L., Sequence Analysis of Oligodeoxyribonucleotides by MS, *Analyt. Biochem., 51*(1973)542-556.

31. Wiebers, J.L., Detection and Identification of Minor Nucleotides in Intact DNA by MS, *Nucl. Acid Res., 3*(1976)2959-2970.

32. Gross, M.L., Lyon, P.A., Dasgupta, A. and Gupta, N.K., MS Studies of Probe Pyrolysis Products of Intact Oligoribonucleotides, *Nucl. Acids Res., 5*(1978)2695-2704.

33. Burgard, D.F., Perone, S.P. and Wiebers, J.L., Sequence Analysis of Oligodeoxyribonucleotides by MS. 2. Application of Computerized Pattern Recognition to Sequence Determination of Di-, Tri- and Tetra-nucleotides, *Biochem., 16*(1977)1051-1057.

34. Franklin, W.E., Direct Pyrolysis of Cellulose and Cellulose Derivatives in a Mass Spectrometer with a Data System, *Analyt. Chem., 51*(1979)992-996.

35. Meuzelaar, H.L.C., Kistemaker, P.G. and Posthumus, M.A., Recent Advances in Py-MS of Complex Biological Materials, *Biomed. Mass Spectrom., 1*(1974)312-319.

36. Kistemaker, P.G., Boerboom, A.J.H. and Meuzelaar, H.L.C., Laser Py-MS: Some Aspects and Applications to Technical Polymers, *Dynamic Mass Spectrom., 4*(1975)139-152.

37. Lum, R.M., Direct Analysis of Polymer Pyrolysis Using Laser Microprobe Techniques, *Thermochim. Acta, 18*(1977)73-94.

38. Lum, R.M., Microanalysis of Trace Contaminants by Laser-probe Pyrolysis, *Amer. Lab., 10*(1978)47-56.

39. Lum, R.M., MoO_3 Additives for PVC: A Study of the Molecular Interactions, *J. Appl. Polym. Sci., 23*(1979)1247-1263.

40. Wechsung, R.F., Hillenkamp, F., Kaufmann, R., Nitsche, R. and Vogt, H., LAMMA--A New Laser Microprobe Mass Analyser, *Scanning Electron Microsc.* (1978)611-620.

41. Unsold, E., Hillenkamp, F., Renner, G. and Nitsche, R., Investigations on Organic Materials Using LAMMA, *Adv. Mass. Spectrom., 7B*(1978)1425-1428.

42. Posthumus, M.A., Kistemaker, P.G., Meuzelaar, H.L.C. and Ten Noever de Brauw, H.C., Laser Desorption-MS of Polar Non-Volatile Bio-Organic Molecules, *Analyt. Chem., 50*(1978)985-991.

43. Horning, E.C., Horning, M.G., Carroll, D.I., Dzidic, I. and
    Stilwell, R.N., New Picogram Detection System Based on a MS
    with an Extended Ionisation Source at Atmospheric Pressure,
    Analyt. Chem., 45(1973)936-943.

44. Beynon, J.H., *Mass and Abundance Tables for Use in MS*, Elsevier,
    Amsterdam, 1963.

45. Djerassi, C., Budziekiewicz, H. and Williams, D.H., *MS of Or-*
    ganic Compounds, Holden-Day, London, 1967.

46. McLafferty, F.W., *Interpretation of MS*, Benjamin, London, 1973.

47. Karni, M. and Mendelbaum, A., The Even-Electron Rule, *Org. Mass*
    Spectrom., 15(1980)53-64.

48. Hughes, J.C., Wheals, B.B. and Whitehouse, M.J., Py-MS of Tex-
    tile Fibres, *Analyst, 103*(1978)482-491.

49. Hickman, D.A. and Jane, I., Reproducibility of Py-MS using
    Three Different Pyrolysis Systems, *Analyst, 104*(1979)334-347.

50. Meuzelaar, H.L.C., Posthumus, M.A., Kistemaker, P.G. and
    Kistemaker, J., Curie-point Pyrolysis in Direct Combination
    with Low Voltage EIMS, *Analyt. Chem., 45*(1973)1546-1549.

51. Maccoll, A. (Ed.), *MTP International Review of Science, Physi-*
    cal Chemistry, Vol. 5, Mass Spectrometry, Butterworths, London,
    1972.

52. Mather, R.E. and Todd, J.F.J., CIMS: A Survey of Instrument
    Technology, *Int. J. Mass Spectrom. Ion Phys., 30*(1979)1-37.

53. Saferstein, R. and Manura, J.J., Py-MS--A New Forensic Science
    Technique, *J. Forensic Sci., 22*(1977)748-756.

54. Hileman, F.D., Johnson, J.O., Songster, J., Futrell, J.H. and
    Petajan, J.H., to be published.

55. Shimizu, Y. and Munson, B., Py-CIMS of Polymers, *J. Polym. Sci.*
    (Chem. Ed.), 17(1979)1991-2001.

56. Beckey, H.D. and Schulten, H.-R., FI- and FD-MS in Analytical
    Chemistry, in C. Merritt, Jr. and C.N. McEwen (Eds.), *Mass*
    Spectrometry, Part A, Marcel Dekker, Inc., New York, 1979, pp.
    145-266.

57. Schuddemage, H.D.R. and Hummel, D.O., Characterisation of High
    Polymers by Pyrolysis within the Field Ionisation MS, *Adv. Mass*
    Spectrom., 4(1968)857-866.

58. Schulten, H.-R. and Gortz, W., Curie-point Pyrolysis and FIMS
    of Polysaccharides, *Analyt. Chem., 50*(1978)428-433.

59. Schulten, H.-R., Beckey, H.D., Meuzelaar, H.L.C. and Boerboom,
    A.J.H., High-Resolution FIMS of Bacterial Pyrolysis Products,
    Analyt. Chem., 45(1973)191-195.

60. Posthumus, M.A., Nibbering, N.M.M., Boerboom, A.J.H. and Schul-
    ten, H.-R., Py-MS Studies on Nucleic Acids, *Biomed. Mass Spec-*
    trom., 1(1974)352-357.

61. Hummel, D.O., Structure and Degradation Behaviour of Synthetic Polymers using Pyrolysis in Combination with FIMS, in C.E.R. Jones and C.A. Cramers (Eds.), *Analytical Pyrolysis*, Elsevier, Amsterdam, 1977, pp. 117-138.

62. Schoppele, S.E., Grizzle, P.L., Greenwood, G.J., Marriott, T.D. and Perreira, N.B., Determination of FI Relative Sensitivities for the Analysis of Coal-derived Liquids and their Correlation with Low-voltage EI Relative Sensitivities, *Analyt. Chem., 48* (1976) 2105-2113.

63. Schulten, H.-R., FDMS and its Application in Biochemical Analysis, in D. Glick (Ed.), *Methods in Biochemical Analysis,* Vol. 24, John Wiley, New York, 1977, pp. 313-448.

64. Schulten, H.-R., Beckey, H.D., Boerboom, A.J.H. and Meuzelaar, H.L.C., Py-FDMS of Deoxyribonucleic Acid, *Analyt. Chem., 45* (1973) 2358-2362.

65. Schulten, H.-R., Py-FI- and Py-FD-MS of Biomacromolecules, Micro-organisms and Tissue Material, in C.E.R. Jones and C.A. Cramers (Eds.), *Analytical Pyrolysis*, Elsevier, Amsterdam, 1977, pp. 17-28.

66. Schulten, H.-R., High-Resolution FI and FD MS of Pyrolysis Products of Complex Organic Materials, in C.-G. Heden and T. Illeni (Eds.), *New Approaches to the Identification of Micro-organisms,* John Wiley, New York, 1975, pp. 155-164.

67. Hunt, O.F., Shabanowitz, J., Botz, F.K. and Brent, D.A., CIMS of Salts and Thermally Labile Organics with FD Emitters as Solid Probes, *Analyt. Chem., 49*(1977)1160-1163.

68. Binks, R., Littler, J.S. and Cleaver, R.L., *Tables for Use in High Resolution MS,* Heyden-Sadtler, London, 1970.

69. Henneberg, D. and Casper, K., Ein einfaches Tabellensystem zur Zuordnung von Bruttoformeln zu Massenwerten aus hochaufgelosten MS organischen Verbindungen, *Z. Analyt. Chem., 227*(1967)241-260.

70. Levsen, K. and Schulten, H.-R., Analysis of Mixtures by CAMS: Pyrolysis Products of DNA, *Biomed. Mass Spectrom., 3*(1976) 137-139.

71. Tuithof, H.H., Boerboom, A.J.H., Kistemaker, P.G. and Meuzelaar, H.L.C., A Magnetic MS with Simultaneous Ion-Detection and Variable Mass Dispersion in Laser-Py and Collision-Induced Dissociation Studies, *Adv. Mass Spectrom., 7B*(1978)838-845.

72. McLafferty, F.W., Wachs, F., Koppel, C., Dymerski, P.P. and Bockhoff, F.M., Ion-Structure Determination from CA Spectra, *Adv. Mass Spectrom., 7B*(1978)1231-1233.

73. Tuithof, H.H., Improvement of Resolution and Transmission in CAMS: Some Basic Considerations, *Int. J. Mass Spectrom. Ion Phys., 23*(1977)147-151.

74. Brenton, A.-G. and Beynon, J.M., MIKE Spectrometry, *Eur. Spectrosc. News.*, *29*(1980)39-42.

75. McLafferty, F.W., An Automated Analytical System for Complex Mixtures Utilising Separation by HR MS and Identification by CAMS, in C.E.R. Jones and C.A. Cramers (Eds.), *Analytical Pyrolysis*, Elsevier, Amsterdam, 1977, pp. 39-48.

76. Russell, D.H., McBay, E.H. and Mueller, T.E., Mixture Analysis by High-Resolution MS-MS, *Int. Lab.*, April(1980)49-61.

77. Beynon, J.H. and Cooks, R.G., Ion-Kinetic Energy MS, *Res. Develop.*, *22*(1971)26-31.

78. Dawson, P.H. (Ed.), *Quadrupole MS and its Applications*, Elsevier, Amsterdam, 1976.

79. Coloff, S.G. and Vanderborgh, N.E., Time-resolved Laser-Induced Degradation of Polystyrene, *Analyt. Chem.*, *45*(1973)1507-1511.

80. Meuzelaar, H.L.C., Kistemaker, P.G., Eshuis, W. and Boerboom, A.J.H., Automated Py-MS: Application to the Differentiation of Micro-organisms, *Adv. Mass Spectrom.*, *7B*(1978)1452-1457.

81. Benninghoven, A. and Sichtermann, W., Secondary Ion MS. A New Analytical Technique for Biologically Important Compounds, *Org. Mass Spectrom.*, *9*(1977)595-597.

82. Anderson, D.M., Biemann, K., Orgel, L.E., Oro, J., Owen, T., Shulman, G.P., Toulmin III, P. and Urey, H.C., MS Analysis of Organic Compounds, Water and Volatile Constituents in the Atmosphere and Surface of Mars. The Viking Mars Lander, *Icarus*, *16*(1972)111-138.

83. Meuzelaar, H.L.C. and Haverkamp, J., *Pyrolysis-Mass-Spectrometry of Biomaterials*, Elsevier, Amsterdam, in press.

84. Levy, E.J., The Application of the Molecular Thermometer Concept to the Standardisation of Polymer Degradation Patterns, 4th International Symposium on Analytical and Applied Pyrolysis, Budapest, 11-15th June 1979, Abstracts, p. 10.

85. Turner, L.P. and Barr, W.R., Py-GC of Ribonucleosides, Ribonucleotides and Dinucleotides, *J. Chromatogr. Sci.*, *9*(1971) 176-181.

86. Caccamese, S., Foti, S., Maravigna, P., Montaudo, G., Recca, A., Luderwald, I. and Przybylski, M., Copolyamides from Adipic and Truxillic Acids: Synthesis and Characterisation by Direct Pyrolysis in the MS, *J. Polym. Sci. (Chem. Ed.)*, *15*(1977)5-13.

87. Sugimura, Y. and Tsuge, S., Fundamental Splitting Conditions for Pyrogram Measurements with Glass Capillary GC, *Analyt. Chem.*, *50*(1978)1968-1972.

88. Cieplinski, E.W., Ettre, L.S., Kolb, B. and Kemner, C., Py-GC with Linearly Programmed Temperature Packed and Open Tubular Columns. The Thermal Degradation of Poly-olefins, *Z. Analyt. Chem.*, *205*(1964)357-365.

89. Medley, E.E., Simmonds, P.G. and Manatt, S.L., A Py-GC-MS Study of the Actinomycete Streptomyces Longisporoflavus, *Biomed. Mass Spectrom.*, *2*(1975)261-265.

90. Maccoll, A., Pyrolysis, Photolysis and Electronolysis, in R.I. Reed (Ed.), *Modern Aspects of MS*, Plenum Press, New York, 1968, pp. 143-168.

91. Dougherty, R.C., The Relationship between MS, Thermolytic and Photolytic Reactivity, *Dynamic Chem.*, *45*(1974)93-138.

92. Posthumus, M.A. and Nibbering, N.M.M., Py-MS of Methionine, *Org. Mass Spectrom.*, *12*(1977)334-337.

Chapter 5

Data Handling

The proper manipulation of pyrogram data is the final phase of analytical pyrolysis, and in many cases it is an essential prelude to the success of the analysis. This is particularly true when Py-GC or Py-MS data are used for fingerprint characterization. Here, quite similar pyrograms require classification, and in taxonomy it is unusual to find unique products which allow definitive identification. In general, unknown samples are compared with standard materials, and a degree of correspondence between the pyrograms is established. Visual inspection has been to the forefront in such comparisons, but fails because no quantitative assessment of the degree of correspondence is available, and no account is taken of the variability inherent between replicate determinations. For all but the most clear-cut differences, numerical data handling is preferred. The more sophisticated methods rely on numerical pattern-recognition techniques [1,2], they are capable of revealing very small differences between pyrograms and provide a numerical assessment of the match. For such methods to be successful, it is necessary for the data to be coded and scaled.

5.1. SCALING AND CODING OF DATA

5.1.1. Retention Time Standardization

Providing tuning and calibration have been properly undertaken, the mass scale of mass pyrograms is exactly reproducible and corresponding peaks in various pyrograms may be readily identified. In contrast, Py-GC data is subjected to some variability in that retention times exhibit small changes, due to progressive changes in column efficiency and operating parameters. For automatic comparisons, a range rather than a specific retention time must be taken to ensure that corresponding peaks are matched correctly. This problem can readily be overcome by a retention time standardization procedure [3-5].

In a typical situation, a number of reference peaks are chosen from among those in the pyrogram. Ideally, these should be present in all pyrograms, be prominant, well resolved, and easily identified and also have approximately the same ratio to each other from sample to sample [4]. If such conditions cannot be met, then the addition of internal standards such as the methyl esters of fatty acids [6] or hydrocarbons [7] may be possible. At least two reference peaks are required [3], but because of the complexity of the pyrogram and the long duration of Py-GC runs, seven peaks spread throughout the pyrogram were used [4]. The proportionating procedure is illustrated in Fig. 5.1.

5.1.2. Response Normalization

Differences in sample size in particular will cause the raw intensity data to show significant variations. Each pyrogram should, therefore, be normalized (scaled) to exclude loading effects. This can be done by expressing the intensity of each peak in the pyrogram (I) as a fraction of the total area, peak height or total ion current (ΣI) to give the normalized value (I^n):

$$I_i^n = \frac{I_i}{\sum_{j=1}^{N} I_j} \tag{5.1}$$

Alternatively, the intensity of each peak may be expressed as a percentage of that of the most intense peak (I_B):

$$I_i'^n = \frac{I_i \times 100}{I_B} \tag{5.2}$$

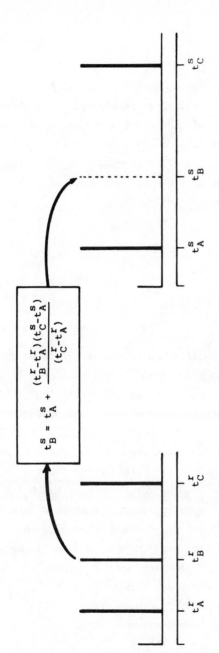

FIGURE 5.1 Standardization of retention times. t^r indicates the raw retention times and t^s the standardized retention times of components A, B and C. A and C are the nominated reference peaks. (From Ref. 4.)

Although of different magnitude, the two values are related by

$$I_i^n = I_i'^n \frac{I_B}{100 \sum\limits_{j=1}^{N} I_j} \tag{5.3}$$

Nevertheless, for comparisons of data sets Eq. (5.1) gives a better basis as variations in intensity may be averaged out, whereas Eq. (5.2) normalizes to one instantaneous measurement only. An alternative normalization procedure suitable for closely similar samples assumes that the summed areas of seven nominated reference peaks in standard (I^S) and unknown samples (I^u) are equivalent [4]. Scaling of the unknown data is performed from

$$I_i^n = I_i \frac{\sum\limits_{j=1}^{7} I_j^s}{\sum\limits_{j=1}^{7} I_j^u} \tag{5.4}$$

5.1.3. Peak Selection

The normalized pyrogram usually consists of a selection of peaks rather than the full data set. Py–GC data is selected from the most intense peaks which show reproducible resolution. The identification of a range of synthetic polymers may be achieved by comparing the three peaks only [8]. In taxonomic applications more information is required and 13 peaks have been shown to be adequate for the characterization of microorganisms [9], although 15 [10] to 24 [11] are perhaps typical with some studies involving comparisons of up to 100 components [4]. Py–MS is more amenable to automated data collection so that most peaks in a mass pyrogram are retained. The mass range is instrumentally determined; early work used an upper mass limit of m/z 50, but m/z 140, 250, or 300 are now more general. Peaks normally excluded from the data set are interfering background signals such as the ion series m/s 28, 32, 40, 44 from air [12]. One other cause for the temporary or permanent omission of peaks is variability.

For effective characterization of an unknown sample by comparison with a reference set, the pyrograms of different samples should be readily distinguishable whereas those from the same sample should be effectively superimposable. Thus, two sources of variation may be identified [13]. The variation between replicates may be classified as the *inner variance* (intrasample variation) which, ideally, should be as low as possible. The variation among different samples, the *outer variance* (intersample variation), should be significantly larger than the inner variance for meaningful conclusions to be drawn. The proper selection of the analytical set clearly requires that preliminary data is to hand. In particular, some peaks are more variable than others in a pyrogram. Exclusion of these decreases the inner variance, but peaks which show a high degree of reproducibility are not necessarily those which are most useful in differentiating between samples [13]. Peaks retained in the normalized pyrogram should thus have a small inner variance and a large outer variance [9]. This may be quantified by the variance ratio or F-test (Table 5.1) [14]. The significance of this value is determined by the number of observations in each set and may be established by reference to tables [15]. The use of stepwise discriminant analysis [16] has been used as a guide to the selection of significant peaks showing a high degree of characteristicity [17]. A value $F > 4.0$ was found to be useful, but this is effectively determined by the number of replicates and independent samples within the data set. With three replicates and 10 samples a significance level is $F(9,2) = 19.4$, whereas with 10 replicates and 30 samples significance is achieved when $F(29,9) = 2.86$ ($P = 95\%$). Using this approach for whole cells and cell fragments of 10 Salmonella serotypes, it was found that three peaks from flagella pyrograms and four peaks from DNA pyrograms enabled confident identification to be made. Whole cells (five peaks with 90% correct identification) and cell walls were less characteristic [17]. Table 5.2 lists some other parameters which are used in describing the statistics of repeat measurements.

TABLE 5.1 Calculation of the Variance Ratio

$\bar{I}_k = \dfrac{\sum\limits_{i=1}^{M_j} I_k^i}{M_j}$	$V_k = \dfrac{\sum\limits_{i=1}^{M_j} (I_k^i - \bar{I}_k)^2}{M_j - 1}$
$F_k(M_o-1, M_i-1) = \dfrac{V_k^o}{V_k^i}$	

I_k^i = Intensity of k th peak in pyrogram (i)

$\bar{I}_k$ = Mean intensity of k th peak in pyrogram (i)

M_j = Number of pyrograms in data set

V_k = Variance of k th peak

V_k^i = Inner variance of k th peak from M_i pyrograms

V_k^o = Outer variance of k th peak from M_o pyrograms

F_k = Variance ratio of k th peak. N.B. The quotient and
degrees of freedom are such that $F_k > 1.0$

Source: Ref. 14.

TABLE 5.2 Statistics of Repeat Measurements

$\sigma_k = (V_k)^{\frac{1}{2}}$	$CV_k = 100\bar{I}_k / \sigma_k$
$Se_k = \pm\sigma_k / (M_j)^{\frac{1}{2}}$	$e_k^{95\%} = \pm t_{M_j-1}^{95\%} Se_k$

σ_k = Standard deviation of k th peak

CV_k = Coefficient of variation (%) of k th peak

Se_k = Standard error about $\bar{I}_k$ (P=68.3%)

$e_k^{95\%}$ = Error limits about $\bar{I}_k$ (P= 95%)

$t_{M_j-1}^{95\%}$ = t-Value for M_j-1 degrees of freedom (P=95%)

The temporary omission of intense peaks from the pyrogram normalization procedure may be required if a significant amount of the inner variance is associated with these fragments. Their inclusion causes considerable variations between normalized pyrograms which can largely be overcome if such peaks are not used to calculate the summed intensity parameter. Very large peaks (e.g., exceeding 6% total ion intensity), or strongly varying peaks (exceeding 1% of total variance) have thus been temporarily eliminated from the calculation [18]. This approach, for example, excludes m/z 18, 31, 32, 43, for glycogen and m/z 17, 18, 34, 43, 44 for BSA. The pyrogram is normalized to the *remaining* total ion intensity and the omitted peaks (similarly adjusted) are returned to the data set. The total ion intensity thus exceeds 100%. The situation is complex because the variance cannot properly be estimated until normalization is effected. It may be that several normalization calculations are necessary to enable the best basis to be established.

5.1.4. Character Weighting

Most comparisons between pyrograms have used the unweighted normalized intensities. The differences between inner and outer variance indicate that some peaks are more characteristic than others for fingerprint identification. Weighting for reproducibility (using inverse mean inner variance) and specificity (using outer mean variance) has been found to increase the reliability of identification and to enable closely related samples to be differentiated [13] (Table 5.10).

5.1.5. Coding

For computer-based calculations, the reduced and normalized pyrograms are stored as pairs of digitized points, e.g., m/z or retention time versus intensity [4,9,11,12,18,19]. For applications which require less sophisticated data processing, pyrograms may be stored and compared as recorder traces [20], as tabulated retention time-intensity data [8], as pyrogram maps [21,22], and as line [23] or bar [24] diagrams. A further data reduction technique has also been proposed for mass pyrograms [25]. The pyrolysis of biological materials gives rise to homologous series of alkenes, ketones, nitriles, amines, and acids. If all peaks occurring within a 14-amu range are summed, a condensed pyrogram (a mixed series of homologous ions) is obtained. Although lacking the detail of the original mass pyrogram, this reduction retains overall chemical specificity of the sample.

5.2. FINGERPRINT COMPARISONS

Numerical comparisons of pyrograms attempt to establish a degree of similarity which enables samples to be grouped and identified. Various indexes have been proposed for this purpose, ranging from simple peak counts to sophisticated multivariate pattern recognition techniques [26,27]. The model chosen is determined by the nature of the study. When gross chemical differences between samples are apparent, pyrograms are characterized by the presence of unique fragments, or by large differences in relative intensities [8]. In these cases, which are exemplified by synthetic polymers and single component systems, a simple similarity coefficient is adequate. With complex pyrograms, exemplified by those obtained from microorganisms, differentiation may well be based upon small intensity differences only. Here, a more careful data-handling procedure is indicated, and multivariate statistical techniques have shown most promise.

5.2.1. Similarity Coefficients

Various similarity coefficients have been used as a basis for numerical taxonomy [26,28]. The divergence of two samples is quantified by summing the differences between various pairs of observations; in pyrolysis, these are usually the intensities of corresponding peaks in the pyrograms of the respective samples. Ideally, the coefficients are normalized so that values range from zero (no peaks are common to both samples) to 100% (identical samples). Unknowns are compared with standards or library data by the calculation of a

TABLE 5.3   Calculation of Similarity Value for Py-GC Data

$$S_{i,j} = \frac{100N_s}{N_s + N_D}$$

$S_{i,j}$ = Similarity value (%) between pyrogram (i) and pyrogram (j)

N_s = Number of peaks common to pyrograms (i) and (j)

N_D = Number of unique peaks in pyrograms (i) and (j)

Source: Ref. 5.

similarity matrix containing similarity coefficients for all possible combinations.

Similarity Value

The presence of unique peaks in fungal pyrograms enabled Vincent and Kulik to calculate similarity by counting the proportion of peaks common to pairs of pyrograms (Table 5.3) [5]. A perfect match is indicated when N_D is zero and S = 100%. Up to 123 peaks were compared in this way and a retention time window of 1% was allowed. A surprising degree of similarity was observed between species, and although this approach enabled relationships between organisms to be proposed, no account was taken of intensity differences, so that much valuable information was neglected. This modification was introduced by Emswiler and Kotula [17], who redefined N_s as the number of peaks common to both pyrograms when a retention time window of 2.5% mean value *and* a peak area window of 25% mean value were applied. The correspondence conditions are now more severe and the conformity index (C) calculated from this data was found to be lower than the similarity value. This allowed greater differentiation between species to be achieved. This is illustrated in Table 5.4 which compares similarity values and conformity indices for whole cells and cell fragments for Salmonella serotypes. It is noteworthy that *S. heidelberg* and *S. saint-pauli* (which are close serotypes)

TABLE 5.4   Similarities of Salmonella Serotypes

Bacteria	Index	Whole Cells	Cell Wall	Flagella	DNA
S. heidelberg:	S	100	90	53	39
S. saint-pauli	C	68	39	18	32
S. heidelberg:	S	71	79	35	63
S. montevideo	C	32	18	2	3

Source: Ref. 17.

match exactly when similarity is used, but differ significantly when
conformity (peak intensities differ) is used. It is also apparent
that flagella and DNA isolates are better than whole cells for positive discrimination between the serotypes [17].

Similarity Coefficient

An alternative approach [9] which has been applied successfully to
Py-GC [10] and Py-MS [29] data uses the quantitative differences in
peak intensities between the two pyrograms. This allows the calculation of a similarity coefficient (Table 5.5) which is suitable for
the comparison of pyrograms when all peaks correspond in retention
time or m/z value. Differences are calculated by measuring the intensity ratio of corresponding peaks in the two pyrograms. The intensities are always arranged so that the largest peak is the denominator and the ratio is less than unity. This also allows a comparison of pyrograms which contain unique fragments ($I_i^k = 0$, $I_j^k > 0$,
ratio = 0) to be made. The similarity coefficient is the mean peak
intensity ratio. This varies in magnitude from 1.0 (a perfect
match, $I_i^k = I_j^k$) to 0.0 which indicates that no peaks are common to
the two pyrograms. In practice, values of $S_{i,j}$ above 0.84, calculated from 13 to 15 peaks, would suggest equivalence of microbiological samples [9,10].

TABLE 5.5   Calculation of a Similarity Coefficient for Py-GC Data

$$S_{i,j} = \frac{\sum_{k=1}^{N} \left[\frac{I_i^k}{I_j^k} \right]}{N}$$

$S_{i,j}$ = Similarity coefficient between pyrogram (i) and pyrogram (j)

I_i^k = Normalised intensity of the k th peak in pyrogram (i)

I_j^k = Normalised intensity of the k th peak in pyrogram (j)

N.B. The quotient is always arranged so that $I_i^k < I_j^k$
i.e. The larger peak is the denominator.

N = Number of unique peaks in the data set

Source: Ref. 9.

TABLE 5.6   Calculation of a FIT Factor for Py-MS Data

$$F_{i,j} = 1000 \left[1 - \frac{\sum_{k=m_1}^{m_2} (I_i^k - I_j^k)^2}{\sum_{k=m_1}^{m_2} [(I_j^k)^2 + (I_j^k)^2]} \right]$$

$F_{i,j}$ = FIT factor between pyrogram (i) and pyrogram (j)

I_i^k = Relative intensity of the k th peak in pyrogram (i)

I_j^k = Relative intensity of the k th peak in pyrogram (j)

m_1 = Lower m/z limit

m_2 = Upper m/z limit

Source: Ref. 30.

TABLE 5.7   Rank Order Comparison of Pyrograms

$$r_{i,j} = 1 - \frac{6 \sum\limits_{k=1}^{N}(R_i^k - R_j^k)^2}{N(N^2-1)}$$	$$d_{i,j} = \frac{100 \sum\limits_{k=1}^{N}|R_i^k - R_j^k|}{(N^2-I)/2}$$

$r_{i,j}$ = Spearman rank order correlation coefficient between pyrogram (i) and pyrogram (j)

$d_{i,j}$ = Dissimilarity (%) between pyrogram (i) and pyrogram (j)

R_i^k = Rank order of intensity of k th peak in pyrogram (i)

R_j^k = Rank order of intensity of k th peak in pyrogram (j)

N = Number of peaks in data set

I = zero when N is even
  = unity when N is odd

Source: Ref. 32.

FIT Factor

A FIT factor has also been proposed to study the reproducibility of Py-MS data (Table 5.6) [12,30,31]. With this parameter, a perfect match results in a zero numerator ($F_{i,j}$ = 1000) and a total mismatch yields a quotient of unity ($F_{i,j}$ = 0). Typical values for replicate analyses are above 975 for a mass range of m/z 25 to 200 [12].

Rank Order Correlation Coefficient

Rank order correlation is one further example of a scaled similarity coefficient. Here, peaks are numbered according to the intensity rank order and differences in rank order between corresponding components in the two pyrograms are used to calculate the parameter. A normal procedure is to use the Spearman coefficient (Table 5.7). The dissimilarity between the two data sets is expressed as the ratio of the sum of the squares of the observed deviations $[\Sigma(R_i^k - R_j^k)^2]$ to the sum of squares of the maximum deviation possible within the data set $[N(N^2 - 1)/3]$. The expression also contains a normalization factor ($x2$), so that the coefficient is maximized ($r_{i,j} = 1$) when there is no difference in rank order ($R_i^k = R_j^k$) and a value $r_{i,j} = -1$ results when there is maximum deviation within the data set $[\Sigma(R_i^k - R_j^k)^2 = N(N^2 - 1)/3]$.

The use of squared deviations means that small variations in rank order within a large data set have little significance. To overcome this, the modulus of the deviations, rather than their squares, has been used to derive a dissimilarity index (Table 5.7) [32]. Pyrograms were also divided into arbitrary sections which also artificially increases the deviation by reducing the magnitude of the denominator. The effect of such manipulations may be seen in that for replicate pyrograms from a fungal sample a dissimilarity of 5.6% was calculated using four subdivisions of the pyrogram. If these data were calculated as a single set, a value of 1.6% results and the Spearman rank correlation coefficient for this set has a value of 0.999. Clearly, a procedure which depends so greatly upon arbitrary subdivisions of a pyrogram cannot be of general applicability. If rank orders are to be used, it would appear more appropriate to use the statistically recognized Spearman coefficient. However, for most data sets comparison may be readily achieved by similarity coefficient or FIT factor calculations and a BASIC computer program which undertakes these calculations is presented in Appendix 3. These methods, however, compare the total variability between pyrograms. Multivariate methods and factor analysis overcome this problem and are clearly better for large, highly similar sets.

t-Test

The use of t-tests (Table 5.8) has also been described in the comparison of the mean intensities of selected peaks from various bitumen samples [33]. This parameter is probability based and uses replicate determinations to reveal if the mean values of two peaks

TABLE 5.8   Calculation of a t Value to compare the Mean Intensities
of Two Peaks

$$V_k = \frac{[\sum_{1=1}^{M_i}(I_{k,1}^i - \bar{I}_k^i)] + [\sum_{1=1}^{M_j}(I_{k,1}^j - \bar{I}_k^j)]}{M_i + M_j - 2}$$

$$t_k = \frac{[\bar{I}_k^i - \bar{I}_k^j]}{[V_k(1/M_i + 1/M_j)]^{\frac{1}{2}}}$$

t_k = t-Value for comparison of mean values of k th peak in pyrogram
        (i) and pyrogram (j). There are $M_i + M_j - 2$ degrees of freedom.

V_k = Variance of data

$I_{k,1}^i$ = Intensity of the 1 th replicate of the k th peak in pyrogram (i)

$I_{k,1}^j$ = Intensity of the 1 th replicate of the k th peak in pyrogram (j)

$\bar{I}_k^i$ = Mean intensity of the k th peak in pyrogram (i)

$\bar{I}_k^j$ = Mean intensity of the k th peak in pyrogram (j)

M_i = Number of replicates (1) in pyrogram (i)

M_j = Number of replicates (1) in pyrogram (j)

Source: Ref. 33.

are significantly different [15]. It does, however, make the im-
plicit assumption that the variances of the data sets are equivalent.
Four peaks were found sufficient to differentiate the samples under
test, but it is probable that a similarity coefficient or a multivariate method would facilitate data handling, particularly if only
small differences between samples were evident.

5.2.2. Distance Functions

Similarity coefficients are effectively a means of expressing the
divergence or distance between members of the analytical set.   An
alternative method is to use a formal distance function [34].   The

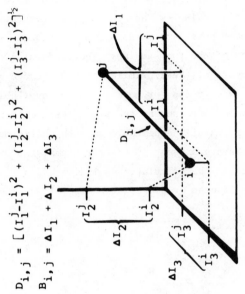

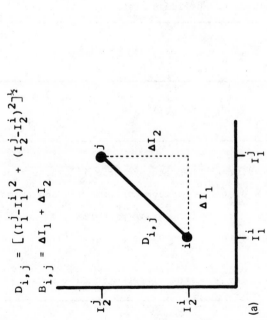

FIGURE 5.2 Distance functions for two samples using two-dimensional (a) and three-dimensional (b) data.

Euclidean distance ($D_{i,j}$) between two points in two- and three-
dimensional space is illustrated in Fig. 5.2, together with the so-
called city block approach ($B_{i,j}$). This is equivalent to displaying
two pyrograms (i and j), having two and three peaks respectively.
The intensities of these peaks are used as the coordinates to plot
the relative positions of the two samples. The calculated distance
is a quantitative measure of the dissimilarity. Distances may be
computed in a similar way if four or more peaks are available.
Here, it is not possible to represent the plot diagrammatically but
the distance formula is equally applicable to such multidimensional
data. Thus, if N peaks are present in the pyrograms, the distance
in N-dimensional space may be calculated as indicated in Table 5.9.
Normalization may be effected by dividing the sum of squares by the
dimensionality:

$$D_{i,j} = \left(\sum_{k=1}^{N} (I_k^j - I_k^i)^2 / N \right)^{\frac{1}{2}}$$

TABLE 5.9   Calculation of Euclidean Distance

$$D_{i,j} = \left[\sum_{k=1}^{N} (I_k^j - I_k^i)^2 \right]^{\frac{1}{2}}$$

$$B_{i,j} = \left[\sum_{k=1}^{N} (\Delta I_k) \right]$$

$D_{i,j}$ = Euclidean distance between pyrogram (i) and pyrogram (j)

$B_{i,j}$ = City block distance between pyrogram (i) and pyrogram (j)

I_k^i = Intensity of the k th peak in pyrogram (i)

I_k^j = Intensity of the k th peak in pyrogram (j)

ΔI_k = $|I_k^j - I_k^i|$

N = Dimensionality - number of peaks in data set

Source: Ref. 34.

Euclidean distances are sensitive to scaling and may be dominated by a few intense contributions. Further, the characteristicity of each peak depends upon the inner and outer variances. To reflect the greater importance of some peaks in the classification system weighting factors may be introduced:

$$D_{i,j} = \left(\frac{\sum_{k=1}^{N} (I_k^j - I_k^i)^2 w_k}{\sum_{k=1}^{N} w_k} \right)^{\frac{1}{2}}$$

where w_k is the weighting factor for the kth peak.

Weighting factors based upon inner and outer variance appear to be the most useful in this context [13]. These are indicated in Table 5.10. Successive weighting by reproducibility, specificity or characteristicity increased the discrimination of the distance function until two serotypes of Listeria, which exhibited minute differences in mass pyrograms, were classified correctly [13].

The *chi-squared* (χ^2) *distribution* has similarities with this approach. The calculation is shown in Table 5.11. By reference to critical values [15], this parameter tests whether groups of observations, normally distributed, show significant dissimilarity. The use of absolute intensities results in domination by intense peaks. This may be overcome if the intensity of each peak is expressed in standard deviation units. In this form the expression may be used to explore reproducibility by testing whether significant deviation between replicates is apparent.

The χ^2 test demands that all observations are independent and uncorrelated, a condition which is unlikely to be fulfilled with pyrolysis data. However, if the statistic is normalized and the square root taken, a distance function (D_i) is generated:

$$D = \left[\frac{1}{N} \sum_{k=1}^{N} \left(\frac{I_k^{'i} - \mu_k}{\sigma_k} \right)^2 \right]^{\frac{1}{2}}$$

This is the euclidean distance function, normalized for dimensionality and weighted for reproducibility (Table 5.10). It estimates

TABLE 5.10 Distance Function Weighting Factors

$$r_k = [(\sum_{i=1}^{M} v_k^i)/M]^{-1} \qquad s_k = [\sum_{i=1}^{M} (\bar{I}_k^i - \bar{I}_k)^2]/M^*$$

$$c_k = r_k \cdot s_k$$

r_k = Reproducibilty of k th peak (Reciprocal mean inner variance)

s_k = Specificity of k th peak (Outer variance term)

c_k = Characteristicity of k th peak

v_k^i = Variance of k th peak from replicates of pyrogram (i)

$\bar{I}_k^i$ = Mean peak intensity of k th peak from replicates of pyrogram (i)

$\bar{I}_k$ = Mean peak intensity of k th peak from all mean pyrograms in data set: $[(\sum_{i=1}^{M} \bar{I}_k^i)/M]$

N = Number of peaks in each pyrogram

M = Number of independent pyrograms in data set

*M-1 for variance

Source: Ref. 13.

the divergence of one replicate from the other determinations, and as such it is a useful guide to the reproducibility of replicate pyrograms. For glycogen and BSA replicates an individual distance fourfold greater than the average distance (2.6 and 3.1, respectively) indicated rejection of the run [18].

A similarity measure may also be calculated from distance:

$$S_{i,j} = 1 - \frac{D_{i,j}}{D_{(max)i,j}}$$

The Mahalonobis D^2 value is a further generalized distance function measuring the dissimilarity between the means of two groups [11]. It is defined as

TABLE 5.11 χ^2 Statistic

$$\chi_i^2 = \sum_{k=1}^{N} \left[\frac{I_k^i - \bar{I}_k}{\bar{I}_k} \right]^2 \qquad\qquad \chi_i^2 = \sum_{k=1}^{N} \left[\frac{I_k^{\prime\,i} - \mu_k}{\sigma_k} \right]^2$$

χ_i^2 = Chi-squared statistic for pyrograms (i)

I_k^i = Intensity of k th peak in replicate pyrogram (i)

$\bar{I}_k$ = Mean intensity of k th peak from all replicates excluding pyrogram (i)

$I_k^{\prime\,i}$ = Intensity of k th peak in replicate pyrogram (i) expressed in units of its standard deviation

μ_k = Mean intensity of k th peak from all replicates excluding pyrogram (i) expressed in units of standard deviation

σ_k = Standard deviation of the k th peak from all replicates excluding pyrogram (i)

N = Number of peaks in pyrogram (i)

$$D_{i,j}^2 = \underline{I}' S^{-1} \underline{I}$$

where $\underline{I}$ is a column vector, and $\underline{I}'$ its transpose (row vector), being derived from the two column vectors $\bar{\underline{I}}_i$ and $\bar{\underline{I}}_j$, which contain the means of the respective intensities, via $\underline{I} = (\bar{\underline{I}}_i - \bar{\underline{I}}_j)$. S^{-1} is the inverse of a square matrix which contains the pooled within-group covariances for the groups (i) and (j). D^2 is essentially a parameter derived from a sum of squares which is normalized for intra-group variability. It is thus quite analogous to the χ^2-derived value above. When distances between each of M pyrograms have been calculated a distance matrix of $M(M - 1)/2$ entries, representing the relationships in M-dimensional space, is available for classification. To facilitate the analysis of this data, particularly when small variations between samples are observed, multivariate statistical techniques are used.

5.2.3. Multivariate Statistical Methods

In contrast to the unidimensional similarity coefficient approach, multivariate methods enable samples to be compared in multidimensional space [35-37]. Relationships between samples may be optimized without the data reduction which characterizes the univariate approach. For this reason, multivariate techniques hold more promise as a means of classifying similar samples and as a basis for the identification of unknown samples via library comparisons.

To illustrate the approach, Fig. 5.3a shows a series of pyrograms from samples r to z. The intensities are recorded in Table 5.12. These data may be pyrograms from sample and library sources (identification of microorganisms); they may be derived from similar sources and classification and interrelationships are sought (geological or soil samples), or else various batches from a production process may be monitored for reproducibility (synthetic polymers). In this data set three peaks are used: the distances may be represented in three-dimensional space using peak intensities as coordinates, and this plot is displayed in Fig. 5.3b. It is seen that

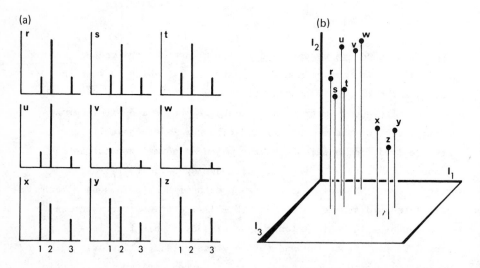

FIGURE 5.3 Three-dimensional partition of samples based upon three peaks in each pyrogram.

TABLE 5.12 Peak Intensity and Reproducibility Data for Pyrograms in Fig. 5.3

INNER	1	2	3		1	2	3		1	2	3	OUTER 1	2	3
r	19	62	19	u	18	71	11	x	44	42	24			
s	21	56	23	v	22	70	8	y	48	38	20			
t	23	58	19	w	23	71	6	z	50	35	25			
$\bar{I}_k$	21.0	58.7	20.3		21.0	70.7	8.3		47.3	38.3	23.0	29.8	55.9	17.22
V_k	4.0	9.3	5.3		7.0	0.3	6.3		9.3	12.3	7.0	231.1	267.3	61.1
V_k^i												6.8	7.3	6.2
F_k	57.8	28.7	11.5		33.0	809.9	9.65		24.8	21.7	8.7	34.1	36.5	9.8

TABLE 5.13   Distance Matrix[a] for Samples in Fig. 5.3

	z	y	x	w	v	u	t	s
r	23.99	21.74	18.71	9.42	8.04	6.97	3.27	4.38
s	20.70	18.82	15.55	13.14	11.86	11.22	2.83	
t	20.77	18.49	15.51	10.61	9.42	9.27		
u	28.96	26.27	23.71	4.05	2.94			
v	27.68	24.79	22.54	1.41				
w	28.20	25.23	23.13					
x	5.35	4.00						
y	3.56							

	uvw	xyz
uvw	25.52	
rst	9.80	19.27

[a]The small table is a reduced distance matrix calculated from the centroids of the three groups.

the samples fall into three groups; the intragroup distances are small, whereas the intergroup distances are significantly larger (Table 5.13). Thus, partition or classification of samples may be achieved. As most pyrograms have more than three peaks, higher dimensional manipulations are necessary to expose the relationships. As partition may not be as clear as in Fig. 5.3, statistical treatment is necessary to evaluate significance. Two-dimensional displays of the distribution also facilitate classification. One such method is to use a d display [38], in which the geometric centres of two groups are used as reference points. The euclidean distances of each point from these reference points is then determined, and the squared distances are used as the coordinates for the two-dimensional representation (Fig. 5.4). This method emphasizes the separation between classes, but the shortcomings of the display must be appreciated; if more than two classes are present, rotation within n-dimensional hyperspace by selection of different reference points for the cartesian plot may be necessary.

The effect of normalizing the data to account for inner and outer variance differences (Table 5.10) using the reproducibility data in Table 5.12 is shown in the upper diagonal of Table 5.14.

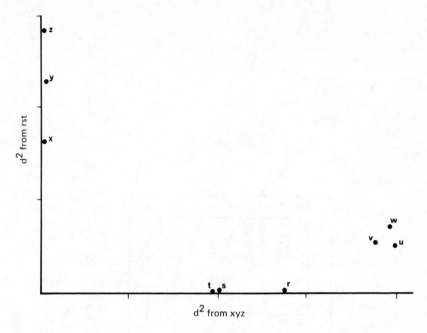

FIGURE 5.4   d-Display for samples in Fig. 5.3.

Peak 3, in particular has a low characteristicity and its influence in the final distance calculation is thus minimized. An alternative normalization is to express intensities $(I_k'^i)$ in terms of standard deviations (σ_k) from the mean intensity $(\bar{I}_k)$ of each dimension (mass number or retention time) calculated from all samples M within the data set. Thus

$$\bar{I}_k = \frac{\sum_{i=1}^{M} I_k^i}{M} \qquad I_k'^i = \frac{I_k^i - \bar{I}_k}{\sigma_k}$$

This procedure (autoscaling) normalizes each dimension to zero mean and unit variance. Distances calculated on this basis are recorded in the lower diagonal of Table 5.14.

TABLE 5.14 Distance Matrix[a] for Samples in Fig. 5.3

	z	y	x	w	v	u	t	s	r	
r	27.25	24.86	21.20	8.01	6.90	6.71	4.81	4.47		r
s	23.61	21.38	17.70	11.79	10.81	11.11	1.91		0.61	s
t	22.66	20.30	16.70	11.08	10.20	10.98		0.57	0.24	t
u	32.35	29.76	26.25	3.70	2.89		1.25	1.78	1.17	u
v	30.39	27.73	24.34	1.17		0.46	1.61	2.15	1.57	v
w	30.68	27.98	24.66		0.28	0.73	1.88	2.44	1.85	w
x	6.13	4.00		2.91	2.67	2.43	1.31	1.15	1.52	x
y	2.97		0.61	2.59	2.39	2.24	1.36	1.44	1.57	y
z		0.71	0.41	3.23	3.01	2.79	1.71	1.54	1.11	z
	z	y	x	w	v	u	t	s	r	

[a]The upper diagonal is normalized to include inner and outer variances (Table 5.11). The lower diagonal has been calculated from data normalized to zero mean and unit variance for each dimension.

TABLE 5.15   K Nearest Neighbors for Samples in Fig. 5.3

Neighbour	r	s	t	u	v	w	x	y	z
First	t	t	s	v	w	v	y	z	y
Second	s	r	r	w	u	u	z	x	x
Third	u	u	u	r	r	r	t	t	s

K-Nearest-Neighbor Technique

The nearest-neighbor technique assigns samples to the class to which
their K nearest neighbors in n-dimensional space belong (Table 5.15).
K is a small, odd number, usually 1 or 3. This technique has been
used to evaluate Py-GC data on 120 synthetic polymers [23]. Pyrograms were coded, using either 20 or 40 peaks with K = 1 or 3. Various structural features within the polymers were studied (e.g.,
presence of benzene ring, CN and CO functions) and classification on
the basis of these features was attempted. It was found that considerable overlap of the classes was apparent which reduced the certainty of classification. An increase in the number of peaks analyzed per pyrogram (i.e., the dimensionality) improved the precision
from 79.4% (average correct classification, N = 20) to 82.2% (N =
40). This work exposes the problem of data preparation. Peaks were
chosen by dividing the pyrogram into zones of equal width and selec-
ting the most intense peak in each zone. The intensities were normalized and the logarithm used for further calculation. This has
the effect of increasing the importance of small peaks at the expense of more intense components and will lead to compression of the
distances. Further, calculations were performed on the integer
values alone, which effectively means that differentiation between
corresponding intensities is dependent upon the base for normalization and requires an order of magnitude difference.

Linear Learning Machine Method

An alternative procedure which involves further calculation is the
linear learning machine [39]. In this technique a plane (hyperplane)

is sought in multidimensional space which separates two classes of
sample. This is achieved by considering each sample to be described
by a vector containing the N peak intensities derived from the pyrogram. A further member is added to the set. This N + 1 component
has the value of unity, and has the effect of forcing the hyperplane
through the origin. Samples (i) and (j) from different classes are
thus described by the vectors

$$\underline{I}^i = (I_1^i, I_2^i, I_3^i, \ldots, I_N^i, 1) \qquad \underline{I}^j = (I_1^j, I_2^j, I_3^j, \ldots, I_N^j, 1)$$

For each element of the array in turn a weighting factor or vector component (w) is sought via an iterative technique. The magnitude of w is selected so that when the product sums (s) are
calculated:

$$s^i = \sum_{k=1}^{N+1} I_k^i w_k \qquad s^j = \sum_{k=1}^{N+1} I_k^j w_k$$

these have opposite signs (i.e., $s^i > 0$; $s^j < 0$). This indicates
that samples (i) and (j) are on opposite sides of the hyperplane and
that partition into classes is effective. The magnitude of the
product sum is a measure of the distance of the sample from the
hyperplane. The larger the magnitude the greater is the confidence
that the samples are classified correctly. The advantage of this
technique is that the vector components may be estimated from a
small set of samples (the training set) and unknown samples may be
classified by reference to these components, rather than to a whole
library of data. It is necessary, however, to ensure that a sufficient number of samples are included in the training set. If N
peaks are used for classification, a minimum of 3N spectra per
class must be available. Further, the technique makes binary deci-
sions and is thus ineffective for classes not represented in the
training sets.

If M classes of sample are present, M - 1 hyperplanes are
necessary to fully classify samples and a multilayered learning
machine is required [39]. The minimal spanning tree [40] is

FIGURE 5.5 Minimal spanning tree for samples in Fig. 5.3.

constructed by joining each class to its nearest neighbor in N-dimensional space, with the restriction that no closed loops appear (Table 5.13, Fig. 5.5).

The vector $\underline{V}_1$ (containing N + 1 vector components w) will enable partition of the samples into classes xyz or rst, uvw. Subsequent application of the second vector ($\underline{V}_2$) allows final differentiation into the rst or uvw classes to be effected (Table 5.16).

The linear learning machine approach has been applied to the classification of structural elements in synthetic polymers, using the same data as before [23] and a further set of 63 pyrograms [41]. The linear learning machine method was found to be less effective than the nearest-neighbor technique (64% correct classifications) with one set of data [23], but better results were obtained with an alternative set [41]. Here, 20- (82.8% correct, 17 iterations to identify vector components) and 40- (82%, 8 iterations) dimensional data were used with a training set of 30 pyrograms and 33 samples to be classified. These results were again used to determine the presence of specific functional groups in various synthetic polymers. Although promising in concept, insufficient partition between classes was again evident. This may well be due to data presentation--intensities were evaluated on a scale from 0 to 9--and the approach is worthy of further study. A FORTRAN program to undertake learning machine calculations is listed in the book by Jurs and Isenhour [1]. Statement 30 should be corrected to 30 IDPR (I) = I + NTRSET. A BASIC translation of this program is given in Appendix 3.

Cluster Analysis

Cluster analysis [42] is a range of techniques which are concerned with the recognition of groups or clusters of points in N-dimensional space. Essentially, the distance matrix is searched for the two

TABLE 5.16 Learning Machine Classification for Samples in Fig. 5.3[a]

| $\dfrac{V_1}{xyz\,V\,rst,uvw}$ | | xyz' | | rst' | | uvw' | | $\dfrac{V_2}{rst\,V\,uvw}$ | | rst' | | uvw' | |
		I_i	$I_i \times W_i$	I_i	$I_i \times W_i$	I_i	$I_i \times W_i$	rst'	$\dfrac{V_2}{V\,uvw}$	I_i	$I_i \times W_i$	I_i	$I_i \times W_i$
W_1	-0.203	47	-9.54	21	-4.26	21	-4.26	W_1	0.135	21	2.84	21	2.84
W_2	0.161	38	6.12	59	9.50	71	11.43	W_2	-0.123	59	-7.26	71	-8.73
W_3	0.004	23	0.09	20	0.08	8	0.03	W_3	0.329	20	6.58	8	2.63
W_4	0.098	1	0.10	1	0.10	1	0.10	W_4	0.100	1	0.10	1	0.10
S			-3.23		5.42		7.30	S			2.26		-3.16

[a]Training was undertaken with a dead zone of ±1 using the raw data in Table 5.11.

points with the smallest distance of separation. These are replaced
by one point at the center of gravity of the pair. A new distance
matrix is now computed and this is again searched for its smallest
element. This process is repeated until all the points have been
assigned to clusters. Clearly, this process will continue until all
points have been assigned to a single cluster unless control parameters are included to establish an acceptable partition. One such
variable is the radius of the hypersphere containing the cluster.
Variation of this between limits should not affect the number of
clusters and indicates that these are compact and well separated.
Various programs [42] and algorithms [43,44] which perform these
analyses have been described, and suitable listings have appeared
[42]. Cluster analysis programs are also part of the BMDP package
[16].

The technique of cluster analysis has also been used as the
basis of the taxometric map (TAXMAP) classification of fungal pyrograms [6,45-48]. Ideally, in this context, each cluster should consist of the replicate analyses from each organism. Independent
clusters should contain the different strains or species which com-
prise the analytical set. The clusters obtained in these analyses
were mapped by connecting each cluster to its nearest neighbors in
N-dimensional space. In this way, relationships between the pyrograms of the various organisms may be exposed. In the event it was
found that the pyrograms did not necessarily indicate genetic similarity as determined by more traditional tests [6]. Nevertheless,
discrete and compact clusters for replicates were obtained and the
method has clear applications in fingerprint identification of
complex samples.

The method has also been applied to the differentiation of various polymers [23]. In this instance, individual clusters were obtained from rubbers, polyacrylonitriles, and polyethylenes, whereas
various condensation polymers showed no tendency to form clusters.

This technique may also be used in monitoring production. In
this instance, the appearance of clusters would indicate that
greater process control is required.

Nonlinear Mapping

One problem associated with multidimensional pattern recognition is that the relationships are not easily represented in two dimensions. Thus, data may be held in a large matrix [i.e., M(M - 1)/2 entries (cf. Table 5.13 for three samples with three replicates each!)] with the inherent difficulty of interpretation. Alternatively, the projection of the distance matrix as a nearest-neighbor map into two dimensions (as with taxonometric maps) causes distortion and hinders visual recognition of patterns. A technique which overcomes this disadvantage to a large extent is the nonmetric multidimensional scaling or nonlinear mapping technique of Kruskal [49,50]. This involves the calculation of a new configuration in two dimensions which best reflects the original configuration in N-dimensional space. The basis for this procedure is that the distances in the two-dimensional representation ($d_{i,j}$) must reflect as nearly as possible the distances in the N-dimensional array ($d_{i,j}$), so that the rank order of distances (monotone relationship) is retained. The quality of the fit is assessed by means of the stress (S) in the configuration such that:

$$S = \left(\frac{\sum\limits_{\substack{i=1 \\ j=i+1}}^{\substack{M-1 \\ M}} (d_{i,j} - \hat{d}_{i,j})^2}{\sum\limits_{\substack{i=1 \\ j=i+1}}^{\substack{M-1 \\ M}} (d_{i,j})^2} \right)^{\frac{1}{2}}$$

The distances ($\hat{d}$) are sought which minimize the stress factors. When a perfect monotonic sequence is obtained (i.e., corresponding d and $\hat{d}$ values have the same rank order) the stress of the configuration is zero. If such a configuration cannot be achieved, the nearest monotonic sequence is generated by sequentially moving pairs of points to distort the monotonicity. The configuration is thus continuously changed until a minimum stress factor is encountered. At

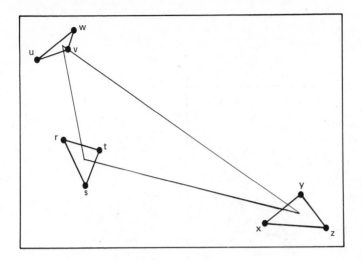

FIGURE 5.6   Nonlinear map for samples in Fig. 5.3.

this point the two-dimensional matrix displays many of the relationships present in multidimensional space. In particular, the nearest
neighbors are illustrated well and are closely associated in the nonlinear map. In contrast, it is often necessary to distort relative
distances so that intergroup separations may be disproportionate.

The data illustrated in Fig. 5.3 is displayed as a nonlinear
map in Fig. 5.6. This retains the overall distances of the original
data and reflects the interrelationships exposed by the three-
dimensional plot. The nonlinear mapping technique is of increased
value with data of higher dimensionality, which cannot be represented simply. In these cases, there is often no better alternative
to display trends and associations within the data. The technique
has been used extensively by Meuzelaar's group to classify microorganisms and bio-organic samples [13], and by Carmichael to represent
taxometric maps [47]. Perhaps the most potent illustration of the
power of modern analytical pyrolysis techniques is the application
of nonlinear mapping to the mass pyrograms obtained from two serotypes (I and IVB) of Listeria [13]. The average spectra of the two
strains displayed minute differences only (Fig. 5.7a), but the

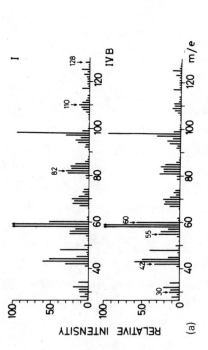

(b)

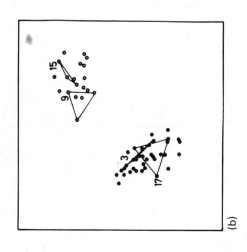

(a)

FIGURE 5.7 Average mass pyrograms (a) and nonlinear map (b) showing partition of Listeria into Sero-types I or IVB. (From Ref. 13.)

nonlinear map, weighted for characteristicity, clearly showed the
partition into the two serotypes (Fig. 5.7b). This elegant mode of
data handling will surely see many more applications in future. The
algorithm for the calculation has been described in detail [50] and
the FOM program packages are in regular use [51,52].

Factor Analysis

Factor analysis [53,54] is a technique in which correlations between
components (peak intensities in a pyrogram, features) of various
classes (samples) are examined in order to expose those factors which
are of significance in describing the classes most effectively. The
rationale is that many features (peaks) are present in the pyrograms of an analytical set. In most instances, however, there is
considerable degeneracy of information and many of the measurements
are interrelated. The variation evident between samples may thus be
explained by a small number of independent properties, usually very
much smaller than the number of features detected. In a mass pyrogram, for example, an ion series produced by electron-impact fragmentation of a single pyrolysis product should correlate highly with a
similar series in a second mass pyrogram. Here, instead of a series
of independent comparisons between the pyrograms, we have a series
of replicate determinations of the same comparison. Factor analysis
identifies such trends in the data--allowing, for example, the determination of the number of distinctive compounds displayed in the
pyrogram--and also allows considerable data reduction to be effected.
The resultant factor pyrograms contain the essential information of
the raw set, but in a much more compact and efficient presentation.

To illustrate,the approach, Fig. 5.8a records a scatter diagram
in which the intensities of two peaks from a series of mass pyrograms
have been plotted. The points are enclosed by an ellipse with major
and minor axes and a regression line is plotted through the points.
The slope of this line gives the correlation between the two variables. This information may be used to construct a vector diagram,
as the cosine of the angle between the two vectors is equal to their

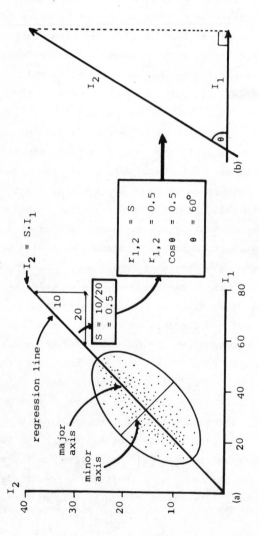

FIGURE 5.8 Correlation of the intensities of two peaks in a series of mass pyrograms: (a) rectilinear plot; (b) vector diagram (s, slope; r, correlation between intensities I_1 and I_2; θ, angle between vectors I_1 and I_2).

correlation (Fig. 5.8b). Thus, highly similar features have a high correlation and the vectors approach coincidence.

For the many peaks in a set of pyrograms, all possible correlations are calculated and hence a vector map of the relationships may be produced. These relationships are again in multi-dimensional space. As even three-dimensional representations of these lose clarity, Fig. 5.9a holds a correlation matrix of three intensities which resolve into a two-dimensional vector map (Fig. 5.9b). The significance of each vector (peak) is assessed by defining a reference vector. In principal components analysis, this reference, or first factor vector, is the centroid or resultant of all vectors. When this axis has been calculated, the angles subtended to it by each component vector are determined. The cosines of these angles represent the correlation to the first factor vector. These correlations are termed the factor loadings and represent the weighting of each peak to the first factor. This analysis is recorded in Table 5.17. A second factor vector, orthogonal to the first, is next defined, and the process is repeated. The loadings of each peak give rise to the components of the second factor. Each factor may be assessed for importance by the latent root, which is the sum of squares of the loadings. The explained variance is the quotient of the latent root and the number of peaks. As each new factor is extracted, the weighting becomes progressively less, and eventually no further factors need be extracted to account for the variance

TABLE 5.17 Factor Analysis of Data in Fig. 5.9

Peak	Factor I		Factor II		Communality (h^2)
	Angle	Loading	Angle	Loading	
1	52.68	0.6063	142.68	-0.7953	1.000
2	7.32	0.9919	82.68	0.1274	1.000
3	47.32	0.6779	42.68	0.7351	1.000
Latent root		1.8110		1.1891	3.000
Explained variance (%)		60.37		39.63	100.000

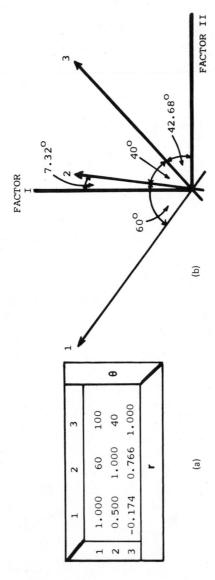

FIGURE 5.9   Correlation matrix and vector map of three intensities.

TABLE 5.18   Factor Analysis of Samples in Fig. 5.3

Peaks	1	2	3	FACTOR I loading
1		157.52	54.27 θ	0.908
2	-0.924		147.14	-0.997
3	ᵣ 0.584	-0.840		0.869
Latent Root				2.574

within the data set. A useful cutoff point is when the latent root falls below unity. If the analysis is undertaken to completion, the sum of squares of the loadings (communality) is equal to the sum of the latent roots. An accepted indication of significance for each loading would be values of >0.3 or <-0.3. For completion, the results of factor analysis on the samples in Fig. 5.3 are given in Table 5.18. Here, over 85% of the variance is accounted for by the first factor and no further factors are sought. Each peak also contributes significantly to the factor. This behavior is hardly surprising in view of the small sample set employed.

In analyses of this type the factor vectors are equivalent to the major and minor axes of the elliptical data distribution (Fig. 5.8), and the factors are principal components. Factor analysis extends this approach by developing factors which are derived from the principal components by rotation. The axes are rotated so that they pass through clusters of experimental points. The coordinates of these points are considerably simplified and individual points depend strongly on one factor only (i.e., partial factors). Thus, the orientation of the coordinate system is chosen so that the independent factors emphasize pure properties (mass spectra), rather than depend upon the mixed factors of principal components. Rotations may involve orthogonal (Varimax) or oblique (Promax) operations.

To give some indication of the power of factor analysis in revealing patterns in mass pyrograms, Figure 5.10 displays a series of mass pyrograms derived from an acrylic paint. This data is presented

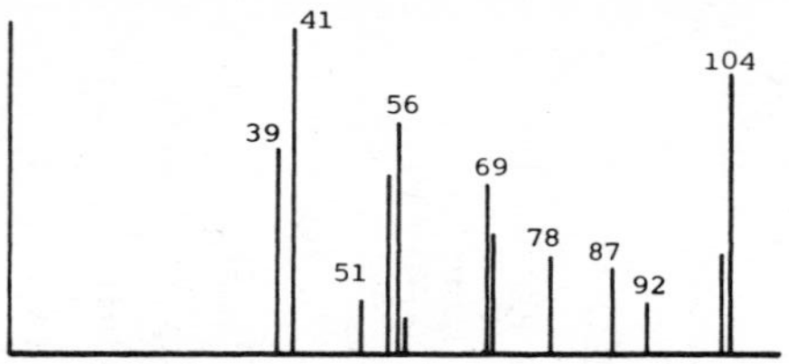

RUN	39	41	51	55	56	57	69	70	78	87	92	103	104
1	0.200	-0.153	0.000	0.000	-0.103	-0.027	0.111	0.551	-0.099	-0.097	-0.348	-0.105	-0.181
2	0.432	0.107	-0.327	-0.873	-0.835	0.435	0.610	1.209	-0.604	0.519	-0.718	-0.592	-0.453
3	0.200	-0.586	0.327	-0.873	-1.274	-0.027	0.012	-0.435	0.238	-0.097	0.392	0.382	0.297
4	0.045	-0.499	0.327	0.000	-0.250	-0.489	-0.188	-0.929	0.743	-0.507	0.762	0.544	0.433
5	0.045	-0.066	0.000	0.873	0.921	-0.489	-0.088	-1.422	0.069	-0.302	0.022	0.220	0.092
6	-0.804	-1.625	1.306	0.873	0.043	-1.413	-1.682	1.867	1.248	-1.533	1.131	1.193	1.388
7	-0.573	-1.019	0.980	1.746	1.652	-1.413	-1.582	0.222	0.743	-1.533	0.762	0.868	0.910
8	-2.426	-0.499	1.306	0.436	-0.103	-0.489	-0.586	1.045	1.248	-0.507	1.131	1.193	1.183
9	-2.040	-0.066	0.980	-0.436	-0.542	-0.027	0.111	0.058	0.743	0.109	0.762	0.868	0.979
10	-0.341	0.107	-0.653	1.309	1.652	-0.489	0.211	1.209	-0.604	-0.302	-0.718	-0.592	-0.521
11	0.586	0.973	-0.980	0.000	0.628	0.435	0.510	-0.271	-1.110	0.929	-1.088	-1.078	-0.931
12	0.432	-0.413	0.000	-1.309	-1.566	-0.027	-0.088	0.551	0.406	0.314	0.392	0.382	0.160
13	0.354	0.800	-0.980	0.873	1.360	0.435	0.111	0.222	-0.941	0.519	-0.718	-1.078	-0.999
14	1.204	1.406	-0.980	-1.746	-1.420	1.821	1.606	-1.586	-1.110	1.545	-0.718	-1.241	-1.203
15	1.127	1.753	-1.633	-0.873	-0.396	1.821	1.606	-1.093	-1.615	1.545	-1.828	-1.565	-1.544
16	0.818	1.406	-1.306	-0.873	-0.103	1.359	1.108	-0.929	-1.110	1.134	-1.088	-1.078	-1.271
17	0.741	-1.625	1.633	0.873	0.336	-1.413	-1.782	-0.271	1.754	-1.738	1.871	1.680	1.660
Mean	568.41	944.76	155.00	93.00	696.71	93.06	465.88	308.65	258.59	232.47	123.94	284.65	775.65
σ	12.947	11.546	3.062	2.291	6.835	2.164	10.037	6.082	5.938	4.875	2.704	6.164	14.667

FIGURE 5.10 Py-MS data from an alkyd paint.

TABLE 5.19 Factor Analysis of Data in Fig. 5.10

m/z	PRINCIPAL COMPONENTS		VARIMAX		PROMAX	
	I	II	I	II	I	II
39	0.632	0.018	0.598	-0.207	0.629	-0.317
41	0.960	0.100	0.933	-0.247	0.965	-0.421
51	-0.960	-0.238	-0.982	0.118	-0.983	0.304
55	-0.634	0.761	-0.323	0.936	-0.525	0.980
56	-0.251	0.947	0.101	0.974	-0.120	0.937
57	0.956	-0.210	0.820	-0.535	0.919	-0.682
69	0.957	-0.147	0.843	-0.477	0.928	-0.629
70	-0.506	0.184	-0.408	0.352	-0.477	0.423
78	-0.956	-0.263	-0.987	0.093	-0.983	0.280
87	0.962	-0.166	0.841	-0.496	0.930	-0.647
92	-0.931	-0.295	-0.975	0.055	-0.962	0.240
103	-0.958	-0.254	-0.986	0.102	-0.984	0.289
104	-0.971	-0.215	-0.984	0.143	-0.991	0.328
EV(%)	72.00	14.89	64.82	22.07		

as autoscaled intensities, with each mass having a mean intensity of zero and a standard deviation of unity. Factor analysis reveals that there are two factors, which together account for over 86% of the variance within the data set (Table 5.19).

A more useful presentation is available. A factor mass pyrogram (factor spectrum) may be calculated by determining, for each factor, the product at each mass of the factor loading, and the standard deviation of the peak intensity. Each factor now represents the contribution of each mass to the total variance of the data. These calculated intensities are analogous to the intensities of certain peaks within the original mass pyrogram, with one difference. The factor loadings are derived with positive or negative contributions. This is reflected in the intensities of the factor mass pyrograms which comprises two series of peaks, one positive and the other negative. These represent two closely correlated components, but with different dependences: as the intensity of one increases, the other decreases. The factor mass pyrogram calculated from the first Varimax

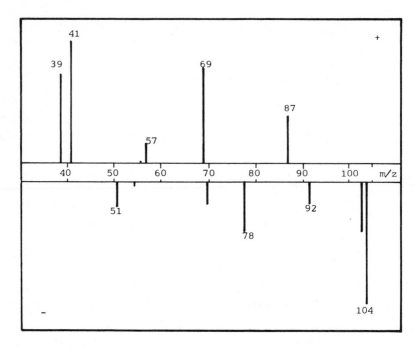

FIGURE 5.11   First Varimax factor mass pyrogram calculated from
Table 5.19 and Fig. 5.10.

factor in Table 5.19 is displayed in Fig. 5.11. It may be recognized
that the factor mass pyrograms show a high degree of similarity to
the mass spectra of pure components, when determined at 70 eV. Under
low-energy ionization conditions the factor mass pyrogram will represent those species (molecular ions) whose formation is controlled by
similar factors. In this particular example the spectra correspond
to butyl methacrylate and styrene, which are the major pyrolysis
products from the paint. Comparison of this data with that displayed in Fig. 4.22 reveals the effectiveness of this approach, although peaks which are common to two or more components may not be
appropriately extracted. Thus, the peak at m/z 39 is common to both
components but is assigned solely to the methacrylate.

A related approach is the soft independent models describing
class analogy (SIMCA) pattern recognition technique [55,56]. In this
method a principal components analysis is performed on the replicates

from each class of a known standard series. Each class may thus be characterized by a unique factor vector. Unknown samples are analyzed similarly and are assessed for similarity to the known standards through a comparison of the principal components. Thus, instead of comparing whole pyrograms peak for peak, a reduced pyrogram is used. This contains the essentials of the raw data by retaining those peaks whose intensities correlate highly (the characteristic peaks), while a reduced emphasis is placed upon widely varying components with no apparent pattern. This is a significant advance on unidimensional approaches such as similarity coefficients and FIT factors, which are based on the total variability between pyrograms. Classification of unknowns can thus be undertaken with greater confidence. The comparison between the various principal components is undertaken by a multiple regression technique in which deviations between the mean values of a class and those of an unknown sample (residuals) are assessed. The sample is classified if the residuals fall within the limits generated from the within-class variation. These techniques also enable outlying samples--those which do not belong to any of the classes in the learning set--to be identified. A series of papers have discussed the reproducibility parameters assumed in this multivariate approach [57-59], and have illustrated the application of SIMCA to fungal taxonomy. Comparison of uni- and multivariate approaches showed that the former succeeded in classifying 45.5% of samples correctly, whereas pattern recognition increased this to 84.8%. As 26 peaks were used in these studies, a fraction of the total number of components produced, it is highly probable that greater certainty of classification is possible using a larger data base.

Canonical correlation analysis is one further related technique which seeks relationships between two sets of variables. It may be considered as a double principal components exercise in which those components of the first variable (peak intensities of class i) which correlate most highly with components of the second variable (peak intensities of class j) are determined. As with principal components

or factor analysis, a series of canonical variate axes are output,
each with a set of loadings for each peak in the data set. These
indicate the importance or contribution of each peak to the canonical correlation. Axes are extracted until most of the variation
within the data is accounted for. Data from the first two canonical
variates has been used to produce a two-dimensional partition of
five bacterial genera [11]. The coefficients of each canonical variate axis were multiplied by the intensities of the corresponding
pyrogram peaks and summed to produce the two coordinates. The rectilinear plot of first vs second canonical variate showed differentiation of the genera when autoscaling, rather than merely normalization, was applied.

To date, factor analysis has been essentially limited to mass
spectra [60,61], and little use of the technique in analytical pyrolysis has appeared, although its value in the sequence determination
of DNA through Py-MS analysis has been demonstrated [62]. Applications have been essentially restricted to principal components, and
surprisingly have been restricted to Py-GC analyses [63,64]. Results
were useful in identifying the more important components in the pyrogram, and plots using the first and second principal components as
coordinates enabled outlying points to be identified and reproducibility to be studied. Clustering into species or generic groups was
not particularly evident, although association dependent upon nutrient media was observed [64]. Factor analysis has also been used
successfully in assessing the structure of four-component copolymers
[65].

It would appear that factor analysis is an elegant technique
which is appropriate for discriminant analysis, but more especially
to reveal patterns in complex multivariate data. Py-MS studies, involving both pattern recognition (taxonomy) and mechanistic (component identification) aspects, would surely benefit from a wider application of this mode of data handling.

Indeed, a recent report confirms the potential of the factor
analysis approach [66], with a reexamination of the reproducibility
of Py-MS [18]. Eight factors were found to account for over 95% of

the variance but usually only the first factor was found useful for physicochemical inference. The variates examined were essentially those discussed earlier (Sec. 4.3) and factor pyrograms representing the positive and negative components of the first factor were presented. The differences in mass distribution between these two components enabled a qualitative description of the variation in reproducibility to be made. For example, the glycogen factor pyrogram reveals that hexose fragments (Table 7.6) characterize the low-temperature pyrograms, whereas many of these undergo further reaction and smaller fragments characterize higher temperature pyrolyses. In contrast, with albumin the characteristic protein residues (Table 7.1) are in greater evidence at higher temperatures, presumably due to the greater energy required for the C-C fissions, which are predominant reactions. Similarly, the effect of expansion chamber temperature reveals that at higher temperatures the transfer of larger molecules is more efficient, but enhanced EI fragmentation results.

Computer programs to perform factor analysis are available with the BMDP package [16], the SPSS package [67,68] and from QCPE [69]. BMDP and SPSS also contain canonical correlation and discriminant analysis routines.

Multiple regression analysis [63] and set theory [70] have also been applied to the classification of pyrograms. In view of the array of mathematical procedures now available for pyrogram comparison, it is to be hoped that numerical analysis of all pyrograms which do not exhibit unique characteristic components will be deemed essential.

5.3. QUANTITATIVE ASPECTS

Although data handling in the fingerprint applications of analytical pyrolysis is of vital importance, adequate computation and control of the results of quantitative analyses is also essential [71,72]. In this mode, the intensity of a peak in a pyrogram is taken to be quantitatively related to the amount of a precursor in the original sample which fragments to produce the volatile component. Unless

considerable experience is available on the performance of the assay,
the usual approach is to construct a calibration curve of response
(y) versus concentration (x). In the event that operating parameters,
particularly the amount pyrolyzed, are not standardized exactly, it is
good practice to introduce an internal standard. The response then
becomes the intensity ratio of the component and internal standard
peaks. The best line through the data points is the regression line,
and unknown samples may be estimated by interpolation, using the
measured peak intensity ratio. The best line is normally calculated
by minimizing the squared residuals (in y) using a least squares procedure. The quality of the line--the extent of the scatter of data
points about the line--is assessed by the correlation coefficient
(r), or more properly its square. Tables of significance for this
parameter, which depends upon the number of data points, are available [15,73], but these values say little about the value of the line
in quantitative analysis. The important features are the limits of
error about the mean interpolated value, which are governed by the
limits of error about the slope and intercept. Error limits allow a
value judgment to be made concerning the random error associated
with the assay. If these are wider than those which may be safely
tolerated by the analyst, further analytical development to increase
control of the variable is required.

The program STAT (Appendix 3) is designed to undertake the calculations entailed in the assessment of error limits. This takes as
input the series of paired response (y) and concentration (x) points
and calculates the various parameters and probability estimates to
describe the correlation between the two variables. Unknown points
are then taken and the interpolated values, together with the associated error limits, are calculated. Two options are available.
One is appropriate if the assay has been designed with replicate
determinations of response for a series of identical concentration
values. Here, reproducibility and regression are assessed by means
of an analysis of variance procedure. If no replicates are available, a second option assesses the data through a simple least-squares

procedure. The program also contains a subroutine which calculates reproducibility parameters (mean, standard deviation, etc.) for repeat measurements. This may be addressed directly if required.

When the assay performance is satisfactory, further manipulation of the data may be required. Typically, quality control procedures require an assessment of long-term trends: Is the product slowly approaching an unsatisfactory level? Although this may be assessed by simply plotting the assay results on a time base, a more appropriate method is to apply the Cusum (cumulative sum) technique [72,74]. This is useful for the study of any variable with respect to time and may be used either as a postmortem device or as a quality control procedure.

The technique may be illustrated with the data in Table 5.20 which records the results of a time series of quality control assays, the residuals from the anticipated or reference value, and the cumulative sum of the residuals (Cusum). The data is plotted as in Fig.

TABLE 5.20   Cusum Table for 30 Successive Process Runs

Run	1	2	3	4	5	6	7	8	9	10	Mean
Assay	103	102	98	100	102	103	104	98	103	97	101
Residuals (assay−100)	3	2	−2	0	2	3	4	−2	3	−3	
Cusum	3	5	3	3	5	8	12	10	13	10	

Run	11	12	13	14	15	16	17	18	19	20	Mean
Assay	99	97	102	98	96	103	97	97	101	100	99
Residuals	−1	−3	2	−2	−4	3	−3	−3	1	0	
Cusum	9	6	8	6	2	5	2	−1	0	0	

Run	21	22	23	24	25	26	27	28	29	30	Mean
Assay	100	97	102	103	100	96	101	100	98	103	100
Residuals	0	−3	2	3	0	−4	1	−1	−2	3	
Cusum	0	−3	−1	2	0	−2	−1	−1	−3	0	

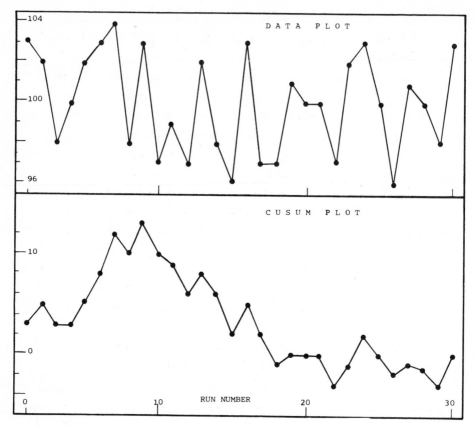

FIGURE 5.12   Plot of data in Table 5.20.

5.12, and shows that the slope of the Cusum curve is positive when
the current mean is greater than the reference value (runs 1 to 10),
it is negative when the current mean is less than the reference
value (runs 11 to 20) and is zero when the reference value is
achieved (runs 21 to 30).   To use this technique to analyze past
production runs, the Cusums are calculated, not from an arbitrary
mean, but from the overall mean of the data.   In Table 5.20 both
values are equivalent at 100%.   The maximum deviation of the Cusum
line from the base line is measured ($|C_{max}|$) and this is divided by
the standard deviation (σ) which has been derived from earlier

satisfactory runs. The critical value of this ratio is available
from tables, or may be estimated from:

$$y = 1.574 + 0.213S - 0.0015S^2 + 4.918 \times 10^{-6}S^3$$

where S is the length of span, i.e., the number of observations in a
set giving rise to the maximum permissible Cusum (y). If $|C_{max}|/\sigma$
> y a significant change in mean level has occurred.

Quality control may be undertaken similarly, but now the Cusum
graph is scaled so that the ratio of the horizontal to vertical
scale is 1:2σ and each new point is plotted as it becomes available.
A V-mask is placed over the points and the distribution of points
within the V are noted. The slope of the Cusum plot indicates the
deviation of the current mean value from the expected result, and
outlying points are eclipsed by the V-mask (Fig. 5.13). Thus lack
of process control may be exposed. The V-mask is characterized by
its angle (2θ) and the distance of the apex from the current point.
The choice of these variables is determined by the level of control
demanded, and is illustrated in Table 5.21. Pyrolysis techniques
may also find a role in acceptance sampling of raw materials and in
manufacture and the statistical principles involved have been
described [72,75,76].

Kinetic profiles and mechanisms of thermal degradation are
further areas of interest which involve quantitative assessments.
The subroutines in program STAT may be used to calculate such data,
provided this is initially transformed into a linear form. Alternatively, short transformation subroutines may be added to the program to undertake this automatically. The transformations required
for a first-order run and for Arrhenius parameters are shown in
Table 5.22, although these are not the only approach to solutions
of these equations [77].

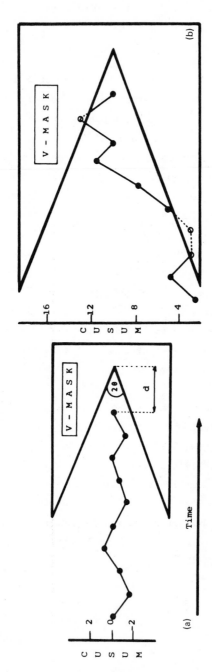

FIGURE 5.13 V-Mask Cusum technique for quality control: (A) runs 21 to 30, $\sigma = 2$; (B) runs 1 to 10, $\sigma = 1$, $2\theta = 43.6°$, $d = 2$.

TABLE 5.21   V-Mask Parameters

d	2θ (°)	Deviation of current mean[a] (σ)	Typical run length required to expose lack of control
1	53.13	0	19
		1	5
		2	2
	77.32	0	700
		1	30
		2	4
2	43.60	0	30
		1	6
	70.00	1	80
		2	6
5	33.40	0	120
		1	7
	48.46	1	25
		2	5
8	28.07	0	200
		1	10
	35.58	1	20
		2	5

[a]$\sigma = 0$, incorrect lack of control indicated.

Source: Calculated from data from Ref. 74.

TABLE 5.22   Linear Transformations of Kinetic Equations for Use with Program STAT

Equation	Linear transformation	x	y	Slope	Intercept
First order $C_t = C_o e^{-kt}$	$\ln C_t = \ln C_o - kt$	t	$\ln C_t$	$-k$	$\ln C_o$
Arrhenius $k = Ae^{-E/RT}$	$\ln k = \ln A - \dfrac{E}{RT}$	$\dfrac{1}{T}$	$\ln k$	$-\dfrac{E}{R}$	$\ln A$

REFERENCES

1. Jurs, P. and Isenhour, T., *Chemical Applications of Pattern Recognition,* John Wiley, New York, 1975.

2. Clifford, H.T. and Stephenson, W., *An Introduction to Numerical Classification,* Academic Press, New York, 1975.

3. Sweeley, C.C., Young, N.D., Holland, J.F. and Yates, S.C., Rapid Computerised Identification of Compounds in Complex Biological Mixtures by GC-MS, *J. Chromatogr., 99*(1975)507-517.

4. Needleman, M. and Stuchbery, P., The Identification of Micro-organisms by Py-GC, in C.E.R. Jones and C.A. Cramers (Eds.), *Analytical Pyrolysis,* Elsevier, Amsterdam, 1977, pp. 77-88.

5. Vincent, P.G. and Kulik, M.M., Py-GC of Fungi: Differentiation of Species and Strains of Several Members of the Aspergillus Flavus Group, *Appl. Microbiol., 20*(1970)957-963.

6. Sekhon, A.S. and Carmichael, J.W., Classification of some Gymnoascaceae by Py-GC using added Marker Compounds, *Sabouraudia, 13*(1975)83-88.

7. Hall, R.C. and Bennett, G.W., Py-GC of Several Cockroach Species, *J. Chromatogr. Sci., 11*(1973)439-443.

8. May, R.W., Pearson, E.F. and Scothern, D., *Pyrolysis-Gas Chromatography,* Chemical Society, London, 1977.

9. Meuzelaar, H.L.C., Kistemaker, P.G. and Tom, A., Rapid and Automated Identification of Micro-organisms by Curie-point Pyrolysis Techniques. 1: Differentiation of Bacterial Strains by Fully Automated Curie-point Py-GC, in C.-G. Heden and T. Illeni (Eds.), *New Approaches to the Identification of Micro-organisms,* John Wiley, New York, 1975, pp. 165-178.

10. Stack, M.V., Donoghue, H.D., Tyler, J.E. and Marshall, M., Comparison of Oral Streptococci by Py-GC, in C.E.R. Jones and C.A. Cramers (Eds.), *Analytical Pyrolysis,* Elsevier, Amsterdam, 1977, pp. 57-68.

11. MacFie, H.J.H., Gutteridge, C.S. and Norris, J.R., Use of Canonical Variates Analysis in Differentiation of Bacteria by Py-GC, *J. Gen. Microbiol., 104*(1978)67-74.

12. Hughes, J.C., Wheals, B.B. and Whitehouse, M.J., Py-MS of Textile Fibres, *Analyst, 103*(1978)482-491.

13. Eshuis, W., Kistemaker, P.G. and Meuzelaar, H.L.C., Some Numerical Aspects of Reproducibility and Specificity, in C.E.R. Jones and C.A. Cramers (Eds.), *Analytical Pyrolysis,* Elsevier, Amsterdam, 1977, pp. 151-166.

14. Scheffe, H., *The Analysis of Variance,* John Wiley, New York, 1967.

15. Fisher, R.A. and Yates, F., *Statistical Tables,* 6th Edition, Longman, Edinburgh, 1963.

16. Brown, M.B. and Dixon, W.J., *BMDP-79 Biomedical Computer Programs*, University of California Press, Berkeley (BMDP, Health Sciences Computing Facility, University of California, Los Angeles).

17. Emswiler, B.S. and Kotula, A.W., Differentiation of Salmonella Serotypes by Py-GC, *Appl. Environ. Microbiol., 35*(1978)97-104.

18. Windig, W., Kistemaker, P.G., Haverkamp, J. and Meuzelaar, H.L. C., The Effects of Sample Preparation, Pyrolysis and Pyrolysate Transfer Conditions in Py-MS, *J. Anal. Appl. Pyrol., 1*(1979)39-52.

19. Menger, F.M., Epstein, G.A., Goldberg, D.A. and Reiner, E., Computer Matching of Pyrolysis Chromatograms of Pathogenic Organisms, *Analyt. Chem., 44*(1972)423-424.

20. Reiner, E., Abbey, L.E. and Moran, T.F., Py-GC of Normal Human Cells and Amniotic Fluid, *J. Anal. Appl. Pyrol., 1*(1979)123-132.

21. Kingston, C.R. and Kirk, P.L., Some Statistical Aspects of Py-GC in the Identification of Alkaloids, *Bull. Narcotics, 17*(1965)19-25.

22. Merritt, Jr., C., Dipietro, C., Robertson, D.H. and Levy, E.J., Characterisation of Amino Acids by Py-GC-MS of their Pheynlthiohydantoin Derivatives, *J. Chromatogr. Sci., 12*(1974)668-672.

23. Kullik, E., Kaljurand, M. and Koel, M., Analysis of Pyrolysis-Gas Chromatograms using Pattern Recognition Techniques, *J. Chromatogr., 126*(1976)249-256.

24. Cox, B.C. and Ellis, B., A Microreactor-GC Method for the Identification of Polymeric Materials, *Analyt. Chem., 36*(1964)90-96.

25. Meuzelaar, H.L.C., Posthumus, M.A., Kistemaker, P.G. and Kistemaker, J., Curie-point Pyrolysis in Direct Combination with Low Voltage EIMS, *Analyt. Chem., 45*(1973)1546-1549.

26. Sneath, P.H.A. and Sokal, R.R., *Numerical Taxonomy*, Freeman, San Francisco, 1973.

27. Pankhurst, R.J. (Ed.), *Biological Identification with Computers*, Academic Press, New York, 1975.

28. Ware, G.C. and Hedges, A.J., A Case for Proportional Similarity in Numerical Taxonomy, *J. Gen. Microbiol., 104*(1978)335-336.

29. Kistemaker, P.G., Meuzelaar, H.L.C. and Posthumus, M.A., Rapid and Automated Identification of Micro-organisms by Curie-point Pyrolysis Techniques. II: Fast Identification of Microbiological Samples by Curie-point Py-MS, in C.-G. Heden and T. Illeni (Eds.), *New Approaches to the Identification of Micro-organisms*, John Wiley, New York, 1975, pp. 179-191.

30. Hughes, J.C., Wheals, B.B. and Whitehouse, M.J., Py-MS. A Technique of Forensic Potential, *Forensic Sci., 10*(1977)217-228.

31. Hickman, D.A. and Jane, I., Reproducibility of Py-MS using Three Different Pyrolysis Systems, *Analyst, 104*(1979)334-347.

32. Seviour, R.J., Chilvers, G.A. and Crow, W.D., Characterisation
     of Eucalypt Mycorrhizas by Py-GC, *New Phytol.*, *73*(1974)321-332.

33. Poxon, D.W. and Wright, R.G., Characterisation of Bitumens us-
     ing Py-GC, *J. Chromatogr.*, *61*(1971)142-144.

34. Sokal, R.R., Distance as a Measure of Taxonomic Similarity,
     Syst. Zool., *10*(1961)70-79.

35. Tatsuoka, M.M., *Multivariate Analysis: Techniques for Educa-
     tional and Psychological Research*, John Wiley, New York, 1971.

36. Demster, A.P., *Elements of Continuous Multivariate Analysis*,
     Addison-Wesley, Reading, Massachusetts, 1969.

37. Bock, R.D., *Multivariate Statistical Methods in Behavioural
     Research*, McGraw-Hill, New York, 1975.

38. Fukunaga, K. and Olsen, D.R., A Two-Dimensional Display for the
     Classification of Multi-Variate Data, *IEEE Trans. Comput.*, *C-20*
     (1971)917-923.

39. Nilsson, N.J., *Learning Machines*, McGraw-Hill, New York, 1965.

40. Kruskal, J.B., On the Shortest Spanning Sub-tree of a Graph and
     the Travelling Salesman Problem, *Proc. Amer. Math. Soc.*, *7*(1956)
     48-50.

41. Kullik, E., Kaljurand, M. and Koel, M., Analysis of Py-GC using
     the Linear Learning Machine Method, *J. Chromatogr.*, *112*(1975)
     297-300.

42. Anderberg, M.R., *Cluster Analysis for Applications*, Academic
     Press, New York, 1973.

43. Wishart, D., An Algorithm for Hierarchical Classifications,
     Biometrics, *22*(1969)165-170.

44. Kowalski, B.R., Measurement Analysis by Pattern Recognition,
     Analyt. Chem., *47*(1975)1152A-1162A.

45. Carmichael, J.W. and Sneath, P.H.A., Taxometric Maps, *Syst.
     Zool.*, *18*(1969)402-415.

46. Carmichael, J.W., *The TAXMAP Classification Program*, University
     of Alberta Mold Herbarium, Edmonton, Canada.

47. Brosseau, J.D. and Carmichael, J.W., Py-GC Applied to a Study
     of Variation in Arthroderma Tuberculatum, *Mycopathalogia*, *63*
     (1978)67-69.

48. Garbary, D. and Mortimer, M., Use and Analysis of Py-GC in
     Algal Taxonomy, *Phycologia*, *17*(1978)105-106.

49. Kruskal, J.B., Multi-dimensional Scaling by Optimising Goodness
     of Fit to a Non-metric Hypothesis, *Psychometrika*, *29*(1964)1-27.

50. Kruskal, J.B., Non-metric Multi-dimensional Scaling: A Numeri-
     cal Method, *Psychometrika*, *29*(1964)115-129.

51. Meuzelaar, H.L.C., Biomaterials Profiling Centre, University of
     Utah, Salt Lake City, Utah.

52. Haverkamp, J., Biomolecular Physics Dept., FOM Institute for Atomic and Molecular Physics, Kruislaan, Amsterdam, The Netherlands.

53. Child, D., *The Essentials of Factor Analysis,* Holt, Rinehart and Winston, London, 1970.

54. Comrey, A.L., *A First Course in Factor Analysis*, Academic Press, New York, 1973.

55. Wold, S., Pattern Recognition by Means of Disjoint Principal Components Models, *Pattern Recogn.*, *8*(1976)127-139.

56. Wold, S. and Sjostrom, M., SIMCA, in B.R. Kowalski (Ed.), *Chemometrics*, ACS Symposium Series No. 52, American Chemical Society, Washington, D.C., 1977.

57. Blomquist, G., Johansson, E., Soderstrom, B. and Wold, S., Reproducibility of Py-GC Analyses of the Mould Penicillium brevicompactum, *J. Chromatogr.*, *173*(1979)7-17.

58. Blomquist, G., Johansson, E., Soderstrom, B. and Wold, S., Classification of Fungi by Py-GC Pattern Recognition, *J. Chromatogr.*, *173*(1979)19-32.

59. Blomquist, G., Johansson, E., Soderstrom, B. and Wold, S., Data Analysis of Py-GC by Means of SIMCA Pattern Recognition, *J. Anal. Appl. Pyrol.*, *1*(1979)53-65.

60. Malinowski, E.R. and McCue, M., Qualitative and Quantitative Determination of Suspected Components in Mixtures by Target Transformation Factor Analysis of their MS, *Analyt. Chem.*, *49* (1977)284-291.

61. Rozett, R.W. and Petersen, E.M., Factor Analysis of MS: Fragmentation Patterns, *Adv. Mass Spectrom.*, *7B*(1978)993-1001.

62. Burgard, D.R., Perone, S.P. and Wiebers, J.L., Sequence Analysis of Oligodeoxyribonucleotides by Py-MS: Application of Computerised Pattern Recognition to Sequence Determination of Di-, Tri- and Tetra-Nucleotides, *Biochemistry*, *16*(1977)1051-1057.

63. Martens, A.J. and Glas, J., Numerical Treatment of Complex Pyrograms: Application to the Analysis of Ethylene-Propylene Copolymers by Chromatography of their Pyrolysates, *Chromatographia*, *5*(1972)508-515.

64. Gutteridge, C.S., MacFie, H.J.H. and Norris, J.R., Use of Principal Components Analysis for Displaying Variation between Pyrograms of Micro-organisms, *J. Anal. Appl. Pyrol.*, *1*(1979)67-76.

65. Eustache, H., Robin, N., Daniel, J.C. and Carrega, M., Determination of Multipolymer Structure using Py-GC, *Eur. Polym. J.*, *14*(1978)239-243.

66. Windig, W., Kistemaker, P.G., Haverkamp, J. and Meuzelaar, H.L. C., Factor Analysis of the Influence of Changes in Experimental Conditions in Py-MS, *J. Anal. Appl. Pyrol.*, *2*(1980)7-18.

67. Nie, N.H., Hull, C.H., Jenkins, J.G., Steinbrenner, K. and
     Bent, D.H., *Statistical Package for the Social Sciences*,
     McGraw-Hill, New York, 1975.

68. Klecka, W.R., Nie, N.H. and Hull, C.H., *SPSS Primer*, McGraw-
     Hill, New York, 1975.

69. Quantum Chemistry Program Exchange, Chemistry Dept., Indiana
     University, Bloomington, Indiana.

70. Merritt, Jr. C. and Robertson, D.H., Qualitative Analysis of
     GC Eluates by Means of Vapour Phase Pyrolysis. II. Classifica-
     tion by Set Theory, *Analyt. Chem.*, *44*(1972)60-63.

71. Grant, E.L. and Leavenworth, R.S., *Statistical Quality Control*,
     McGraw-Hill, New York, 1972.

72. Davies, O.L. and Goldsmith, P.L., *Statistical Methods in Re-
     search and Production*, Oliver and Boyd, Edinburgh, 1972.

73. Rohlf, F.J. and Sokal, R.R., *Statistical Tables*, W.H. Freeman,
     San Francisco, 1969.

74. Chamberlain, J.D., Cumulative Sum Techniques, *Analyt. Proc.*, *17*
     (1980)172-176.

75. Caulcutt, R., Acceptance Sampling, *Analyt. Proc.*, *17*(1980)166-
     172.

76. Dodge, F.H. and Romig, H.G., *Sampling Inspection Tables*, John
     Wiley, New York, 1959.

77. Davies, O.L. and Budgett, D.A., Accelerated Storage Tests on
     Pharmaceutical Products: Effect of Error Structure of Assay
     and Errors in Recorded Temperature, *J. Pharm. Pharmacol.*, *32*
     (1980)155-159.

PART **B**

APPLICATIONS

Chapter 6

Synthetic Polymers

6.1. INTRODUCTION

The thermal degradation of synthetic polymers has proved to be an
extremely important analytical technique for revealing composition,
structure, and stability profiles. Techniques such as thermogravimetry [1] and differential thermal analysis [2] have been used in
parallel with analytical pyrolysis to produce a firm foundation for
thermal analyses [3-7]. The pyrolysis approach has many attributes
to recommend it. Minimal sample preparation is required, and the
technique may be applied alike to simple polymers and to complex
mixtures of formulated and treated products such as rubbers. The
small amounts of sample required, the high-resolution analysis, and
the qualitative and quantitative information available from this
technique have made pyrolysis an essential tool in polymer chemistry
and production. The mechanistic and kinetic information which may
also be obtained from such studies enhances its utility further.

Special methods have evolved which extend the range of applications. Pyrolysis-hydrogenation-GC is a technique in which the complexity of a pyrogram is mediated by the reduction of unsaturated
pyrolysis products to the parent hydrocarbons [8-10]. This simplification enables the resolution of geometrical isomers to be achieved.
Such products depend upon the stereochemistry of the polymer and thus
allow tacticity to be assessed. Sample introduction techniques have
also received attention [11,12]. Particularly useful for the quality

control of compounded rubbers is a one-step, two-shot procedure
(chromatopyrography) which enables both the formulation and the
polymers to be identified. The first stage is analysis of the volatile fragments (heated injection port) which characterize the formulation, and this is followed by a second step which pyrolyzes the
polymeric residue. Inorganic residues may be studied off-line.
Stepwise pyrolysis, in which the same sample is subjected to a series of identical or increasing temperatures has also been described
[13-15]. These procedures allow degradation profiles to be established which are useful for characterization purposes, and to enable
kinetic measurements to be undertaken. A series of papers has compared traditional Curie-point pyrolysis with a photo-irradiation
system which has proven to be useful in exposing polymer microstructure [16-20].

Perhaps the most important advance in technology, however, is
that associated with mass spectrometry and its computer backup. The
current state of the art enables high reliability of Py-MS analyses
to be achieved and the interfacing of gas chromatographs to spectrometers is well understood. Thus, new dimensions in analytical pyrolysis are achievable using Py-GC-MS-DS (data system) and Py-MS-DS. The
data system not only facilitates data handling, but in many cases the
combination allows unique studies to be undertaken. The application
of Py-GC-MS-DS to the identification of fibers, filled rubbers,
paints, and other polymers through computer reconstructed pyrograms
and mass spectra well illustrates the potential of available methodology [21].

The applications of analytical pyrolysis to polymer analysis
have been reviewed extensively [22-26]. In many of these studies an
appreciation of the mechanisms involved in the pyrolysis process has
led to the design of elegant and imaginative experiments which have
exposed subtle aspects of microstructure and stability.

6.2. PYROLYSIS MECHANISMS

The mechanisms of polymer degradation have been discussed in detail
[27]. The pyrolysis processes occur through well-defined pathways
and may be classified according to polymer type (vinyl or condensation).

6.2.1. Vinyl Polymers

Three main degradation pathways may be distinguished. The importance
of each depends upon the substituents along the polymer backbone, and
all three pyrolysis processes may be identified in some polymers.

Depolymerization

This process corresponds to depropagating polymer chain radicals
which results in the unzipping of the polymer chain to yield sequential monomer units. The low-temperature fission of poly α-methylstyrene is an example which follows uniquely this pathway [28]. Polymethyl methacrylate, polytetrafluorethylene, and polystyrene also
largely fragment in this way and pyrograms are characterized by intense monomer components [29]. Depolymerization may also be identified in polypropylene and polybutadiene. Variations in pyrolysis
temperature influence degradation pathways (see Fig. 2.1), and small
structural changes may modify decomposition processes (Figs. 1.4 and
1.5). With proper control of the analytical parameters dimers,
trimers, and even tetramers may be recognized (Table 3.9), although
polymethyl methacrylate does not display this behavior. The detection of vinyl cyclohexene and butadiene from polybutadienes and dipentene and pentene from rubbers illustrates this trend.

   Unzipping is also observed during the pyrolysis of copolymers.
The quantitative assessment of the monomer ratio enables the composition of the polymer to be established (Sec. 3.2.3). In these cases,
care to ensure appropriate calibration standards is essential. It
has been shown that the rate of formation of monomer units is dependent upon the nature of neighboring units, the *boundary effect*. For
instance, in a copolymer of units A and B, monomer $\bar{A}$ in the pyrogram

may result from triad sequences -AĀA-, -AĀB-, or -BĀB-. Thus three
formation probability constants for each monomer may be identified
[30]. If these three constants differ significantly from one another,
the implication is that the monomer yield of a copolymer is dependent
upon sequence and not merely overall composition. Thus, pyrograms
from a mixture of homopolymers will correspond to those from block
copolymers, but will differ significantly if random copolymers are
studied. The various formation constants have been measured using
three standard polymers with different compositions. That these constants vary with the pyrolysis temperature is not surprising.

The formation of oligomers from copolymers proceeds in an analogous manner and, depending upon the distribution of monomer units,
various pure and mixed oligomers are obtained (Table 3.12). The
proportions of these products in the pyrolysate enable the microstructure of the copolymer to be studied. The Boundary Effect is
also important. When monomer units are the sole fragment of a poly-
mer--as with polymethyl methacrylate--it may be expected that oligomeric units, from two or more adjacent residues in a copolymer, would
not be detected. This has been shown to be true for a methyl methacrylate (M) -styrene (S) copolymer in which MS, SS, MSM, MSS, SMS, and
SSS oligomers were identified. The microstructure, however, may be
revealed by recognition of boundary effects, i.e., that the monomer
yields are determined by the microstructure. Thus, the pyrogram
enables the *run number* [31], a measure of randomization, to be calculated. This is recorded in Table 6.1 for component A (PMA).
Similar data is available for component B (St) [30]. Sequence information is also available through Py-MS studies of isotope clusters
of halogenated polymers [29].

Elimination

The elimination of small, neutral molecules is a common fragmentation
pathway in molecules with suitable leaving groups. Thus, whereas
polyhalogenated compounds undergo unzipping, polymers such as polyvinyl chloride initially fragment by loss of HCl. The polyolefin

TABLE 6.1 Calculation of Run Number for a Copolymer

Formation Probability Constant	$450^\circ C$	$500^\circ C$	$550^\circ C$
$k_1 = K(AAA)A$	124	158	179
$k_2 = K(AAB)A$	200	230	227
$k_3 = K(BAB)A$	204	229	234

$$R = \frac{200z}{r_A z^2 + 2z + r_B}$$

$$R_A = 200F_A \frac{(k_1-k_2)\pm[k_2^2-k_1k_3+Y_A(k_1-2k_2+k_3)]^{\frac{1}{2}}}{k_1-2k_2+k_3}$$

R   = Run number

R_A = Run number from Py-GC for Component A

z   = Initial monomer feed ratio

r   = Monomer reactivity ratios for A and B

F_A = Mole fraction of Component A in polymer: $[f_A/(f_A+f_B)]$

Y_A = Normalised yield of monomer A from pyrogram:
      Observed yield of A/f_A

Source: Ref. 30.

thus produced undergoes "kick back" fission through a six-membered transition state to yield benzene, the major pyrolysis product from PVC [24]. In the case of polyvinyl acetate the eliminated molecule (acetic acid) is the major product, and has been used for quantitative assessment [32]. This mode of fragmentation may also be used to provide information on the sequence distribution of copolymers, in addition to quantitative assessment of monomer ratios. The detection of inverse monomers through chlorination of polypropylene has already been discussed (see Sec. 3.2.3), and this technique has also been found useful to determine the structure of other chloro-polymers. Vinyl chloride-vinylidene chloride polymers, for example, have been analyzed by this approach [33]. PVC on pyrolysis yields principally benzene, whereas polyvinylidene chloride gives 1,3,5-trichlorobenzene and some monomer. Triad sequences in the copolymer undergo similar fragmentation. Thus, chlorobenzene and 1,3-dichloro-benzene are additional products which enable microstructure to be elucidated. The excellent correlation between the theoretical and pyrolysis results confirms the validity of the approach.

Random Chain Cleavage

When all bond dissociation energies within the polymer are similar and no rearrangement reactions are possible, random chain cleavage results. Usually this process competes with the other fragmentation modes and results in small hydrocarbon and terminal olefin and diene components which arise through intramolecular H-transfer reactions, disproportionation and radical combination. Typical polymers to ex-hibit random cleavage are polyethylene and polypropylene, while the degradation may also be identified in polybutadiene, polyisoprene, and polystyrene. All three main fragmentation modes are discernable in polyacrylonitrile. It is interesting to note that low-voltage Py-MS [29] or Py-FDMS pyrograms of polyethylene [34] show consider-able complication, although many components have even masses, show-ing them to be pyrolysis, rather than MS fragmentation, products. Py-GC of polyethylene is similar and a series of triplet peaks (hydrocarbon, α-olefin, α,ω-diene) extending beyond C_{30} are observed (Fig. 3.8). The intensity of the peaks above C_{21} may be used to distinguish the grade of polythene. With low density samples, the n-hydrocarbon is the most intense component, while the olefin pre-dominates in the high density grade [35]. Temperature effects are also important with this fragmentation process [36]. Short branch-ings within the polymer may result in a range of isoalkanes, depend-ing upon the bonds cleaved [10,37].

Random chain cleavage allows stereoregularity and chemical inversion of residues to be studied. Hydrogenation to yield saturated products simplifies the pyrogram and enables geometrical isomers (diasteromers) to be separated. These allow the regularity of the polymer to be probed [8,10] (Fig. 3.11) and agree well with chlorination studies [10,38].

6.2.2. Condensation Polymers

Pyrolysis of condensation polymers may be accompanied by unzipping, elimination of small molecules and random or directed chain cleavage. Unzipping is a feature of the low-voltage Py-MS mass pyrogram from Nylon-6, which is characterized by a major peak at m/z 113 (ε-caprolactam) [29]. Such processes are also observed with polymers such as poly(ε-caprolactone), in which cleavage of the ester function occurs. A ketene and an ω-hydroxy residue are formed:

$$\cdots O-(CH_2)_5-CO-O-(CH_2)_4-CH_2-CO-O-(CH_2)_5-CO \cdots$$

$$\cdots O-(CH_2)_5-CO-O(CH_2)_4-CH=C\overset{+\cdot}{=}O \ + \ \overset{+\cdot}{H}O-(CH_2)_5-CO \cdots$$

Fragmentation yields the carboxonium ion series

$$m/z\ 457 \rightarrow m/z\ 343 \rightarrow m/z\ 229 \rightarrow m/z\ 115$$

Each loss, confirmed by metastable transitions, corresponds to 114 mass units (the caprolactone monomer). Py-GC studies of aromatic polyesters confirm the probe pyrolysis results (cf. Figs. 4.7 and 4.8), but significant amounts of decarboxylation accompanied the elimination reactions initiating chain fission (Fig. 6.1) [39].

Nylon further yields cyclopentanone (Py-GC) and quantitative assessment of nylon copolymers and mixtures based upon the relative yields of ε-caprolactam and cyclopentanone is possible [40].

Random chain cleavage is revealed in Py-MS studies of polyethers and polythioethers [41,42]. Figure 6.2 illustrates this for poly(oxy-1,4-phenylene) and shows that the phenyloxyphenol (m/z 186) should be about twice as intense as the diphenylether (m/z 170) or dihydroxyphenylether (m/z 202) [42]. In contrast, the corresponding

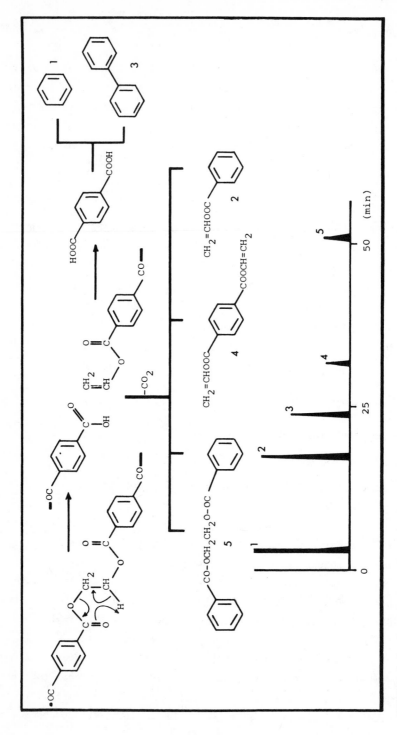

FIGURE 6.1 Py-GC analysis of polyethyleneglycol terephthalate. (Adapted from Ref. 39, from the *Journal of Chromatographic Science*, by permission of Preston Publications, Inc.)

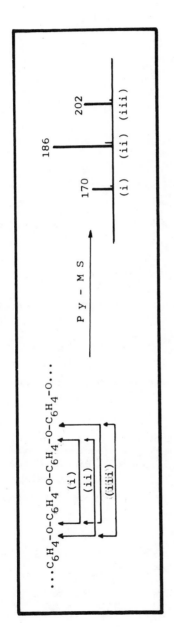

FIGURE 6.2 Partial Py-MS fragmentation of poly(oxy-1,4-phenylene). (From Ref. 42, G. Montaudo, M. Przybylski, H. Ringsdorf, *Makromol. Chem.* *176*:1763(1975), Hüthig and Wepf Verlag, Basel.)

thioether showed directed cleavage with each fragment being characterized by a terminal phenyl and thiophenyl residue (e.g., Ph-S-C$_6$H$_4$-
SH, m/z 218). A second series of peaks containing three or more
residues were probably cyclic products [43].

An understanding of the origins of the pyrolysis products of
condensation polymers has also been used effectively to study composition [44,45], copolymer structure [46,47], and stereochemistry
[48].

6.3. APPLICATIONS

The application of analytical pyrolysis to the study of synthetic
polymers accounts for a large proportion of the pyrolysis literature.
This reflects the diversity and utility of the information available
through pyrolysis, and the wide range of samples which are amenable
to study.

6.3.1. Fingerprint Identification

Fingerprint identification involves a comparison of a pyrogram from
an unknown sample with those from standard materials. The standard
pyrograms may be held in a library or else determined alongside the
unknowns. The former option requires a high degree of long-term reproducibility to ensure adequate performance, whereas the latter demands a small number of samples or else a short analysis time, as
with Py-MS. Although unique fragments may speed identification, it
is more usual that a final identification depends on a comparison of
peak intensities. To date, this has often been an uncritical, visual
approach, undertaken by the reproduction of typical pyrograms.
Clearly, now that data handling (Chap. 5) is far advanced, a more
thorough approach is anticipated.

Pyrograms libraries have been described [24,49] and one system
enables polymer identification on the basis of three intense peaks
to be undertaken [24]. Identification of the components was also
reported. Fingerprint pyrograms have been reported for polyurethanes
[50,51] and vinyl and condensation polymers using a dual-flow system

[52]. This latter system also enabled functional group analysis to be undertaken. Other reports have included coumarone-indene and cyclopentadiene polymers [53], polybutadiene [54-56], phenolic resins [57] and fluoropolymers using polystyrene as an internal standard [58]. More complex samples have also been studied and these include plastics [59,60], paints [61], rubbers [21,29,62,63], fibers [64,65], adhesives [66], and paper [67]. This latter study revealed the presence of synthetic polymers such as PVC and butadiene-styrene co-polymers in paper samples. Brake linings [68], mineral oil additives [69], and flame retardants using a phosphorus detector [70] have also been studied. The power of Py-MS in this field was demonstrated by the detection of chloroprene as a component of a sample of butyl rubber [29].

Recently, a comprehensive scheme for the identification of poly-mers has been described [71]. Rather than relying upon fingerprint pyrograms, however, specific pyrolysis products were used as the basis for assessment. Recommended pyrolysis and analysis conditions, dependent on the nature of the problem, were recommended; sample pyrograms were presented, and the pyrolysis products expected from various polymers were reported. Tables 6.2 and 6.3 summarize the approach.

This emphasis on the structure of pyrolysis products is a wel-come development and has certain advantages over the traditional fingerprint approach. In particular, the fingerprint method re-quires a very high degree of reproducibility. This demand is sub-stantially reduced if product identification is undertaken. Also, the standardized conditions used for the measurement of library pyrograms may not be optimum for each polymer. This is particularly so for complex mixtures, when the presence of small amounts of com-ponents may be totally undetected. In addition, fingerprinting gives no direct indication of structure. Clearly, product identification should be undertaken whenever possible to enhance the value of the analysis still further.

TABLE 6.2 Py-GC Conditions for Polymers in Table 6.3

PYROLYSIS	COLUMN	STATIONARY PHASE	SUPPORT	TEMPERATURE			FLOW RATE (Argon)
				INITIAL	PROGRAM	FINAL	
770°C 10s	3mx3mm ID	10-15% Reoplex 400	Chezasorb	40-45°C 5-6 min	4-6°C min^{-1}	160°C	20-25 ml min^{-1}
770°C 10s	3mx3mm ID	10-15% Polyphenylether	Chromosorb W	40°C 5-6 min	4-6°C min^{-1}	180°C	20-25 ml min^{-1}
770°C 10s	2mx3mm ID	10-15% Polyphenylether or Reoplex 400	Chromosorb W	100°C			20-25 ml min^{-1}
	5mx3mm ID	15% Tricresyl phosphate	Chezasorb	60°C			
770°C 10s	50-100m x 0.5mm	SE-30, OV-101 or OV-17	WCOT	40°C 5 min	2°C min^{-1}	200°C 35 min	4.5 ml (He)min^{-1}

Source: Ref. 71.

TABLE 6.3 Pyrolysis Products from Various Polymers

Polymer	Products
Acrylate rubbers	Methyl, ethyl, or butyl acrylate
Butyl rubbers	Isobutylene
Me-Styrene-butadiene rubbers	Butadiene, vinylcyclohexene, styrene
Nitrile rubbers	Butadiene, vinylcyclohexene, acrylonitrile
Poly(butadienes)	Butadiene, vinylcyclohexene
Poly(chloroprenes)	Chloroprene
Poly(isobutylenes)	2-Methylbut-1-ene, 2-methyl-1,3-butadiene
Poly(isoprenes)	Isoprene, dipentene
Poly(urethanes)	Various

Source: Ref. 71.

6.3.2. Analysis of Composition

Analysis of polymer composition is concerned with the detection, identification, and quantitative determination of components in mixed or copolymer systems. This type of analysis may be used widely as an industrial quality control technique, particularly in conjunction with a Cusum V-mask analysis of trends. Applications may be found, particularly in the manufacture or acceptance sampling of plastics, paints, rubbers, insulating materials, and textiles. The minimal sample preparation and the high resolution available makes the technique equally applicable to the determination of major components or minor contaminants in a vast range of products. Polymer mixtures may be distinguished from copolymers by the different effects of pyrolysis temperature (Fig. 3.10).

One noteworthy example of analytical pyrolysis in quality control is the assessment of carboxyl-terminated butadiene-nitrile (CTBN) polymer in epoxy resins [72]. This material is added to toughen and improve the flexibility of epoxy resins used in a structural or adhesive capacity. In aerospace applications in particular, cresol-novalac resins, modified by CTBN are used extensively.

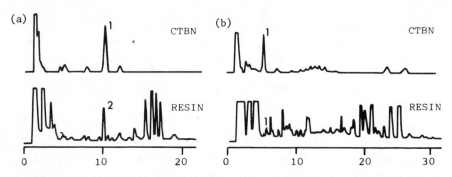

FIGURE 6.3 Pyrograms from CTBN and a cresol-novalac epoxy resin:
(a) SE-30 column and (b) polyphenyl ether-Carbowax 20M column (1,
vinylcyclohexene; 2, vinylcyclohexene + N-containing component).
(Adapted from Ref. 72, from the *Journal of Chromatographic Science*,
by permission of Preston Publications, Inc.)

Py-GC-MS of CTBN yielded vinylcyclohexene as a well-resolved fragment, and butadiene. The epoxy resin, with no CTBN additive,
yielded peaks coincident with butadiene, but was clear in the vinylcyclohexene region. A synthetic, treated resin also produced vinylcyclohexene (from CTBN) and showed that this residue was appropriate
for qualitative and quantitative study. However, when a manufactured
sample of an amine-cured resin was analyzed considerable differences
were observed. The most serious was that a nitrogen-containing fragment (possibly a methylpyrrole or pyrazine) was now coincident with
the analytical peak. A column change from SE-30 to a more polar
polyphenyl ether; Carbowax 20M stationary phase enabled resolution
of vinylcyclohexene to be achieved (Fig. 6.3). This clearly illustrates the value of mass spectrometry in developmental work.

Quantification of CTBN was achieved through calibration standards prepared by spiking synthetic blends of resin with known
amounts of the additive. Good linearity was obtained and a reproducibility of ±0.2% w/w (>10%) was determined. The minimum detection level of CTBN was estimated to be 1% w/w. The analytical pyrolysis approach was found to be superior to methods such as IR and HPLC.

Quality control of rubbers is a further area where important
pyrolysis applications are found. The problems associated with
these analyses, with particular reference to tyre rubber, have been

discussed [73,74] and an automated system has been described (Fig. 3.15). Pyrograms from rubbers are essentially characterized by the monomers or dimers from unzipping reactions (Fig. 3.9), and quantitative assessment of the blend may be achieved using characteristic fragments [73,75]. Additives may also be assessed [11,21,76] and the degree of curing [77] or quantification of cured products may be readily achieved [78]. Pyrolysis and ozonolysis GC have also been applied to polybutadiene analysis [54].

Other applications have included the determination of monomer ratios in copolymers [79,80], sulfur additives in polypropylene [81], fire-retardant materials containing PVC [82], and nonvolatile deposits in used catalysts [83]. The weathering of fiberglass epoxy resin composites has been studied by laser-Py-GC [84], laser-probe pyrolysis has revealed trace contaminants in polymers [85] and various textile fibers have been characterized [86]. Photolysis GC enables identification and quantification of various additives [87] and polymers [20] to be achieved, and the application of pyrolysis-molecular weight GC has been reviewed [88]. The use of internal standards (e.g., polymers such as polystyrene or polymethyl methacrylate which yield well-characterized peaks) has been discussed [89,90], but is clearly limited to soluble samples.

6.3.3. Determination of Structure

The determination of polymer structure usually necessitates the identification and quantification of the larger pyrolysis fragments. These yield information on the sequence arrangement of subunits, enable random and block copolymers to be distinguished and also enable stereochemical assignments to be made. Pyrolysis of random and block copolymers is mediated by boundary effects and these may be distinguished by the different dependence of monomer yields on composition [30]. This has been used to good effect to determine diad sequences in styrene-glycidyl methacrylate copolymers [91]. Typical pyrograms are shown in Fig. 6.4 and reveal monomer, dimer, and co-dimer fragments. The recovery of monomer and dimers from a series

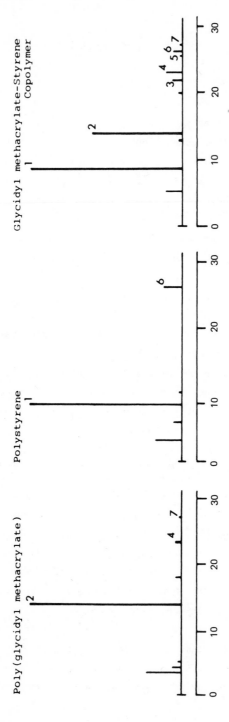

FIGURE 6.4 Major components from the pyrolysis of styrene and glycidyl methacrylate homo- and copolymers (1, Styrene; 2, Glycidyl methacrylate; 3, GMA-St; 4, GMA dimer; 5, GMA-St; 6, Styrene dimer; 7, GMA dimer). Adapted from Ref. 91.)

TABLE 6.4   Pyrolysis of Styrene-Glycidyl Methacrylate Copolymers

GMA in feed	GMA in copolymer	Monomer yield	Dimer yield	Other products
0	0	77.0	9.0	14.0
20	35.7	84.9	5.0	10.1
30	43.8	87.4	4.5	8.1
40	50.4	85.9	4.3	9.8
50	56.4	84.8	3.1	12.1
60	62.6	85.4	2.9	11.7
70	69.3	91.0	2.5	6.5
80	77.3	91.9	1.5	6.6
90	87.1	91.7	0.8	7.5
100	100	93.2	1.8	5.0

Source: Ref. 91.

of polymers containing various ratios of styrene and the methacrylate (GMA) varies because of the boundary effect (Table 6.4), although the proportion of styrene:styrene dimer and GMA:GMA dimer remain constant. The standard polymers allowed the probability constants for the production of the various dimers from different tet-
rameric subunits (e.g., $\overline{AA}$ from $\overline{AAAA}$, $\overline{AAAB}$, $\overline{BAAB}$; $\overline{AB}$ from $\overline{AABA}$, $\overline{AABB}$,
$\overline{BABA}$, $\overline{BABB}$; $\overline{BB}$ from $\overline{ABBA}$, $\overline{ABBB}$, $\overline{BBBB}$) to be measured. Correction
factors based upon these constants were used to calculate the diad
content of the copolymer from pyrolysis yield. Excellent agreement
between theoretical and experimental values was obtained, although
it was pointed out that conversion and monomer feed ratio are required for this approach. It is therefore inappropriate for a
totally unknown sample.

The structure of cationic [92] and anionic [93,94] exchange
resins have been studied by Py-GC-MS. Pyrolysis of styrene-divinylbenzene copolymers yielded over 50 determined products, including
methyl chloride (anion), trimethylamine (functional group) and p-
divinylbenzene (substitution pattern) from an anionic exchanger.

Identification and quantification of these enables the degree of
cross-linking in the matrix, the type and structure of the functional
group and counterion, and the position of substitution to be obtained.
Py-MS is revealed to be a reliable and rapid alternative [95]. The
quality control of styrene–divinylbenzene copolymers has also been
described [96].

Other work concerned with structure elucidation has involved
stepwise pyrolysis to distinguish homo- and copolymers [97], elemental analysis [98], and studies on the crystallinity of polyethylene
[99]. Rubbers [100], acrylonitrile–methacrylic acid copolymers [5],
butadiene copolymers [101], siloxanes [102,103], and plasma–polymer-
ized hydrocarbons [104] have also been studied. More recently,
styrene–acrylate copolymers [105] and polyalkenes [106,107] have
also received attention. This last work [107] is noteworthy in that
pyrolysis products from well-studied polymers were reported, e.g.,
for PVC, polystyrene, poly(vinyl acetate), poly (α-methylstyrene)
and a styrene–acrylonitrile copolymer. Despite this, many new products were identified, including 8 out of 15 from polystyrene and 15
out of 24 from poly(α-methylstyrene). Again MS aided by a data
system was instrumental.

Stereochemical studies have also been undertaken [8,10,48,108,
109].

6.3.4. Kinetic and Mechanistic Studies

The kinetic and mechanistic aspects of pyrolysis processes are now
under intensive study. In addition to providing a firm basis for
the application of analytical pyrolysis techniques, such information
is of value in exposing thermal stability profiles and degradation
routes of polymeric materials. Their widespread use, ranging from
textiles and foams, through paints and lacquers to insulating and
structural components, makes these properties of vital concern.
Thus, such diverse areas as the evolution of toxic gases, flame-
retardation studies, and the recovery of raw materials from polymer
wastes have much to gain from the analytical pyrolysis approach.

In view of the importance of mechanistic and kinetic information, it is gratifying to note that an ever increasing number of contributions now include interpretation of data in these terms. This trend demonstrates the current awareness of the importance of basic pyrolysis processes and confirms analytical pyrolysis as a powerful molecular probe.

One of the earliest reports of pyrolysis kinetics, and certainly one of the most quoted papers in pyrolysis literature is the calculation of the half-lives of polymers at various temperatures undertaken by Farre-Ruis and Guiochon [110]. Assuming first-order kinetics and extrapolating known rate constants to pyrolysis temperatures, the Arrhenius equation was used to calculate half-lives. For the first time, it was shown that pyrolysis may well be complete before the pyrolysis wire has reached the equilibrium temperature. The heating rate was also shown to have a marked effect upon the true pyrolysis temperature--an effect later verified experimentally (Table 2.1). Contemporary studies by Lehrle and Robb were investigating the potential of Py-GC for kinetic studies [111-113]. Modern pyrolysis units with well-characterized TRT and T_{eq} parameters now enable measurements to be undertaken with confidence.

A typical experiment involves the pyrolysis of a sample at a range of temperatures for different periods of time. The time series enables each isothermal rate of pyrolysis to be calculated. The Arrhenius equation then allows the activation energy and the preexponential factor to be determined (Table 5.22). This approach has been reported for *cis*-1,4-polybutadiene [13]. Butadiene and vinylcyclohexene were the major pyrolysis products, and the formation of the monomer was monitored in the kinetic runs. The major disadvantage with this approach is that each point requires the pyrolyzer to be loaded with a new sample. Considerable delay is experienced while the system stabilizes (ca. 30 min); so a full kinetic profile is a slow process. To overcome this problem the technique of sequential pyrolysis has been described [13,14]. Here, the same sample is pyrolyzed repeatedly for fixed time periods at a preselected temperature.

At each subsequent pyrolysis the product yield decreases and the process is repeated until no further products are evolved. This time series of results enables the rate constant to be calculated with only one loading of the pyrolyzer. The process is repeated at a range of other temperatures to enable Arrhenius parameters to be calculated. In the case of *cis*-1,4-polybutadiene (7.4 µg), up to 10 sequential pyrolyses were performed, the initial pyrogram was recorded over 30 min and subsequent runs lasted about 5 min each. First-order kinetics were indicated and the Arrhenius equation was valid over the range 450 to 530°C. Deviations above this range were due to the temperature rise time (8 msec) having a significant effect, due to the greater rate of degradation. A typical isothermal time series is shown in Fig. 6.5 and results are illustrated in Table 6.5. The precision is significantly better than that obtained with conventional pyrolysis. The importance of ensuring that the sample and the pyrolysis wire are at the same temperature must be emphasized. This has been discussed in detail [14] and it was concluded that sample sizes of less than 5 µg should be avoided.

Other kinetic work has included the calculation of kinetic parameters for styrene-methyl acrylate copolymers using Py-GC [114, 115]. The rate constants were found to depend upon the nature and sequence of the monomer units, and thus sequences were determined. Thermogravimetry and DTA have also been used [116,117] and in combination with Py-GC [118]. In conjunction with MS analysis, the thermal degradation of PTFE and copolymers in helium and air has been studied by Py-GC and thermogravimetry [119]. Temperature decomposition profiles were determined by monitoring products such as C_2F_4, C_3F_6 and C_4F_8 (pyrolysis 450 to 800°C), or ions from C_2F_4 (m/z 100, 81, 50) and C_3F_6 (m/z 150, 131, 69). The data were suitable for quantitative interpretation and the calculation of Arrhenius parameters was undertaken. Different decomposition stages were apparent for copolymer decomposition (E:-Stage 1: C_3F_6 evolution, 295 kJ mol^{-1}, Stage 2: C_3F_6 evolution, 358 kJ mol^{-1}), whereas PTFE was found to evolve C_2F_4 (the main degradation product) with an activation

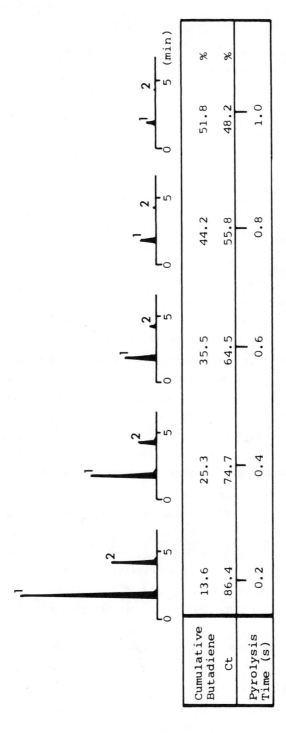

Cumulative Butadiene				
Ct	13.6	25.3	35.5	44.2	51.8	%
	86.4	74.7	64.5	55.8	48.2	%
Pyrolysis Time (s)	0.2	0.4	0.6	0.8	1.0

FIGURE 6.5 Sequential pyrolysis of *cis*-1,4-polybutadiene at 450°C (1, butadiene; 2, vinylcyclohexene). (Adapted from Ref. 13, from the *Journal of Chromatographic Science*, by permission of Preston Publications, Inc.)

TABLE 6.5   Results from Sequential Pyrolysis of *cis*-1,4-Polybuta-
diene[a]

Pyrolysis temperature (°C)	450	470	490	510	532	552	575
Rate constant (sec^{-1})	0.73	1.26	3.01	4.62	10.35	23.11	46.21
Minimum pyrolysis time (sec)	7.60	4.40	1.84	1.20	0.54	0.24	0.12

[a]$E = 157$ kJ/mol; $A = 1.26 \times 10^{11}$/sec).

Source: Ref. 13.

energy of 360 kJ mol^{-1}. These compare favorably with the value of
322 kJ mol^{-1} taken by Farre-Rius and Guiochon in their earlier work.

Cross-correlation Py-GC, in which time- and temperature-resolved
pyrograms are recorded, allows continuous monitoring of degradation
processes. The system is suitable for polymers undergoing rapid degradation, and continuous calculation of chromatograms, without completion of the previous analysis, is possible [120-122]. Samples are
introduced automatically onto the column after short, predetermined
time intervals (1 to 2 sec). The recorder trace is a composite of
all superimposed runs, and the chromatograms are computed by correlation of the input sequence with detector output. Thus, the behavior of each component throughout the degradation may be followed
(Fig. 6.6). However, analyses must be undertaken isothermally, and
as no net separation of products is achieved, component identification must be undertaken separately. Nevertheless, this technique
holds considerable promise in mechanistic and kinetic studies. For
example, the energy of activation for styrene release from polystyrene (320 kJ/mol) compares favorably with other work (318 kJ mol^{-1})
[122].

The rate constants for the polymerization of ethylene and propylene oxides have been calculated on the basis of the distribution of
fragments determined by Py-MS [123].

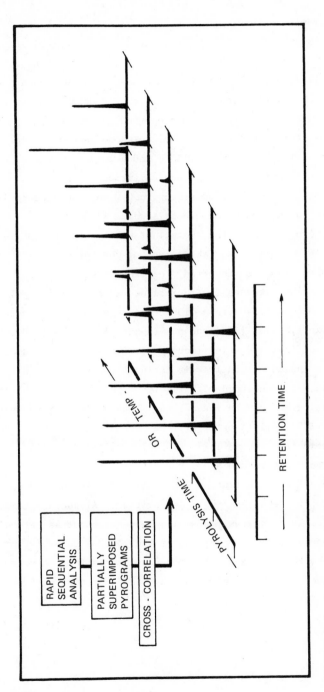

FIGURE 6.6 Cross-correlation chromatography.

Mechanistic analyses have been undertaken by Py-MS and Py-GC techniques, although the former, with its direct chemical information content, has been to the forefront, particularly direct probe techniques [43,124-129]. Py-GC-MS has been used to identify many of the fragments from methyl methacrylate-methacrylic acid copolymers [130]. A free radical route to various hydrocarbon species was proposed and confirmed by deuterium labeling. Polysulphone decomposition was followed by similar procedures [131], and a random cleavage decomposition process via free radicals was confirmed for polyethylene and poly(iso butylene), with fragments up to C_{55} being monitored [132]. Fractionated polymers have also been studied [133,134].

The determination of the thermal stability of PVC [135] using time-resolved Py-GC [4], and the effect of morphology on PVC pyrolysis [136] have also been reported, together with work on the influence of stereochemistry on the thermal degradation of 1,4-polybutadienes [137] and polyisoprenes [138], bis-p-toluene sulphonates of 2,4-hexadiyne-1,6-diol [139] and poly(α-methylstyrene) [140]. This last work describes radiation-induced pyrolysis which yields monomer, dimer and an indane derivative via cationic chain reactions. Other reports have been concerned with fluoro polymers [141], rubbers [142, 143], branching in polyethylenes [144], highly alternating copolymers of styrene [145], poly(α-methylstyrene) copolymers [146], cross-linked styrene copolymers [147], and polyacrylonitrile [148].

A thorough understanding of the kinetic and mechanistic factors involved in polymer degradation would allow prediction and explanation of pyrogram peaks and their intensities. Although some progress has been made in this area [149-151], success cannot as yet be claimed. However, the recognition of the role of molecular processes in pyrolysis, the advanced methodology available for their study and the importance of such results in many fields of application will surely spur further endeavor.

6.3.5. Other Applications

The reports of other pyrolysis studies are extensive. Many have not used analytical pyrolysis techniques (i.e., small sample size and integrated pyrolysis-analysis-data handling system) and have been concerned with physical as well as chemical effects. Nevertheless, useful conclusions have been drawn [152-155], although it is possible that even more information on the primary processes involved could be obtained using the advantages afforded by analytical pyrolysis, particularly with time- and temperature-resolved studies.

The pyrolysis of thermally resistant polymers is characterized by significant amounts of carbonaceous residues and is an important field of study [43,129,156]. The effect of additives on pyrolysis yields and degradation pathways is readily studied by analytical pyrolysis, and reports in this area are increasing [76,82,157-162]. Typical are studies on poly(isoprene) and polypropylene which have involved determination of product distribution, degradation rates at various temperatures under different atmospheres, and Arrhenius parameters [163-166]. Chromyl chloride was used to incorporate low levels of chromium into the polymer. Between 388 and 438°C pure and treated polypropylene was pyrolyzed at similar reaction rates. This, however, disguised the fact that the activation energy for the decomposition of the treated polymer fell from 213 to 184 kJ mol^{-1}. Only small differences were observed in the distribution of the major pyrolysis products. Significant differences were observed when oxidative pyrolysis was undertaken. In these studies, the presence of 1.5% chromium was shown to *increase* the activation energy (by 42 kJ mol^{-1}) and char formation was also observed. This is in contrast to polypropylene which normally leaves no residue. The limiting oxygen value (26.4 from 17.4) and self-ignition temperature (400°C from 250°C) for the treated material also revealed the fire-protecting ability of the additive. It was also found that the detailed pyrolysis differed, depending upon whether in vacuo, static or

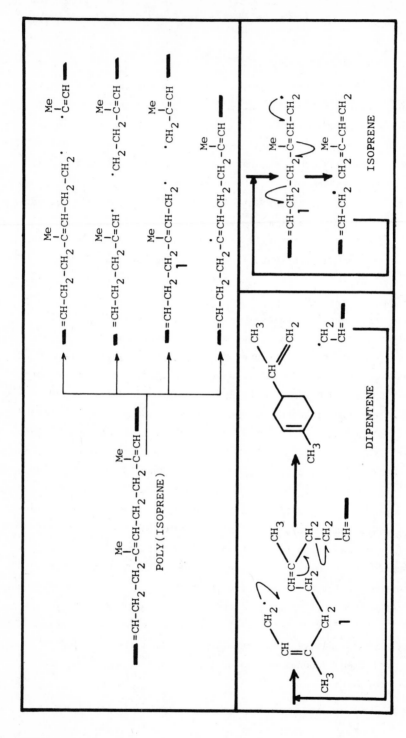

FIGURE 6.7 Pyrolysis pathways of poly(isoprene) showing the depolymerization of one fragment radical to dipentene or to isoprene.

318

flowing carrier gas conditions were employed. This approach has also allowed a full mechanistic description of the pyrolysis events to be established. The scheme proposed for poly(isoprene) is depicted in Fig. 6.7. The pyrolysis of organometallic compounds is noteworthy in this context [167-174].

The evolution of flammable or toxic gases from structural polymers is also of intense interest and is under active investigation [175-179]. Studies on the reclamation of waste organic matter through pyrolysis are also progressing [180-182], and a more thorough approach to mechanistic and kinetic treatments of hydrocarbon pyrolysis is underway [183]. A new severity function has enabled the evaluation of feedstock for olefin function by means of bench scale pyrolysis and GC. Although of relatively minor application, Py-IR work has also been reported [184-186].

REFERENCES

1. Keattch, C.J. and Dollimare, D., *An Introduction to Thermogravimetry*, Heyden, London, 1975.

2. Pope, M.I. and Judd, M.D., *Differential Thermal Analysis*, Heyden, London, 1977.

3. Crighton, J.S., Characterisation of Textile Materials by Thermal Degradation: A Critique of Py-GC and Thermogravimetry, in C.E.R. Jones and C.A. Cramers (Eds.), *Analytical Pyrolysis*, Elsevier, Amsterdam, 1977, pp. 337-349.

4. Liebman, S.A., Ahlstrom, D.H. and Foltz, C.R., Thermal Degradation Studies of PVC with Time-Resolved Py-GC and Derivative TGA, *J. Polym. Sci. (Polym. Chem. Ed.), 16*(1978)3139-3150.

5. Guyot, A., Bert, M., Hamoudi, A., McNeill, I. and Grassie, N., Pyrolysis of Acrylonitrile-Methacrylic Acid Copolymers and Derivatives, *Eur. Polym. J., 14*(1978)101-107.

6. Ordoyno, N.F. and Rowan, S.M., Interactions Observed During Pyrolysis of Binary Mixtures of Textile Polymers, *Thermochim. Acta, 23*(1978)371-385.

7. Grassie, N., New Developments in the Thermal Degradation of Polymers, *J. Appl. Polym. Sci. (Symp.), 35*(1979)105-121.

8. Dimbat, M., Py-Hydrogenation Capillary GC. A Method for the Determination of the Isotacticity and Isotactic and Syndiotactic Block Length of Polypropylenes, in R. Stock (Ed.), *Gas Chromatography 1970*, Institute of Petroleum, London, 1971, pp. 237-246.

9. Sugimura, Y. and Tsuge, S., Fundamental Splitting Conditions for Pyrogram Measurements with Glass Capillary GC, *Analyt. Chem., 50*(1978)1968-1972.

10. Tsuge, S., Sugimura, Y. and Nagaya, T., Structural Characterisation of Polyolefins by Py-Hydrogenation-Glass Capillary GC, *J. Anal. Appl. Pyrol., 1*(1980)221-229.

11. Hu, J.C-A., Py-GC Analysis of Rubbers and Other High Polymers, *Analyt. Chem., 49*(1977)537-540.

12. Hu, J.C-A., Liquid Sample Introduction in GC, *Analyt. Chem., 51* (1979)2395-2397.

13. Ericsson, I., Sequential Py-GC Study of the Decomposition Kinetics of cis-1,4-Polybutadiene, *J. Chromatogr. Sci., 16*(1978) 340-344.

14. Andersson, E.M. and Ericsson, I., Determination of the Temperature-Time Profile of the Sample in Py-GC, *J. Anal. Appl. Pyrol., 1*(1979)27-38.

15. Sugimura, Y., Tsuge, S. and Takeuchi, T., Characterisation of Ethylene-Methacrylate Copolymers by Conventional and Stepwise Py-GC, *Analyt. Chem., 50*(1978)1173-1176.

16. Hausler, K-G., Schroder, E. and Hartwich, B., Light-Radiation and Curie-Point Pyrolysis. I. Light Radiation Pyrolysis, *Plast. Kautsch., 23*(1976)481-484.

17. Hausler, K-G., Schroder, E., Grosskreuz, G. and Hube, H., Py-GC Charakterisierung von Vernetzen Polymeren, *Plast. Kautsch., 25* (1978)691-696.

18. Hausler, K-G. and Hartwich, B., Lichtstrahl-und Curiepunkt-pyrolyse von Styren-Butadien Copolymeren, *Z. Chem., 19*(1979) 141-142.

19. Hausler, K-G., Popov, G. and Schroder, E., Light Beam and Curie-point Pyrolysis of Polymers. VIII. Study of the Structure of Styrene-divinylbenzene Block Copolymers by Means of Py-GC, *Plast. Kautsch., 26*(1979)76-79.

20. Hausler, K-G., Schroder, E., Huster, B. and Jobst, K., Light Radiation and Curie-point Pyrolysis of Polymers. X. Py-GC Characterisation of Radiation-Chemical Crosslinked Polybutadienes, *Plast. Kautsch., 27*(1980)19-22.

21. Wuepper, J.L., Py-GC-MS Identification of Intractable Materials, *Analyt. Chem., 51*(1979)997-1000.

22. Wolf, C.J., Grayson, M.A. and Fanter, D.L., Py-GC of Polymers, *Analyt. Chem., 52*(1980)348A-358A.

23. Irwin, W.J., Analytical Pyrolysis--An Overview, *J. Anal. Appl. Pyrol., 1*(1979)3-25, 89-122.

24. May, R. W., Pearson, E.F. and Scothern, D., *Pyrolysis-Gas Chromatography*, Chemical Society, London, 1977.

25. Berezkin, V.G., Alishoev, V.R. and Nemirovskaya, I.B., *GC of Polymers*, Elsevier, Amsterdam, 1977.

26. Brauer, G.M., Py-GC Techniques for Polymer Identification, in P.E. Slade, Jr. and L.T. Jenkins (Eds.), *Thermal Characterisation Techniques*, Marcel Dekker, New York, 1970, pp. 41-105.

27. Bamford, C.H. and Tipper, C.F.H. (Eds.), *Comprehensive Reaction Kinetics, Vol. 14*, Elsevier, Amsterdam, 1975.

28. Wall, L.A., Florin, R.E., Aldridge, M.H. and Fetters, C.J., Pyrolysis of Monodisperse Poly-α-Methylstyrene, *J. Res. Nat. Bur. Stand. (Sect. A.), 83*(1978)371-380.

29. Kistemaker, P.G., Boerboom, A.J.H. and Meuzelaar, H.L.C., Laser Py-MS: Some Aspects and Applications to Technical Polymers, *Dynamic Mass Spectrom., 4*(1975)139-152.

30. Tsuge, S., Kobayashi, T., Nagaya, T. and Takeuchi, T., Py-GC Determination of Run Numbers of Methyl Methacrylate-Styrene Copolymers Using the Boundary Effect, *J. Anal. Appl. Pyrol., 1* (1979)133-141.

31. Harwood, H.J., The Characterisation of Sequence Distribution in Copolymers, *J. Polym. Sci., Part B., 2*(1964)601-607.

32. Lehrle, R.S. and Robb, J.C., The Quantitative Study of Polymer Degradation by GC, *J. Gas Chromatogr., 5*(1967)89-95.

33. Tsuge, S., Okumoto, T. and Takeuchi, T., Py-GC Studies on the Sequence Distribution of Vinylidene Chloride-Vinyl Chloride Copolymers, *Makromol. Chem., 123*(1969)123-129.

34. Beckey, H.D. and Schulten, H-R., FI- and FD-MS in Analytical Chemistry, in C. Merritt, Jr. and C.N. McEwen (Eds.), *Mass Spectrometry, Part A*, Marcel Dekker, New York, 1979.

35. Cieplinski, E.W., Ettre, L.S., Kolb, B. and Kemner, G., Py-GC with Linearly Programmed Temperature Packed and Open Tubular Columns. The Thermal Degradation of Poly-olefins, *Z. Anal. Chem., 205*(1964)357-365.

36. Willmot, F.W., Py-GC of Polyolefins, *J. Chromatogr. Sci., 7* (1969)101-108.

37. Michajlov, I., Zugenmaier, P. and Cantow, H.J., Structural Investigations on Polyethylenes and Ethylene-Propylene Copolymers by Reaction-GC and X-Ray Diffraction, *Polymer, 9*(1968)325-343.

38. Senoo, H., Tsuge, S. and Takeuchi, T., Estimation of Chemical Inversions of Monomer Placement in Polypropylene by Py-GC, *Makromol. Chem., 161*(1972)185-193.

39. Sugimura, Y. and Tsuge, S., Studies on Thermal Degradation of Aromatic Polyesters by Py-GC, *J. Chromatogr. Sci., 17*(1979) 269-272.

40. Senoo, H., Tsuge, S. and Takeuchi, T., Py-GC Analysis of 6-66-Nylon Copolymers, *J. Chromatogr. Sci., 9*(1971)315-318.

41. Hummel, D.O., Dussel, H.J., Rosen, H., and Rubenacker, K., FI-
    and EI-MS of Polymers and Copolymers. 4. Aromatic Polyethers,
    Makromol. Chem. Suppl. 1, *1*(1975)471-484.

42. Montaudo, G., Przybylski, M., and Ringsdorf, H., Untersuchungen
    von Polymeren im MS. 6. Poly(oxy-1,4-phenylen), Poly(thio-1,4-
    phenylen) und Poly(dithio-1,4-phenylen), *Makromol. Chem., 176*
    (1975)1763-1776.

43. Hummel, D.O., Structure and Degradation Behaviour of Synthetic
    Polymers Using Pyrolysis in Combination with FIMS, in C.E.R.
    Jones and C.A. Cramers (Eds.), *Analytical Pyrolysis*, Elsevier,
    Amsterdam, 1977, pp. 117-138.

44. Aguilera, C. and Luderwald, I., Strukturuntersuchung von Poly-
    estern durch direkten Abbau im MS. 5. Polyester aus Terephthal-
    saure und/oder Bernsteinsaure und Hydrochinon, *Makromol. Chem.,
    179*(1978)2817-2827.

45. Jacobi, E., Luderwald, I. and Schultz, R.C., Uber die MS Bestim-
    mung des Einbauverhaltnisses in Copolymeren. 2. Copolyester der
    Milchsaure und Glykolsaure, *Makromol. Chem., 179*(1978)277-280.

46. Jacobi, E., Luderwald, I. and Schultz, R.C., Strukturuntersuch-
    ung von Polyestern durch Direkten Abbau in MS. 4. Polyester
    und Copolyester der Milchsaure und Glykolsaure, *Makromol. Chem.,
    179*(1978)429-436.

47. Luderwald, I. and Kricheldorf, H.R., Uber den Thermischen Abbau
    von alternierenden Copolyamiden im MS, *Angew. Makromol. Chem.,
    56*(1976)173-191.

48. Luderwald, I., Przybylski, M. and Ringsdorf, H., Py-MS of Trux-
    illic and Truxinic Polyamides and Related Copolyamides, in C.E.
    R. Jones and C.A. Cramers (Eds.), *Analytical Pyrolysis*, Elsevier,
    Amsterdam, 1977, pp. 297-308.

49. Coakley, J.E. and Berry, H.H., Identification of Elastomers by
    Py-GC, U.S. Tech. Nat. Info. Service (15:8:1971), NAFITR-1713,
    pp. 115-136.

50. Burns, D.T., Johnson, E.W. and Mills, R.F., Rapid Identification
    of Solid Polyurethane Elastomers, *J. Chromatogr., 105*(1973)43-48.

51. Zorina, N.I., Tsarfin, T.A. and Karnishin, A.A., Py-GC Study of
    Polyurethane Composition, *J. Anal. Chem. USSR, 32*(1977)936-939.

52. Iglauer, N. and Bentley, F.F., Py-GC for the Rapid Identifica-
    tion of Organic Polymers, *J. Chromatogr. Sci., 12*(1974)23-33.

53. Luke, B.G., Py-GC of Coumarone-Indene and Cyclopentadiene
    Resins, *J. Chromatogr. Sci., 11*(1973)435-438.

54. Host, M. and Deur-Siftar, D., Characterisation of Polybutadiene
    Type Polymers by Reaction GC, *Chromatographia, 5*(1972)502-507.

55. Braun, D. and Canji, E., Py-GC von Dien-Polymeren, *Angew. Mak-
    romol. Chem., 29-30*(1973)491-505.

56. Shono, T. and Shinra, K., Determination of the Microstructure of Polybutadiene by Py-GC, *Anal. Chim. Acta,* *56*(1971)303-307.

57. Martinex, J. and Guiochon, G., Identification of Phenol-Formaldehyde Polycondensates by Py-GC, *J. Gas Chromatogr.,* *5*(1967) 146-150.

58. Kretzschmar, H-J., Gross, D. and Kelm, J., Py-GC and Spectroscopic Identification of Fluorine Polymers, in C.E.R. Jones and C.A. Cramers (Eds.), *Analytical Pyrolysis*, Elsevier, Amsterdam, 1977, pp. 373-391.

59. Groten, B., Application of Py-GC to Polymer Characterisation, *Analyt. Chem.,* *36*(1964)1206-1217.

60. Nelson, D.F., Yee, J.L. and Kirk, P.L., The Identification of Plastics by Py-GC, *Microchem. J.,* *6*(1962)225-231.

61. Wheals, B.B. and Noble, W., The Py-GC Examination of Car Paint Flakes as an Aid to Vehicle Characterisation, *J. Forens. Sci. Soc.,* *14*(1974)23-32.

62. Cole, M.H., Petterson, D.L., Sljaka, V.A. and Smith, D.S., Identification and Determination of Polymers in Compounded Cured Rubber Stocks by Py-Two Channel GC, *Rubber Chem. Technol.,* *39*(1966)259-277.

63. Foxton, A.A., Hillman, D.E. and Mears, P.R., Identification of Rubber Vulcanisates by Py-GC, *J. Inst. Rubb. Ind.,* *3*(1969)179-183.

64. Derminot, J. and Rabourdin-Belin, C., Possibility for Gas Phase Chromatography in Identification Problems of Textile Fibres, *Bull. Inst. Text. Fr.,* *25*(1971)721-734.

65. Focher, B., Seves, A. and Bollini, M., Py-GC of Textile Materials, *Tinctoria,* *69*(1972)411-420.

66. Leukroth, G., Analysis of Adhesives by Combining GC and IR Spectroscopy using a GC-IR Analyser, *Adhasion* (1970)457-458.

67. Chene, M., Martin-Borret, O., Bollon, A. and Perret, A., Identification of Synthetic High Polymers in Paper by Py-GC, *Papeterie,* *88*(1966)1587-1590.

68. Fisher, G.E. and Neerman, J.C., Identification of Brake Lining Constituents by Py-GC, *Ind. Eng. Chem.,* *5*(1966)288-292.

69. Quigley, D.A., Davies, D.G. and Evans, H.L., Separation and Characterisation of Ashless Adhesives in Additive Concentrates, *Lab. Practice,* *18*(1969)421-424.

70. Cope, J.F., Identification of Flame-Retardant Textile Finishes by Py-GC, *Analyt. Chem.,* *45*(1973)562-564.

71. Alekseeva, K.V., GC Identification of Polymers Using Individual Pyrolysis Products, *J. Anal. Appl. Pyrol.,* *2*(1980)19-34.

72. Maynard, J.B., Twichett, J.E. and Walker, J.Q., Determination of Carboxyl-Terminated-Butadiene-Nitrile Polymer in Commercial Cresol-Novalac Epoxy Resin Systems by Py-Capillary GC-MS, J. Chromatogr. Sci., 17(1979)82-86.

73. Coulter, G.L. and Thompson, W.C., Automatic Analysis of Tyre Rubber Blends by Computer-linked Py-GC, in C.E.R. Jones and C. A. Cramers (Eds.), Analytical Pyrolysis, Elsevier, Amsterdam, 1977, pp. 1-15.

74. Coulter, G.L. and Thompson, W.C., Automatic Analysis of Tyre Rubber Blends by Computer-Linked Py-GC, Proc. Anal. Div. Chem. Soc., 14(1977)212-215.

75. Tsuge, K., Ando, J. and Okubo, N., Quantitative GC Analysis of NR-SBR-BR Three Component Blended Rubbers, J. Soc. Rubber Ind. (Jap.), 42(1969)851-855.

76. Krull, M., Kogerman, A., Kirret, O., Kutyina, L. and Zapolski, D., Py-GC of Capron (Nylon 6) Fibre Stabilised with Ethers of 4-oxydiphenylamine, J. Chromatogr., 135(1977)212-216.

77. CDS Applications Note 081673, Application of Py-GC to the Determination of Degree of Cure of Rubber, Chemical Data Systems, Oxford, Pa.

78. Krishen, A. and Tucker, R.G., Quantitative Determination of the Polymeric Constituents in Compounded Cured Stocks by Curie-point Py-GC, Analyt. Chem., 46(1974)29-33.

79. Evans, D.L., Weaver, J.L., Mukherji, A.K. and Beatty, C.L., Compositional Determination of Styrene-Methacrylate Copolymers by Py-GC, Proton-NMR and C-Analysis, Analyt. Chem., 50(1978) 857-860.

80. Blackwell, J.T., Quantitative Determination of the Monomer Composition in Hexafluoropropylene-Vinylidene Fluoride Copolymers by Py-GC, Analyt. Chem., 48(1976)1883-1885.

81. Sinclair, J.W., Schall, L. and Crabb, N.T., Quantitative Determination of Aliphatic Sulphur-Containing Additives by Py-GC, J. Chromatogr. Sci., 18(1980)30-34.

82. Derby, J.V. and Freeman, R.W., Vapour-Phase Py-GC of Fire-Retardant Materials Containing PVC, Amer. Lab., 6(1974)10-14.

83. Shen, J. and O'Kane, J.L., Monitoring and Identification of Nonvolatile Organic Deposits in Used Catalysts, Analyt. Chem., 49(1977)2374-2375.

84. Merritt, Jr., C., Sacher, R.E. and Petersen, B.A., Laser Py-GC-MS Analysis of Polymeric Materials, J. Chromatogr., 99(1974) 301-308.

85. Lum, R.M., Microanalysis of Trace Contaminants by Laser-Probe Pyrolysis, Amer. Lab., 10(1978)47-56.

86. Gunther, W., Koukoudimos, K. and Schlegelmilch, F., Characterisation of Textile Fibrous Materials by Py-Capillary GC, *Melliand Textilber.*, 60(1979)501-503.

87. Juvet, Jr., R.S., Smith, J.L.S. and Li, K-P., Polymer Identification and Quantitative Determination of Additives by Photolysis-GC, *Analyt. Chem.*, 44(1972)49-56.

88. Kiran, E. and Gillham, J.K., Py-Molecular Weight-GC Vapour Phase IR Spectrophotometry: An On-Line System for Analysis of Polymers, *Dev. Polym. Degradation*, 2(1979)1-33.

89. Esposito, G.G., Quantitative Py-GC by Internal Standard, *Analyt. Chem.*, 36(1964)2183-2185.

90. Gross, D., Verwendung von Standards in Polymer-pyrogrammen, *Z. Anal. Chem.*, 253(1971)40-42.

91. Kalal, J., Zachoval, J., Kubat, J. and Svec, F., Application of Py-GC in the Analysis of Diad Sequence Distribution in Styrene-Glycidyl Methacrylate Copolymers, *J. Anal. Appl. Pyrol.*, 1(1979)143-157.

92. Blasius, E., Lohdet, H. and Hausler, H., Characterisierung von Ionenaustauschern auf Kunstharzbasis durch Py-GC, *Z. Anal. Chem.*, 264(1973)278-286, 290-292.

93. Blasius, E. and Hausler, H., Characterisierung von Anionenaustauchern auf Kuntsharzbasis durch Py-GC, *Z. Anal. Chem.*, 276 (1975)11-19.

94. Blasius, E. and Hausler, H., Ermittlung des Vernetzungsgrades und Beladungszustandes von Ionenaustauschern auf Kunstharzbasis durch Py-GC, *Z. Anal. Chem.*, 277(1975)9-17.

95. Blasius, E., Hausler, H. and Lander, H., Characteristerung von Ionenaustauschern auf Kunstharzbasis durch Py-MS, *Talanta*, 23 (1976)301-307.

96. Sellier, N., Jones, C.E.R. and Guiochon, G., Py-GC as a Means of Determining the Quality of Porous Polymers of Styrene Crosslinked with Divinylbenzene, in C.E.R. Jones and C.A. Cramers (Eds.), *Analytical Pyrolysis*, Elsevier, Amsterdam, 1977, pp. 309-318.

97. McCormick, H., Quantitative Aspects of Py-GC of Some Vinyl Polymers, *J. Chromatogr.*, 40(1969)1-15.

98. Meade, C.F., Keyworth, D.A., Brand, V.T. and Deering, J.R., Determination of Total Oxygenates in Organic Materials for Levels Down to 5 ppm by GC, *Analyt. Chem.*, 39(1967)512-516.

99. Deur-Siftar, D., Application of Py-GC for Characterisation of Polyolefins, *J. Gas Chromatogr.*, 5(1967)72-76.

100. Alexeeva, K.V., Khramova, L.P. and Solomatina, L.S., Determination of the Composition and Structure of Polymers by Py-GC, *J. Chromatogr.*, 77(1973)61-67.

101. Shimono, T., Tanaka, M. and Shono, T., Py-GC of Butadiene Co-
     polymers, *Analyt. Chim. Acta, 96*(1978)359-365.

102. Szekely, T. and Blazso, M., Analytical Pyrolysis of Inorganic
     and Thermostable Polymers, in C.E.R. Jones and C.A. Cramers
     (Eds.), *Analytical Pyrolysis,* Elsevier, Amsterdam, 1977, pp.
     365-372.

103. Blazso, M., Garzo, G. and Szekely, T., Py-GC Studies on Poly
     (Dimethyl Siloxanes) and Poly(Dimethyl Alkylene Siloxanes,
     Chromatographia, 5(1972)485-492.

104. Seeger, M., Gritter, R.J., Tibbitt, J.M., Shen, M. and Bell,
     A.T., Analysis of Plasma-Polymerised Hydrocarbons by Py-GC, *J.
     Polym. Sci. (Chem. Ed.), 15*(1977)1403-1411.

105. Shimono, T., Tanaka, M. and Shono, T., Py-GC of Methyl Methac-
     rylate-α-Methylstyrene Copolymers, *J. Anal. Appl. Pyrol., 1*
     (1979)77-84.

106. Shimono, T., Tanaka, M. and Shono, T., Py-GC of Poly(3-methyl-
     1-alkenes), *J. Anal. Appl. Pyrol., 1*(1980)189-196.

107. Alajberg, A., Arpino, P., Deur-Siftar, D. and Guiochon, G.,
     Investigation of Some Vinyl Polymers by Py-GC-MS, *J. Anal.
     Appl. Pyrol., 1*(1980)203-212.

108. Van Schooten, J. and Evenhuis, J.K., Py-Hydrogenation-GC of
     α-olefin Copolymers, *Polymer, 6*(1965)561-577.

109. Toader, M., Chivulescu, E., Bader, P. and Boborodea, M., Deter-
     mining Polypropylene Isotacticity by IR Spectroscopy and Py-GC,
     Mater. Plast., 10(1973)151-155.

110. Farre-Rius, F. and Guiochon, G., On the Conditions of Flash
     Pyrolysis of Polymers as Used in Py-GC, *Analyt. Chem., 40*(1968)
     998-1000.

111. Barlow, A., Lehrle, R.S., Robb, J.C. and Sunderland, D., Poly-
     (Methyl Methacrylate)Degradation, *Polymer, 8*(1967)537-545.

112. Bagby, G., Lehrle, R.S. and Robb, J.C., Kinetic Measurements
     by Micro-Py-GC, *Polymer, 10*(1969)683-690.

113. Bell, F.A., Lehrle, R.S. and Robb, J.C., Polyacrylonitrile De-
     gradation Kinetics Studied by the Micro-Py-GC Technique, *Poly-
     mer, 12*(1971)579-599.

114. Varhegyi, G. and Blazso, M., Thermal Degradation and Microstruc-
     ture of Vinyl Copolymers. A Mathematical Model, *Eur. Polym. J.,
     14,*(1978)349-352.

115. Blazso, M. and Varhegyi, G., Calculation of Kinetic Parameters
     and Sequence Distribution from Py-GC Data of Styrene-Methyl
     Acrylate Copolymers, *Eur. Polym. J., 14*(1978)625-630.

116. Bouster, C., Vermande, P. and Veron, J., Study of the Pyrolysis
     of Polystyrenes. I. Kinetics of Thermal Decomposition, *J. Anal.
     Appl. Pyrol., 1*(1980)297-313.

117. Kishore, K., Pai Verneker, V.R. and Gayathri, V., Kinetic Studies on Thermal Decomposition of Polystyrene Peroxide, *J. Anal. Appl. Pyrol., 1*(1980)315-322.

118. Farre-Rius, F. and Guiochon, G., Kinetics of Decomposition of Polyesters. II. GC Analysis of the Products of Thermal Decomposition of Polyethylene Glycol Adipate, Poly-1,4-Butanediol Adipate and Poly-1,8-Octanediol Adipate, *J. Gas Chromatogr., 5* (1967)457-463.

119. Morisaki, S., Simultaneous Thermogravimetry-MS and Py-GC of Fluorocarbon Polymers, *Thermochim. Acta, 25*(1978)171-183.

120. Kaljurand, M. and Kullik, E., Application of the Hadamard Transform to GC of Continuously Sampled Mixtures, *Chromatographia, 11*(1978)328-330.

121. Kaljurand, M. and Kullik, E., Continuous Thermal Volatilisation Analysis of Polymers by GC with Pseudo-random Injection of Samples, *J. Chromatogr., 171*(1979)243-247.

122. Urbas, E., Kaljurand, M. and Kullik, E., Study of the Thermal Decomposition of Polymers by On-Line Cross-Correlation-GC, *J. Anal. Appl. Pyrol., 1*(1980)213-220.

123. Lee, A.K. and Sedgwick, R.D., Application of MS to Copolymers of Ethylene and Propylene Oxides. II. Kinetic Analysis, *J. Polym. Sci. (Chem. Ed.), 16*(1978)997-1003.

124. Luderwald, I., Recent Results and New Techniques in Py-EIMS of Synthetic Polymers, *Proc. 5th. Eur. Symp. Polym. Spectrosc., 1978*, Verlag Chemie, Weinheim, 1979, pp. 217-255.

125. Bottino, F., Foti, S., Montaudo, G., Poppalardo, S., Luderwald, I. and Przybylski, M., Synthesis of Polymers Containing Double Bridged Phenylether Units and Characterisation by Direct Pyrolysis in the MS, *Angew. Makromol. Chem., 67*(1978)203-211.

126. Bottino, F., Foti, S., Montaudo, G., Pappalardo, S., Luderwald, I. and Przybylski, M., Direct Pyrolysis in the MS of Aromatic Polysulphonates and Polythiosulphonates, *J. Polym. Sci. (Chem. Ed.), 16*(1978)3131-3137.

127. Chatfield, D.A., Hileman, F.D., Voorhees, K.J., Einhorn, I.N. and Futrell, J.H., Characterisation of Polymer Decomposition Products by EI- and CI-MS, *Appl. Polym. Spectrosc., 1978*:241-256.

128. Kabilov, Z.A., Muinov, T.M., Shibaev, L.A., Suzanov, Y.N., Korzhavin, L.N. and Prokopchuk, N.R., Investigations of Imidization of Polypyromellitamido Acids and Thermal Degradation of Polypyromellitimides by MS Thermal Analysis, *Thermochim. Acta, 28*(1979)333-347.

129. Hummel, O., Decomposition Behaviour of Thermo Stable Polymers as Studied by Py-FI-MS, *Adv. Chem. Therm. Stable Polym., 1977*: 99-118.

130. Gritter, R.J., Seeger, M. and Johnson, D.E., Hydrocarbon Formation in the Pyrolytic Decomposition of Methacrylate Copolymers, *J. Polym. Sci. (Chem. Ed.), 16*(1978)169-177.

131. Gritter, R.J., Seeger, M. and Gipstein, E., Study of the Mechanism of Polysulphone Decomposition by Py-GC and Py-GC-MS, *J. Polym. Sci. (Chem. Ed.), 16*(1978)353-360.

132. Seeger, M. and Gritter, R.J., Thermal Decomposition and Volatilisation of Poly(α-olefins), *J. Polym. Sci. (Chem. Ed.), 15*(1977)1393-1402.

133. Tsuge, S., Okumoto, T., Sugimura, Y. and Takeuchi, T., Py-GC Investigations of Fractionated Polycarbonates, *J. Chromatogr. Sci., 7*(1969)253-256.

134. Tsuge, S., Okumoto, T. and Takeuchi, T., Study of Thermal Degradation of Fractionated Polystyrenes by Py-GC, *J. Chromatogr. Sci., 7*(1969)250-252.

135. Deur-Siftar, D. and Mitrovic, V., Determination of the Thermal Stability of PVC by Reaction GC, *Chromatographia, 5*(1972)573-575.

136. Leisztner, L., Gal, S., Szanto, J. and Kovacs, L., Influence of the Morphology of Samples in the Pyrolysis of PVC, *J. Thermal Anal., 13*(1978)141-147.

137. Tamara, S. and Gillham, J.K., Py-M.Wt. Chromatography--Vapour Phase IR Spectrophotometry. An On-Line System for Analysis of Polymers. IV. Influence of Cis-Trans Ratio on the Thermal Degradation of 1,4-Polybutadienes, *J. Appl. Polym. Sci., 22*(1978)1867-1884.

138. Vacherot, M., Determination of the Microstructure of the 1,4- and 3,4-Polyisoprenes by Py-GC, *J. Gas Chromatogr., 5*(1967)155-156.

139. Ghotra, J.S., Stevens, G.C. and Bloor, D., MS and Py-GC Studies of the Monomer and Polymer of Bis(p-toluenesulphonate) of 2,4-Hexadiyne-1,6-diols, *J. Polym. Sci. (Chem. Ed.), 15*(1977)1155-1167.

140. Yamamoto, Y., Himei, M. and Hayashi, K., Radiation-Induced Pyrolysis of Poly(α-methylstyrene). Degradation by Cationic Chain Reactions, *Macromolecules, 10*(1977)1316-1320.

141. Cascaval, C.N. and Florin, R.E., Py-GC of some Fluorine-Containing Polymers, *J. Fluorine Chem., 14*(1979)65-70.

142. Naveau, J., Parametric Sensitivity in Py-GC. Application to cis-1,4-polyisoprenes, *J. Chromatogr., 174*(1979)109-122.

143. Sidek, B.D., Formation of 2,4-Dimethyl-4-ethenylcyclohexene from the Pyrolysis of Natural Rubber and its Relevance to the Depolymerisation Mechanism of 1,4-Polyisoprenes, *J. Rubber Res. Inst. Malays., 27*(1979)40-45.

144.   Sugimura, Y. and Tsuge, S., Py-Hydrogenation-Glass Capillary
       GC Characterisation of Polyethylenes and Ethylene-α-Olefin
       Copolymers, *Macromolecules, 12*(1979)512-514.

145.   Tsuge, S., Kobayashi, T., Sugimura, Y., Nagaya, T. and Takeu-
       chi, T., Py-GC Characterisation of Highly Alternating Copoly-
       mers Containing Styrene and Tetracyanoquinodimethan Methyl
       Acrylate, Acrylonitrile or Methyl Methacrylate Units, *Macro-
       molecules, 12*(1979)988-992.

146.   Gritter, R.J., Gipstein, E. and Adams, G.E., Py-GC-MS of Poly-
       mers: Poly(isopropenylcyclohexane) and Copolymers with its
       Aromatic Counterpart α-Methylstyrene, *J. Polym. Sci. (Chem.
       Ed.), 17*(1979)3959-3967.

147.   Cascaval, C.N., Scheider, I.A., Poinescu, I.C. and Butnaru, M.,
       Py-GC of Some Cross-Linked Copolymers of Styrene, *Eur. Polym.
       J., 15*(1979)661-666.

148.   Chaigneau, M., MS Analysis of Compounds Formed by Pyrolysis of
       Polyacrylonitrile and Copolymers, *Analusis, 5*(1977)223-227.

149.   Fanter, D.L., Levy, R.L. and Wolf, C.J., Laser Pyrolysis of
       Polymers, *Analyt. Chem., 44*(1972)43-48.

150.   Walker, J.Q. and Wolf, C.J., Py-GC: A Comparative Study of
       Different Pyrolysers, *J. Chromatogr. Sci., 8*(1970)513-518.

151.   Levy, E.J. and Paul, D.G., Application of Controlled Partial
       Gas Phase Thermolytic Dissociation to the Identification of GC
       Effluents, *J. Gas Chromatogr., 5*(1967)136-145.

152.   Still, R.H. and Whitehead, A., Thermal Degradation of Polymers.
       XV. Vacuum Pyrolysis Studies on Poly(p-methoxystyrene) and
       Poly(p-hydroxystyrene), *J. Appl. Polym. Sci., 21*(1977)1199-
       1213.

153.   Ordoyno, N.F. and Rowan, S.M., Interactions Observed During
       the Pyrolysis of Binary Mixture of Textile Polymers, *Thermo-
       chim. Acta, 23*(1978)371-385.

154.   Ferguson, J. and Mahapatro, B., Pyrolysis Studies on Polyacry-
       lonitrile Fibres. Influence of Conditions and M.Wt. on Tensile
       Property Changes During the Initial Stages of Pyrolysis, *Fibre
       Sci. Technol., 11*(1978)55-66.

155.   Camino, G. and Costa, L., Thermal Degradation of a Highly
       Chlorinated Paraffin Used as a Fire-Retardant Additive for
       Polymers, *Polym. Degradation Stab., 2*(1980)23-33.

156.   Mol, G.J., Gritter, R.J. and Adams, G.E., MS of Thermally
       Treated Polymers, *Appl. Polym. Spectroscop., 1978*:257-277.

157.   Lum, R.L., Thermal Decomposition of Poly(butylene Terephth-
       alate), *J. Poly. Sci. (Chem. Ed.), 17*(1979)203-213.

158.   Lum, R.L., MoO$_3$ Additives for PVC: A Study of the Molecular
       Interactions, *J. Appl. Polym. Sci., 23*(1979)1247-1263.

159. Lum, R.L., Investigation of Polycarbodiimide Additive Effects
      on the Acid-Catalysed Hydrolysis of PBTs, *J. Polym. Sci.
      (Chem. Ed.)*, *17*(1979)3017-3021.

160. Starnes, Jr., W.H. and Edelson, D., Mechanistic Aspects of the
      Behaviour of Molybdenum (VI) Oxide as a Fire-Retardant Additive
      for PVC. An Interpretive Review, *Macromolecules*, *12*(1979)797-
      802.

161. Ballistreri, A., Foti, S., Montaudo, G., Pappalardo, S.,
      Scamporrino, E., Arnesano, A. and Calgari, S., Thermal Decom-
      position of Flame-Retardant Acrylonitrile Copolymers Inves-
      tigated by Direct Py-MS, *Macromol. Chem.*, *180*(1979)2843-2849.

162. Varma, I.K. and Sharma, K.K., Thermal Degradation of PVC in
      Presence of Additives. III. Studies in Molecular Weight Dis-
      tribution, *Angew. Makromol. Chem.*, *79*(1979)147-155.

163. Chien, J.C.W. and Kiang, J.K.Y., Polymer Reactions--X. Thermal
      Pyrolysis of Poly(isoprene), *Eur. Polym. J.*, *15*(1979)1059-1065.

164. Kiang, J.K.Y., Uden, P.C. and Chien, J.C.W., Polymer Reactions.
      Part VII. Thermal Pyrolysis of Polypropylene, *Polym. Degrad.
      Stab.*, *2*(1980)113-127.

165. Chien, J.C.W. and Kiang, J.K.Y., Polymer Reactions. 8. Oxida-
      tive Pyrolysis of Poly(propylene), *Makromol. Chem.*, *181*(1980)
      45-57.

166. Chien, J.C.W. and Kiang, J.K.Y., Polymer Reactions. 9. Effect
      of Polymer-Bound Chromium on Oxidative Pyrolysis of Poly(pro-
      pylene), *Macromolecules*, *13*(1980)280-288.

167. Mischer, G., MS of High Polymers, *Adv. Mass Spectrom.*, *7B*
      (1978)1444-1451.

168. Bratspies, G.K., Smith, J.F., Hill, J.O. and Magee, R.J., A
      Thermogravimetry-DTA, EGA and Py-GC-MS Study of Dihalotin
      (IV) bisdiethyldithiocarbamates, *Thermochim. Acta*, *19*(1977)
      335-348.

169. Bratspies, G.K., Smith, J.F. and Hill, J.O., A Thermogravim-
      etry-DTA and Py-GC-MS Study of Several Tin (IV) Dithiocarbamate
      Complexes in an Air Atmosphere, *Thermochim. Acta*, *19*(1977)373-
      382.

170. Bratspies, G.K., Smith, J.F., Hill, J.O. and Magee, R.J., A
      Py-GC-MS and Thermogravimetry-DTA Study of Bis(diethyldithio-
      carbamate)diphenyl Tin(IV), *Thermochim. Acta*, *19*(1977)349-360.

171. Bratspies, G.K., Smith, J.F., Hill, J.O. and Magee, R.J., A
      Thermogravimetry-DTA, EGA and Py-GC-MS Study of Tetrakis(die-
      thyldithiocarbamate) Tin (IV), *Thermochim. Acta*, *19*(1977)361-
      371.

172. Uden, P.C., Henderson, D.E. and Lloyd, R.J., An Interfaced
      Vapour-Phase Instrumental System for Thermal Analysis and Py-
      rolysis, in C.E.R. Jones and C.A. Cramers (Eds.), *Analytical
      Pyrolysis*, Elsevier, Amsterdam, 1977, pp. 351-363.

173. Margitfalvi, J. and Koltai, L., Investigation of the Decomposition of Several Organometallic Compounds by Py-GC, *J. Thermal Anal.*, *15*(1979)251-255.

174. Bratspies, G.K., Smith, J.F. and Hill, J.O., MS Investigation of the Thermal Decomposition of Tetrakis(N,N-diethyldithiocarbamate) Tin (IV) in Vacuum, *J. Anal. Appl. Pyrol.*, *2*(1980)35-44.

175. Urbas, E. and Kullik, E., Py-GC Analysis of Some Toxic Compounds from N-Containing Fibres, *J. Chromatogr.*, *137*(1977)210-214.

176. Freedman, A.N., Gaseous Degradation Products from the Pyrolysis of Insulating Materials Used in Large Electricity Generators, *J. Chromatogr.*, *157*(1978)85-96.

177. Hilado, C.J., Screening Materials for Toxicity of Pyrolysis Gases, *J. Combust. Toxicol.*, *6*(1978)248-255.

178. Cagliostro, D.E. and Zermer, D.C., Modelling the Toxicity of Burning of Poly(methyl methacrylate), *Fire Technol.*, *15*(1979)85-101.

179. Hilado, C.J. and Brauer, D.P., Upholstery Fabrics: Ignitability, Flash-fire Propensity, Toxicity, *Proc. Int. Conf. Fire Safety*, *4*(1979)2-18.

180. Kaminsky, W., Janning, J. and Sinn, H., Pyrolysis--A Path for the Future?, *Eur. Rubber J.*, *161*(1979)15,18,58.

181. Kaminsky, W., Sinn, H. and Janning, J., Industrial Prototypes for the Pyrolysis of Used Tyres and Waste Plastic Materials, *Chem-Ing. Tech.*, *51*(1979)419-429.

182. Buckens, A.G., Mertens, J.J., Schoeters, J.G.E. and Steen, P.C., Experimental Techniques and Mathematical Models in the Study of Waste Pyrolysis and Gasification, *Conserv. Recycling*, *3*(1979)1-23.

183. Szepesy, L., Feedstock Characterisation and Prediction of Product Yields for Industrial Naphtha Crackers on the Basis of Laboratory and Bench-Scale Pyrolysis, *J. Anal. Appl. Pyrol.*, *1*(1980)243-268.

184. Luigart, F., Contribution to the Py-IR Spectroscopy of Polymers-Microanalysis, *Kunstoffe*, *70*(1980)66-67.

185. McCall, E.R., Morris, N.M. and Berni, R.J., IR Analysis of the Vapour-Phase Pyrolysis Products of Flame-Retardant Fabrics, *Text. Res. J.*, *49*(1979)288-292.

186. Truett, W.L., The Characterisation of Organic Polymers via Py-IR, *Adv. Chem. Ser.*, *174*(1979)81-86.

Chapter 7

Biological Molecules

7.1. INTRODUCTION

The rapid extension of analytical pyrolysis into diverse fields such
as taxonomy and soil chemistry has been largely due to progress in
the pyrolysis of biological molecules. Such work has shown that the
various classes of molecules, with the possible exception of nucleic
acids, give highly characteristic pyrograms, and has enabled the
origin of fragments in more complex samples such as bacteria [1,2]
and soil [3] to be proposed. Interest in the pyrolysis of biological molecules has many facets, each relating to the commercial importance of thermal degradation processes. Foremost among these are
combustion [4-9], flame retardation [10-11], and the use of cellulosic wastes for raw material or food production [12-14]. The simulation of smoking conditions [15-18], thermal conversions during cook-
ing [19-22], and the production of mutagenic principals via the
thermal degradation of amino acids and proteins [23,24] are also
under active investigation. Many of these studies have used large-
scale, furnace-type pyrolysis and off-line analysis. This has been
done to mimic the applicable thermal processes. The product distribution is thus determined by a variety of secondary reactions,
and differs significantly from that obtained using a strict analytical pyrolysis approach. The variation in pyrolysis and analytical
conditions makes the detailed comparison of literature data difficult. An extreme, but illustrative case is that of D-glucose, which

yields 2-furaldehyde and 5-hydroxymethyl-2-furaldehyde with Curie-
point Py-GC [25], but gives over 100 products with furnace methods
[26,27]. Reviews in this area have appeared [28-31].

7.2. AMINO ACIDS AND PROTEINS

The pyrolysis of α-amino acids is characterized by decarboxylation
to yield the corresponding amine fragment. This is frequently the
major component in the pyrograms of aliphatic compounds. Bimolecular processes lead to dipeptides and diketopiperazines, which yield
nitriles on fragmentation. These, too, are characteristic of individual amino acids and have been frequently reported. Aldehydes
which contain one carbon atom less than the parent compound are also
diagnostic of aliphatic amino acids and arise via decarbonylation of
the α-lactone produced by an $S_N i$ deamination process. Side chain
stripping involving chain homolysis also leads to saturated and unsaturated residues. These processes have been illustrated for phenylalanine in Fig. 1.6. Larger scale pyrolysis may yield an entirely
different product distribution. For example, stilbene, naphthalene,
phenylnaphthalene, and anthracene have been identified in phenylalanine pyrolysates [37]. Distinctive pyrograms have been reported for
28 different amino acids [32] and detailed studies have identified
many products and have proposed mechanistic events to account for
their formation [33-38]. Differences between published work are
largely due to variations in the pyrolysis and analytical conditions.
Polar aromatic amino acids essentially yield products derived via
sidechain stripping. Thus, tyrosine yields phenol, p-cresol and p-
ethylphenol, together with some rearrangement products such as o-
and m-cresol, o- and m-ethylphenol, as large fragments. Similarly,
tryptophan gives indole, 3-methylindole and vinylindole [34] and
histidine gives imidazole [39]. Major amine fragments, however, may
be detected from tyrosine and tryptophan with different analytical
conditions (Carbowax 20M-KOH) [40]. It is not clear whether the absence of nitriles in these analyses is due simply to transfer conditions, scale, or to more fundamental differences.

The imino acids proline and hydroxyproline are characterized by the appearance of pyrrole as the major fragment, together with smaller amounts of methyl- and ethylpyrroles. Trifunctional amino acids undergo complex reactions with glutamic acid yielding pyrrole and 2-methylpyrrole, together with N-ethylpyrrole and pyrrolid-2-one. Lysine also underwent cyclization and 3,4-dihydroazepinone was the sole large fragment detected [40]. Sulfur-containing amino acids yield H_2S (cystine) or MeSH (methionine). Methionine also yields the nitrile and ethylene as major fragments [38]. No evidence for the intermediacy of diketopiperazines was found, and an [^{15}N]alanine label was not incorporated into the methionine products. A ruby laser Py-GC study of aliphatic amino acids yielded hydrocarbons, olefins, and a characteristic aldehyde fragment [41], whereas infrared ruby pyrolysis yielded complex pyrograms with alanine giving benzene, toluene, and picoline among the products [42]. With β-amino acids deamination predominates over decarboxylation as the major fragmentation pathway [36].

Amino acids have also been characterized through the phenylthio-hydantoin derivative formed during the Edman degradation of polypeptides [43]. Phenylisocyanate, phenylisothiocyanate, and benzonitrile were characteristic of the derivative and each amino acid showed a distinctive pattern. Unique fragments:- phenylalanine-toluene, tryptophan-indole and methylindole, tyrosine-*p*-cresol, and proline-pyrrole were frequently observed.

Despite the fact that pyrolysis processes for amino acids are well understood, there has been comparatively little application of this knowledge to the analysis of polypeptides or proteins. Pyrograms from dipeptides differed from those of mixtures of the individual amino acids [32] and products characteristic of the component monomers were obtained [33]. The sequence arrangement of amino acids has also been shown to affect the product distribution [44], but no comprehensive studies have been undertaken in this area. One clearly documented example of structure elucidation is, however, available. This relates to the peptide sequences in the antibiotic Actinomycins

[45]. The Py-GC of 25 µg of Actinomycin D, for example (Fig. 7.1) yielded four diketopiperazines each derived from adjacent peptide residues. The identification of these products, together with knowledge of the identity of the COOH- terminal residue, enabled the peptide sequence to be established.

The sensitivity and specificity of Py-GC-MS has enabled octapeptides in infusion fluids to be identified. Two closely related hormones, lypressin and felypressin, were detected in ng amounts in aqueous solution in the presence of large amounts of buffers and stabilizers [46]. Felypressin contains two adjacent phenylalanine residues, whereas lypressin has a tyrosine-phenylalanine sequence (Fig. 7.2). The pyrolysis products reflect this difference and, in contrast to felypressin, lypressin evidences p-cresol and phenol as characteristic peaks, together with a significant decrease in the intensity of toluene and styrene fragments (from phenylalanine). Pyrrole, from proline, was a further significant peak in both pyrograms.

Such behavior suggests pyrograms and mass pyrograms derived from proteins should be characterized by the components of amino acid pyrolysis. This is indeed so [44,47-49], although it would appear that side-chain stripping, rather than depolymerization, is the main fragmentation mechanism. Thus, although some nitriles (e.g., from phenylalanine) are produced, and the intense pyrrole contribution from imino acid residues points to some rupture of the backbone, most pyrograms are characterized by the presence of fragments from the sulfur containing or aromatic amino acids. The Py-GC of enzymes, including chymotrypsin, acetylcholinesterase, lactate dehydrogenase, and urease, has been shown to proceed essentially this way and all of the five identified peaks were found to be due to phenylalanine, tyrosine and tryptophan fragments [50]. Indeed, this pyrolysis enabled a method for the quantitative determination of tryptophan residues in proteins to be established [51]. Quantitative studies have also been undertaken using the pyrrole peak which characterizes gelatin and collagen. This has been of use in dental

studies [52] and has enabled the degree of mineralization of the
calcified collagenous tissue, such as tooth enamel and dentin, to
be assessed [53]. In particular, the logarithm of the peak area of
the pyrrole fragments was directly related to the specific gravity
of the developing enamel [28,54,55].

Py-MS studies have also shown the applicability of this technique for producing fingerprint mass pyrograms. In addition, the
presence of characteristic ion series mirror the Py-GC results and
allow some direct chemical inferences to be made. Thus, the peak
at m/z 34 (H_2S) is significantly larger in albumin than in pepsin
and reflects the larger proportion of sulfur-containing amino acids
in the former sample [47]. Trypsin, which contains methionine,
shows the presence of methanethiol (m/z 48), whereas this peak is
absent from insulin, which is methionine-free. Further, the peak
at m/z 117 is also much reduced in insulin. This reflects the absence of tryptophan, which yields indole (m/e 117), although benzyl
cyanide (from phenylalanine) also contributes to this peak [29,55,56].

In addition, other work has shown the value of a nitrogen detector in enhancing pyrolysis fragments from proteins [50,57] and fingerprint pyrograms on Adeno virus structure and other proteins [58] and
enzymes have also appeared [59]. Significant differences between
various reports, due to pyrolysis and analysis conditions reduce the
value of purely fingerprint reports. Pyrolysis of hair [60,61] and
larger scale work has also been reported [62-65]. Characteristic
ions from the Py-MS of proteins are listed in Table 7.1

7.3. POLYSACCHARIDES

Polysaccharides are components of foodstuffs, textiles, and woods,
and their thermal decomposition reactions have a long history. This
reflects the importance of such processes, ranging from combustion
to food preparation. In practice, many of the reactions are concerned with the degradation of large amounts of material with slow
heating, and research work has tended to follow these conditions in
many instances [66]. Thus, furnace pyrolysis and off-line analysis

TABLE 7.1 Characteristic Ions Found in Protein Mass Pyrograms

m/z	Component	m/z	Component
17	Ammonia	81	Alkylpyrroles
28	Ethylene	83	Methylbutanenitrile
30	$CH_2=\overset{+}{N}H_2$	92	Toluene
34	Hydrogen sulfide	94	Phenol
41	Acetonitrile	97	--
42	Propylene	104	Styrene
44	Carbon dioxide	108	Cresol
48	Methanethiol	109	C_3-Alkylpyrroles
55	Propionitrile	117	Indole
56	Butene	117	Phenylacetonitrile
67	Pyrrole	120	C_3-Alkylbenzenes
68	--	131	Methylindole
69	Butanenitrile	145	C_2-Alkylindoles
70	Methylbutene	153	--

has been a major technique, but the proliferation of secondary reactions which occur during such conditions has tended to obscure the primary mechanistic events. The volatile products obtained from heated glucose are recorded in Table 7.2 and reflect the complexity of the product distribution [22,66,67]. It is of interest to note that oligomers may also be isolated [68]. More recent studies undertaken at 400°C and 1000°C with either an N_2 or an H_2 atmosphere discuss the degradation in detail and confirm the overall product distribution with the presence of furans being characteristic [69-71].

Despite such diversity, Curie-point Py-GC studies on glucose (at 460°C) indicate effectively the nature of the initial degradation products [25]. The pyrogram is characterized by two main components, 2-furaldehyde (furfural) and 5-hydroxymethyl-2-furaldehyde, which result from dehydration of the carbohydrate. Levoglucosenone, levulinic acid, and 1,4-3,6-dianhydro-D-glucopyranose [72] were the other identified fragments (Fig. 7.3). The simplicity of the pyrogram has

TABLE 7.2   Thermal Degradation Products of D-Glucose

Methanol	2-Me-butanoic acid	2-Me-5-n-Pr-furan
Ethanol	4-OH-2-pentenoic acid	2,5-di-Et-furan
Methanal	(lactone)	2-(1-Propenyl)furan
Ethanal	Levulinic acid	2-isopropenylfuran
Propanal	Succinic acid	2-Me-3(fur-2-yl)-2-
Butanal	Tartaric acid	propenefuran
2-Me-propanal	Pyruvic acid	2(1-Propenyl)-5-Me-
Pentanal	Benzene	furan
2-Me-butanal	Toluene	2-Furfural
2-Propenal	o-Xylene	3-Furfural
2-Me-propenal	m-Xylene	2-Acetylfuran
2-Butenal	p-Xylene	3-Me-2-furfural
2-Pentenal	Et-benzene	5-Me-2-furfural
1,3-Pentadienal	Styrene	5-Me-2-acetylfuran
Acetone	1,3,5-tri-Me-benzene	1-(Fur-2-yl)propan-
2-Butanone	1,2,4-tri-Me-benzene	1,2-dione
2-Pentanone	1,2,3-tri-Me-benzene	2-Me-tetrahydrofuran-
3-Pentanone	Naphthalene	3-one
2-Hexanone	Phenol	5-hydroxymethyl-2-
2,3-Butandione	Resorcinol	furfural
2,3-Pentandione	Hydroquinone	4-OH-2,5-di-Me-furan-
3-Buten-2-one	Pyrocatechol	3(2H)-one
3-Penten-2-one	Furan	Me-furoic acid
3-Me-3-buten-2-one	2-Me-furan	2-Furoic acid
Cyclopentanone	3-Me-furan	1,6-anhydro-β-D-gluco-
3-Me-cyclopentan-	2-Et-furan	pyranose
1,2-dione	2,5-di-Me-furan	1,4:3,6-dianhydro-β-D-
2-Cyclopenten-1-one	2-Vinylfuran	glucopyranose
Formic acid	2-n-Pr-furan	2,3-Benzofuran
Acetic acid	2-iso-Pr-furan	Maltose
Propanoic acid	2-Et-5-Me-furan	iso Maltose
Butanoic acid	2,3,5-tri-Me-furan	Cellobiose
2-Me-propanoic acid	2-iso-Pr-5-Me-furan	Gentiobiose
Pentanoic acid	2-Me-5-Vinylfuran	

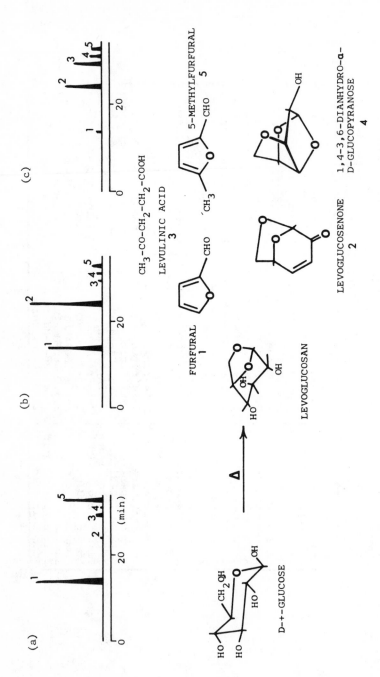

FIGURE 7.3 Curie-point pyrolysis of D-glucose, cellulose, and levoglucosan: (a) glucose, (b) cellulose, and (c) levoglucosan. (Adapted from Ref. 25.)

been confirmed (at 700°C) [72], although no anhydrosugars were reported, and many other lower boiling fragments (benzene, acetone, furan, 2-methylfuran) were also found.

The pyrolysis of other monosaccharides shows essentially the same fragmentation, although the product distribution varies. Disaccharides and oligosaccharides follow analogous pathways [25,73,74] and cellahexose exhibits essentially the pyrolysis depicted in Fig. 7.3 for glucose, although it must be remembered that only volatiles are detected.

Pyrolysis of cellulose, the β-1,4 polymer of glucose, is more complex, and although this also proceeds through dehydration, various other processes occur and initial products are anhydro sugars rather than furan derivatives. At temperatures above 300°C cellulose decomposes endothermically until most of the material has volatilized. The products may be separated into three groups: a carbonaceous residue (char), a highly viscous syrup (tar), and a mixture of volatile gases. These arise through a variety of complex reactions which may occur either sequentially or competitively.

The formation of levoglucosan (1,6-anhydro-β-D-glucopyranose) appears to be one of the main primary degradation routes (Fig. 7.4). This material is found in the tar fraction, together with smaller amounts of its furanose isomer, other anhydro sugars and some randomly linked oligosaccharides. *Depolymerization* and a series of inter- and intramolecular transglycosylation reactions are the main events during this initial stage. Indeed, these reactions may be controlled finely to enable a high yield of D-+-glucose (via hydrolysis of levoglucosan) to be obtained through pyrolysis of cellulosic materials [75].

Dehydration reactions accompany and follow transglycosylation. These result in the production of unsaturated compounds such as levoglucosenone, 3-deoxyglucosone and the furans, which constitute the bulk of the volatiles. At higher temperatures *fission* of sugar residues becomes important and other volatiles such as acrolein, glyoxal and acetaldehyde are produced. *Condensation* of the various

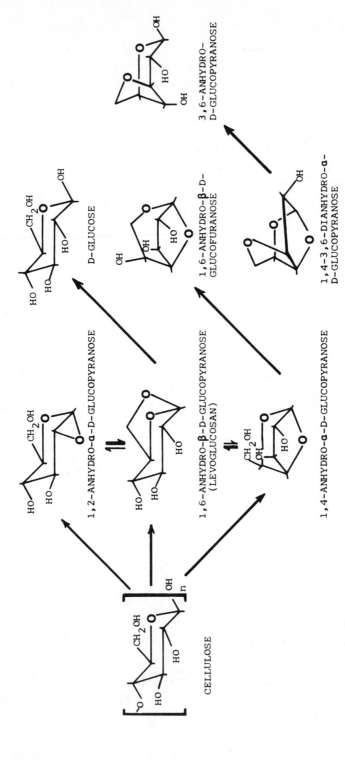

FIGURE 7.4 Pyrolysis of cellulose: initial reactions.

D-GLUCOSE

1,2-ANHYDRO-α-D-GLUCOPYRANOSE

1,6-ANHYDRO-β-D-GLUCOPYRANOSE (LEVOGLUCOSAN)

1,4-ANHYDRO-α-D-GLUCOPYRANOSE

CELLULOSE

1,6-ANHYDRO-β-D-GLUCOFURANOSE

1,4-3,6-DIANHYDRO-α-D-GLUCOPYRANOSE

3,6-ANHYDRO-D-GLUCOPYRANOSE

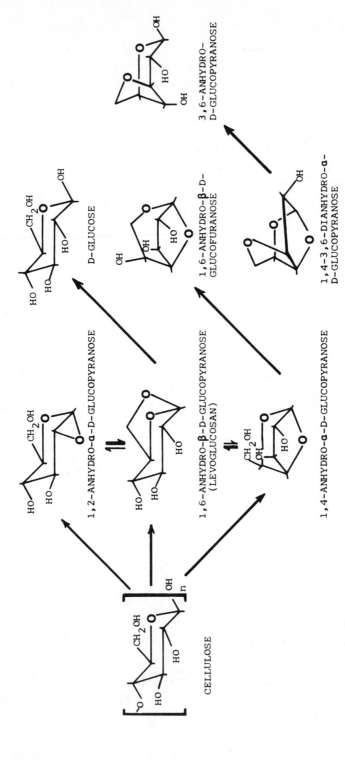

TABLE 7.3 Yields (%) from Cellulose Pyrolysis at 300°C

Pressure	Char	Tar	Levoglucosan	Hydrolysable glucose
760 torr	34.2	19.1	3.6	10.1
1.5 torr	17.8	55.8	28.1	54.6

unsaturated and carbonyl-containing products and free-radical cleavage of the side chains lead to the formation of char. The proportions of char, tar, and volatiles are highly dependent upon the pyrolysis conditions. Typical yields for furnace conditions are shown in Table 7.3. The low pressure conditions allow the distillation of anhydro sugars before further degradation occurs, and hence the yield of char is significantly reduced. This is also noted when increased pyrolysis temperatures are used. For example, at 400°C under vacuum, char formation accounts for about 5% of the product and the tar yield approaches 80%. This clearly indicates the necessity for low loadings and high heating rates in analytical pyrolysis of carbohydrates, although high temperatures are contraindicated.

Curie-point Py-GC again reveals its potential in exposing initial events in the analysis of cellulose (Fig. 7.3). This contrasts with that of glucose in that the anhydro sugars now predominate. Levoglucosan was not detected in the pyrogram--perhaps due to analytical conditions, as this compound can be demonstrated to be the major furnace product by GC [76]--but pyrolysis of this product was shown to produce levoglucosenone, together with smaller amounts of levulinic acid, 5-hydroxymethylfurfural and the 1,4-3,6-dianhydro sugar. The Py-GC results suggest that two major pathways are followed--one via levoglocosan to levoglucosenone and the other via dehydration of monomeric units to yield furfural. In contrast to these results, Curie-point analysis at a higher temperature gives a much greater abundance of lower volatiles [72] and the GC analysis of the furnace pyrolysis products of cellulose reveals a vast abundance of products, many similar to those from glucose [77,78].

TABLE 7.4   Furnace Pyrolysis Products (%) from Treated Celluloses
at 550°C

Product	Cellulose	Treated with 5%		
		H_3PO_4	$(NH_4)_2HPO_4$	$ZnCl_2$
Tar	66	16	7	31
Water	11	21	26	23
Char	5	24	35	31

The combustion of cellulosic compounds proceeds on two fronts:
a rapid flaming combustion of the flammable volatiles and a slower
glowing combustion of residual char.   Mechanistic studies of combustion and flame retardation are under active consideration [79-83].
Retardant additives may act by increasing the proportion of char and
nonflammable volatiles such as water and CO_2 at the expense of the
flammable volatiles or by suppressing the combustion of organic
volatiles in the gas phase.   Retardants are frequently acids or
Lewis acids, and these lower the degradation temperature by the pro-
motion of the various elimination and condensation reactions which
favor char formation.   The effects are illustrated in Table 7.4.

Probe Py-MS of cellulose and treated samples confirm these
results [84,85].   The estimated product distribution is recorded in
Table 7.5.   These values were obtained by the quantitative analysis
of the mass pyrogram using standard spectra and mixture analysis
arithmetic.   Mass fragmentograms also enabled the evolution of individual components to be monitored.

Clearly, the variation in pyrolysis which occurs under these
conditions should be borne in mind when impure carbohydrates are
analyzed.   Although not studied in detail to date, it may well be
expected that differences may be anticipated.   This is further indicated by the fact that neutral (NaCl) or basic (NaOH) additives
also affect the pyrolysis processes.   Conceivably, bases promote
aldol and retro-aldol reactions, together with various elimination
processes.

TABLE 7.5 Probe Py-MS of Cellulose (values are % of total ionization recorded)

FIBRE	Levoglucosan	1,6-Anhydro-β-D-glucofuranose	5-Hydroxymethyl-furfural	CO	CO_2	H_2O
Cotton	30	8	9	10	4	7
Flame-retardant cotton	6	1	6	8	13	33

Source: Ref. 85.

TABLE 7.6 Typical Py-MS Ion Series from Hexoses

m/z	Component	m/z	Component
18	Water	74	Hydroxypropanone
28	Carbon monoxide	84	$C_4H_4O_2$
28	Ethylene	86	Diacetyl
30	Formaldehyde	90	Glyceraldehyde
31	$CH_2{=}\overset{+}{O}H$	90	Dihydroxypropane
32	Methanol	96	2-Furaldehyde
42	Ketene	98	Furfuryl alcohol
42	Propene	98	1,5-Anhydro-2,3-dideoxy-β-D-pent-2-enofuranose
43	$CH_3{-}C{\equiv}O^+$		
44	Carbon dioxide	102	$C_4H_6O_3$
44	Acetaldehyde	110	5-Methyl-2-furaldehyde
46	Formic acid	112	
55	C_3H_3O	114	
56	Acrolein	116	Levulinic acid
57	C_4H_7	120	[Erythrose]
58	Acetone	126	Levoglucosenone
60	Hydroxyacetaldehyde	126	5-Hydroxymethyl-2-furaldehyde
61	$C_2H_5O_2$		
68	Methylbutadiene	128	1,4-3,6-Dianhydro-α-D-glucopyranose
70	Pentene	162	Levoglucosan
72	Pyruvaldehyde		

The low-voltage Py-MS of cellulose has also been reported [86]. The mass pyrogram is characterized by many even mass peaks showing a significant number of molecular ions. The fragmentation mirrors the Py-GC results with peaks at m/z 126 (Levoglucosenone, 5-hydroxymethylfurfural), m/z 110 (5-methylfurfural), and m/z 96 (furfural). Considerable similarity between the smaller pyrolysis products from hexoses has been noted [87]. Thus, the products derived from glycogen pyrolysis, identified by high-resolution FIMS, may be used as a guide to peak identification. These are recorded in Table 7.6.

The importance of cellulose as a prime source of renewable organic carbon has meant that the pyrolysis of this material and its
derivatives is under constant study [88-90]. Starch, the α-1,4 + α-
1,6 polymer of glucose, is also important commercially and thermal
conversions result in the formation of dextrins. Pyrolysis of
starch also yields levoglucosan and volatile products are analogous
to those of other glucose polymers [91-93]. The Py-FDMS of glycogen
[87,88] reveals the most informative fragments of currently reported
studies, for in addition to the usual products (Table 7.6) dimeric
subunits may be identified (Fig. 4.16) [94,95]. This contrasts
markedly with analogous studies on dextrans which exhibit a higher
yield of volatiles and a greater preponderance of proton-transfer
reactions which prevent the appearance of fragments higher than
(monomer-OH). The fingerprint capabilities of low voltage Py-MS
have been amply demonstrated and α- and β-1,3- glucans and glycogan,
glucose polymers differing in linkage and branching, are easily distinguishable from each other and from cellulose [56].

Xylan, a D-xylose polymer but often containing other sugar
residues, is perhaps next in importance after cellulose. This
material, together with cellulose, constitutes the major source of
combustible volatiles in wood. Further, it has less thermal stability and thus plays an important role in the initiation and propagation of pyrolysis and combustion reactions. The pyrolysis of xylan
under furnace conditions gives a range of volatile products with
furfural and acetaldehyde predominating [96], while the tar fraction
contains saccharide residues equivalent to about 50% D-xylose.
Flame retardants such as zinc chloride or NaOH increased the water
and char yields. Curie-point analysis mirrored the simplicity of
the glucose-cellulose results [25] and only two major peaks, arising
from depolymerization and dehydration reactions, were obtained.
D-xylose yielded principally furfural. This product was also found
in the pyrogram of xylan, but a new product, identified as 3-hydroxy-
2-penteno-1,5-lactone was the major component (Fig. 7.5) [97,98].
In contrast to glucose polymers, a continuous modification of the

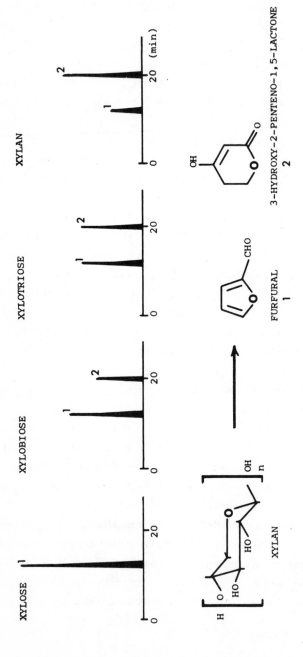

FIGURE 7.5 Py-GC of D-xylose, oligomes, and xylan. (Adapted from Refs. 97 and 98.)

TABLE 7.7   Main Ions in the Mass Pyrogram of Crustacean Chitin

m/z	167	151	137	125	109	97	85	73	59	43	32	18
I(%)	-	8	38	24	100	31	32	61	53	59	99	53

Source: Ref. 29.

product distribution with chain length is observed and lactone formation increases with the number of nonreducing residues in the
chain.   Curie-point pyrolysis at a higher temperature yields a pyrogram very similar to that of glucose [73], but without 5-methylfur-
fural, the six-carbon product.   The pyrolysis of pectins has been
shown to give little information on the degree of polymerization,
but correlation with the degree of methylation was possible [99] and
radiolysis and pyrolysis of glucose have been compared [100].

In view of the role of amino sugars as structural elements
(chitin and the muramic acid--glucosamine polymers in bacterial cell
walls) little work has appeared on the detailed mechanistic events
of pyrolysis.   Work on some amino sugars has shown that substantial
dehydration and charring occurs [101].   Furnace pyrolysis of chitin
(an N-acetyl-D-glucosamine polymer), proposed as a possible tobacco
substitute, has also been undertaken [102].   A considerable range of
products, including neutrals (e.g., toluene), bases (e.g., picolines),
phenols (e.g., phenol), and acids (e.g., acetic acid) were identified.
Py-MS analysis of chitin does not exhibit this complexity, but is
characterized by direct fission and dehydration reactions.   The main
ions are recorded in Table 7.7 [29].   Rapid assessment of the purity
of samples of chitin was also possible with m/z 34 (H_2S) showing
protein contamination and even mass series such as m/z 128, 84, and
60, indicating the presence of other polysaccharides.

The advantages of Py-MS in giving rapid analysis coupled with
the availability of direct chemical information combine to provide a
powerful structural probe for the study of cell polysaccharides.
Although in these applications the technique is still in its infancy, current applications suggest an immense potential.   In

particular, characteristic fragments arise from the various types of
sugar found in these systems and distinction between O- and N-acetyl
compounds may be made [103]. These studies are of particular importance, due to the known importance of surface oligosaccharides as
immunogenic agents.

The partial structures of the B, C and Y capsular polysaccharides from Neisseria meningitidis are shown in Fig. 7.6. The B and
C types are homopolymers of N-acetylneuraminic acid, whereas Type Y
consists of repeating glucose-N-acetylneuraminic acid disaccharide
units. Pyrolysis of B and C fractions shows ion series (m/z: 43,
59, 60, 67, 72, 73, 80, 97, 109, 135, 151) with similarity to that
of chitin (Table 7.7), while pyrolysis of the Y type also showed
characteristic hexose peaks (Table 7.6). The variable amounts of
O-acetylated residues--dependent upon strain and growth conditions--
were assessed by the appearance of characteristic peaks at m/z 60
(acetic acid), 43 ($CH_3-C\overset{+}{\equiv}O$) and 42 (ketene). The B polysaccharide
also revealed more intense peaks at m/z 32 and 31 (methanol) from
the nonacetylated $-CH_2OH$ end group. Other work on bacterial polysaccharides has been briefly reported [55] and Py-GC has been used
to study whole cells and isolated cell fractions and has enabled
serotypes to be distinguished on the basis of antigenic polysaccharide units [104]. Py-MS confirmed this analysis [105].

7.4. LIGNINS

Lignin is the principal nonpolysaccharide polymer found in wood. It
is important as a raw material and also as a source of bioorganic
material in the geosphere. It is largely composed of phenol polymers and although comparatively little work has been reported, both
Py-GC and Py-MS data has appeared [106]. The pyrogram is characterized by the presence of methoxyphenols and furans with some
smaller volatiles, such as acetone and acetaldehyde. The mass pyrograms contain many even mass peaks, which essentially correspond to
products in the Py-GC analysis. The data is summarized in Table
7.8.

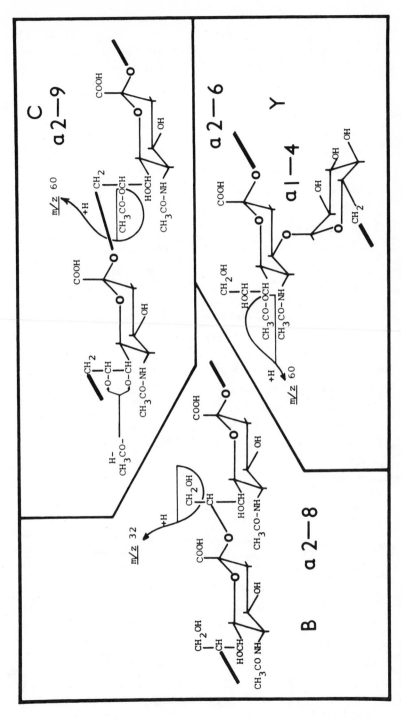

FIGURE 7.6 Partial structures of capsular polysaccharides from *Neisseria meningitidis*. (From Ref. 103.)

TABLE 7.8   Curie-Point Pyrolysis Products from Lignin

m/z	Component	m/z	Component
16	Methane	84	Cyclohexane
18	Water	94	Phenol
44	Carbon dioxide	96	2,5-Dimethylfuran
44	Acetaldehyde	108	Cresols
46	Ethanol	120	C_3-Alkylbenzenes
58	Acetone	124	Methoxyphenols
60	Acetic acid	138	Dimethoxybenzenes
64	Sulfur dioxide	150	Methoxyacetophenone
68	Furan	150	Vinyl-methoxyphenols
82	Cyclohexene	152	Dimethoxytoluene
82	2-Methylfuran	164	Allylmethoxyphenols

Source: Ref. 106

Probe pyrolysis-CIMS has also been proven to be of value in distinguishing between lignins of different origin [107]. This technique allows larger fragments to be detected--albeit as pseudomolecular ions with isobutane as the reagent gas--enabling a detailed comparison of structure to be undertaken (Table 7.9).

7.5. NUCLEIC ACIDS

The detectable pyrolysis products from nucleic acids are highly dependent upon the experimental conditions (Fig. 4.21). This reflects the significant differences in reactivity, polarity and volatility associated with the sugar, phosphate and heterocyclic units. This heterogeneity has complicated the nucleic acid pyrolysis literature, but under appropriate conditions extremely useful information concerning the identity of sugar and base residues and their sequences may be extracted.

Py-GC has proven to be of limited value in this field. In particular, the purine and pyrimidine bases found in nucleic acids are highly polar and cannot be eluted from a GC column without prior

TABLE 7.9 Probe Py-CIMS Products from Lignins

PROPOSED STRUCTURE R^+	SYRINGYL m/z	Intensity Pine	Birch	CONIFERYL m/z	Intensity Pine	Birch	COUMARYL m/z	Intensity Pine	Birch
-CHOH-CHOH-CHOH	243		6						
-CHOH-CH$_2$-CHOH	227		12	197	20	23			
-CHOH-CH-CHO	225		15	195	3	10			
-CO-CH-CHO	223	2	9						
-CH$_2$-CH$_2$-CH$_2$OH$_2$				183	4	7	153	2	
-CH$_2$-CH$_2$-CH$_2$O	211		55						
-CH=CH-CH$_2$-OH$_2$				181	20	25	151	7	5
-CH=CH-CHOH$_2$				180	15	11			
-CH=CH-CHOH	209	4	67	179	100	31	149	15	5
-CH=CH-CHO	208		7	178	5				
-COOH$_2$	199		10	169	3	10			
-CH$_2$-CH$_2$=CH$_2$	195		26	165	10	8	135	2	
-CH$_2$-CH=CH	193	7	100	163	92	67	133	11	
-CH=C=CH							131	2	
-CHOH	183		22	153	18	19			
-CH=CH$_2$ + H				151	7	5	121	3	5
-C=CH + H				149	15	5			
-CH$_3$ + H				139	6	16	109	3	
-CH$_2$	167	17	22	137	15	15	107	4	6
-H + H	155		19	125	5	13	95	2	
-	153	18	19	123	5	16			

Source: Ref. 107

derivatization. Thus, early studies were concerned with the pyrolysis of the bases themselves, and although only very volatile products were detected, these were sufficient to permit differentiation. Isocytosine was found to yield acetone, whereas cytosine did not. Acetone was also a product from thymine and was suitable for quantitative assessment of this pyrimidine [108]. Nucleosides (base-sugar) and nucleotides (base-sugar-phosphate) have also been studied with

pyrolysis at 800 to 850°C [109,110]. These temperatures were chosen
to maximize the production of low-mass volatiles and would clearly
lead to extensive fragmentation, particularly of the labile sugar
residues. Distinction could be made between the members of each
group using fingerprint techniques, but close similarities were evident, for example, between the purines adenosine and guanosine. A
change of GC column and temperature range enabled better resolution
to be achieved and a Py-GC-MS study enabled 17 products to be identified. Particularly useful fragments for identification purposes
were cyanogen, which was produced from the purines only, and aceto-,
propio- and acrylonitriles. The production of these products was
related to the pyrolysis of the purine and pyrimidine bases and to
D-ribose. Ethylene was the major component from all samples, apart
from cytosine (acetonitrile) and thymine (propene nitrile), where it
was the second most intense product. The absence of characteristic
sugar components reflects the vigorous pyrolysis conditions used.
Under different conditions furfuryl alcohol, furfural and hydroxy-
acetone may be detected from D-ribose [73]. The bases in some dinucleotides could be distinguished by the relative proportions of
the nitrile products in the pyrogram, but no sequencing information
was apparent.

More structural information is available using Py-MS. The an-
ticipation was that the more effective transfer conditions would
allow the more polar and larger fragments, necessary for chemical
interpretation, to be detected. Curie-point pyrolysis (at 610°C),
however, was found to yield pyrograms characterized essentially by
sugar residues [111]. Laser pyrolysis-MS [29] and field-ionization
studies [112] confirmed the absence of purines and pyrimidines in
the pyrogram. This was surprising because pyrolysis of mixtures of
these bases yielded intense pyrograms due mainly to the molecular
ions of the intact bases released by evaporation. The low-voltage
EIMS and FIMS pyrograms were found to be essentially similar and
high resolution MS enabled the identity of many of the fragments to
be proposed. These were found to be characteristic of ribose (RNA)

TABLE 7.10 Curie-Point Pyrolysis Products from DNA and RNA

DNA	RNA	Component	DNA	RNA	Component
17	17	Ammonia	81		C_5H_5O
27	27	HCN	82		2-Methylfuran
29	29	CHO		84	Hydroxyfuran
32	32	Methanol	86		$C_4H_6O_2$
41	41	Acetonitrile	95		C_5H_5NO
43	43	CHNO	96		4-Pyrone
44	44	Acetaldehyde	98	98	Furfuryl alcohol
46	46	Formic acid	98		α-Angelicalactone
53		Propenenitrile	99	99	$C_5H_7O_2$
55		$HC{\equiv}C{-}CH{=}\overset{+}{O}H$	103	103	Benzonitrile
56		Propenal	112		$C_6H_8O_2$
60	60	Acetic acid		114	Hydroxyfurfuryl alcohol
68	68	Furan	116		$C_4H_4O_4$
70	70	C_4H_6O	116		Dehydroxyribose
72	72	C_4H_8O	117		$C_5H_9O_3$

Source: Refs. 111, 112.

or deoxyribose (DNA) and are recorded in Table 7.10. The structure of the component at m/z 116 ($C_4H_4O_4$) was not proposed [112], but this was not levulinic acid ($C_5H_8O_3$)--a fragment proposed for hexoses [25], although angelicalactone, the dehydration product of this compound, has been detected [112]. Confirmation of the proposed structures has been obtained via CAMS (Fig. 4.18) [112]. Various elimination and rearrangement reactions, initially involving the phosphate moeity, were proposed to account for the products (Fig. 7.7) [111]. It is conceivable that a phosphate-base ion pair or complex is established. This will be significantly less volatile than the free bases and is possibly the cause of the absence of base fragments in the pyrograms. Indeed, under FDMS conditions such complexes may be detected and are clearly implicated in the degradation pathway [113].

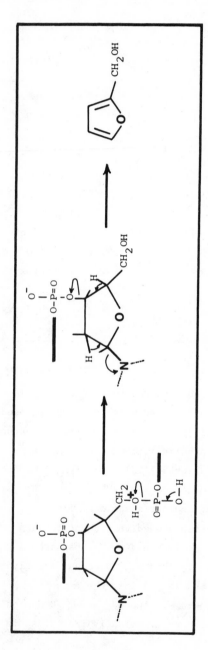

FIGURE 7.7 Pyrolytic formation of furfuryl alcohol from DNA. (From Ref. 111.)

In contrast, Py-FD-MS has proven capable of detecting large
fragments from DNA (Fig. 4.21) [113]. No organic ions were detected
below m/z 112 and the parent bases were readily identified (as M + 1
ions). Similar patterns at higher mass numbers indicated the presence of related, but more complex units. Cluster ions of inorganic
phasphate (e.g., acids and monosodium salts of phosphorus oxyacids)
were also detected, again as M + 1 ions, and these were found to be
useful as internal reference peaks. Dinucleotide ions with two or
three phosphate residues were also present and were identified as
double-charged M + 2 ions. This is a significant result and clearly
is of importance in the determination of nucleic acid sequence distributions. The components and proposed structures are summarized
in Table 7.11. Clearly, the majority of these products are too polar
for detection with normal pyrolysis procedures, although some success
is apparent with laser methods [29].

Probe Py-MS has also proven to be useful in the structure elucidation of nucleic acids. Here, it would seem appropriate to assume
that the base-phosphate complexes undergo further pyrolysis or electron-impact fragmentation to yield the observed product distribution.
Earliest studies [114] using this technique for DNA analysis did not
recognize the pyrolytic degradation prior to ionization [115], but
masses up to m/z 391 were reported. Elemental compositions were
determined for 17 fragments and no nucleotides or other phosphate-
containing residues were found. Nevertheless, base fragments are
much in evidence and may readily reveal the presence of minor nucleotides in DNA [116].

The major fragmentation products of DNA result from cleavage
and/or elimination at each phosphodiester linkage. This leads to
characteristic m/z values for the purine and pyrimidine bases and
for the nucleosides [117,118]. Seven main fragments for each base,
some containing phosphate, may be identified and these were also detected in oligomers of up to 25 residues. These are represented,
together with the m/z values for each base, in Fig. 7.8. Measurement of the intensities of some or all of these fragments allows

TABLE 7.11 Py-FDMS Products from Herring DNA (m/z)

COMPONENT	B	CYTOSINE	5-METHYL CYTOSINE	THYMINE	ADENINE	GUANINE
B + H		112	126	127	136	152
B + H_3PO_4		210	224		234	250
B + (methylfuran - 2H)$_2$		272			296	312
B + (methylfuran - 2H)$_3$					376	392
B + (methylfuran - 2H)$_4$				447	456	472
Nucleoside - $2H_2O$		192	206		216	232
Nucleotide - H_2O		290				314
Nucleotide		308		323		

(Deoxycytidine diphosphate + 2H)$^{2+}$	390	(Cytidine-thymidine triphosphate + 2H)$^{2+}$	693
(Deoxythymidine diphosphate + 2H)$^{2+}$	404	(Dithymidine triphosphate + 2H)$^{2+}$	708
(Na deoxycytidine diphosphate + 2H)$^{2+}$	411	(Na cytidine-thymidine triphosphate+2H)$^{2+}$	715
(Na deoxythymidine diphosphate + 2H)$^{2+}$	426	(Na dithymidine triphosphate + 2H)$^{2+}$	730
(Na dithymidine triphosphate + 2H)$^{2+}$	650	(Na ^{13}C-dithymidine triphosphate + 2H)$^{2+}$	731

Source: Ref. 113.

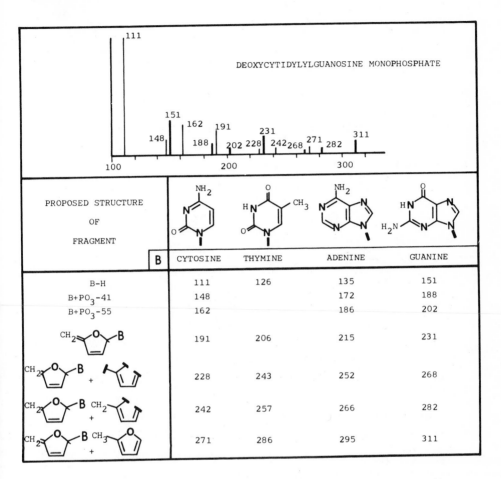

FIGURE 7.8 Probe pyrolysis products of DNA: application to a dinucleotide monophosphate (m/z). (Adapted from Ref. 117, adapted
with permission from J. L. Wiebers and J. A. Shapiro, Sequence Analysis of oligodeoxyribonucleotides by MS. I. Dinucleoside monophosphate, *Biochemistry* *16*:1044-1050. Copyright 1977 American Chemical
Society.)

quantitative estimation of the base composition of the nucleic acids
[119] and their sequence [117,120]. Two approaches were used for the
sequencing studies. Initially, dinucleoside monophosphates were
studied [117]. The intensities of the characteristic ions were found
to vary with sequence. Thus, the most intense fragment from deoxy-
thymidylyladenosine monophosphate was found at m/z 126 (from thymine)

whereas the base peak from deoxyadenylylthymidine monophosphate occurred at m/z 135 (from adenine).   These patterns are still present
with tri- and tetranucleotides, although the amount of data necessary
to determine the sequence of bases is considerably greater.   Factor
analysis, nearest neighbor and nonlinear mapping approaches were used
to reduce the data and to enable dinucleotide sequences to be revealed.   Prediction of the terminus (3' or 5') (also from the pyrolysis data) enabled the tetranucleotide sequence to be determined [120].

The probe pyrolysis results in RNA have shown less promise for
sequence determination [121] although, again, the base composition is
readily revealed.   Homogeneous RNA polymers--poly(uradylic acid),
poly(cytidylic acid), poly(adenylic acid), and poly(guanylic acid)--
were used initially as test samples.   The pyrograms from each essentially reflected the mass spectra of the pure bases and poly(cytosine)
was the only sample to evidence peaks at a mass higher than the component base.   Synthetic RNA samples with mixed nucleotides were also
examines.   Poly(A,U,U) was encouraging in that the uracil and adenine
peaks were clearly observed, and in a ratio of 2:1, the proportion of
bases in the polymer.   Adenine and uracil were again quantitatively
represented in the pyrogram of poly(A,G,U) consisting of equimolar
amounts of the bases.   However, guanine was an extremely weak component of the pyrogram.   The greater thermal stability of poly(G),
decomposition at 250°C, compared to the other polymers, decomposition
at 130 to 150°C, is the cause of this deviation.   This problem was
further accentuated with natural RNA when only small amounts of cytosine and guanine were detected.   Clearly, further work concerning
volatilization techniques is required before routine techniques are
available for RNA analysis.

7.6.   LIPIDS AND SMALLER BIOMOLECULES

Lipids may be classed as the long-chain fatty acid esters of glycerol
(glycerides).   Such materials are not composed of the repeating units
which characterized the biopolymers which have been discussed earlier,
but nevertheless, they are important biological molecules which also

may be found complexed to other biopolymers: lipoproteins and lipopolysaccharides. The dietary and culinary importance of lipids has
ensured considerable interest in thermal fragmentation, polymerization, and autoxidation reactions [122-124].

The primary pyrolysis process is probably the elimination of a
fatty acid residue (Fig. 7.9). The products may undergo further
fragmentation to yield acrolein and homologues derived from the fatty
acid, but the observed product distribution appears highly dependent
upon pyrolysis and analysis conditions. In contrast to MS fragmentation, the principal products are of high mass (up to that of the
parent acid) and are characterized by three main homologous series.
These are: the carboxylic acids (produced by disproportionation of
the alkyl substituent and consisting of -anoic and -enoic components),
the hydrocarbons (produced by decarboxylation of the saturated acids)
and the olefins (produced by disproportionation or decarboxylation)
[125]. These processes have been verified for fatty acid esters [126]
and the study of hydrocarbons has shown that chain branching may be
revealed [127]. The extent to which these reactions occur and the
transfer of polar products to the analytical segment will have a considerable influence on the appearance of the pyrogram. Indeed, diglycerides may be analyzed without apparent decomposition by GC, and
pyrolysis of phosphoglycerides (at 350 to 400°C using the heated injection port) was found to yield the corresponding diglyceride as the
only retained component of the pyrogram [128]. Thus, dipalmitoyl
phosphatidyl choline (lecithin) gave a product with a mass spectrum
identical to 1,3-dipalmitin. Pyrolysis, silylation and GC analysis
has also been undertaken [129]. In contrast, probe pyrolysis of the
lecithin was accompanied by much fragmentation of the fatty acid
residues.

Laser pyrolysis of saturated (methyl stearate, myristate, and
palmitate) and unsaturated (methyl oleate and olaidate) fatty acid
esters, diglycerides (dipalmitin) and phospholipids (lecithin) was
characterized by homologous olefin and hydrocarbon series, although
complexity was indicated by the appearance of benzene and toluene

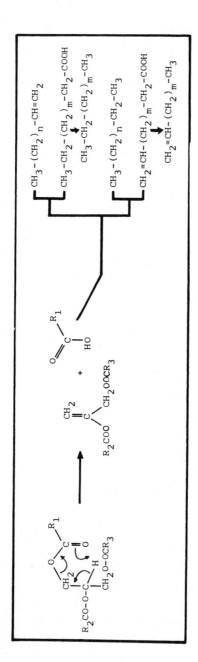

FIGURE 7.9 Pyrolysis of a triglyceride.

TABLE 7.12   Characteristic Ions Found in Lipopolysaccharide Mass
Pyrograms (m/z)

72	73	81	82	84	85	96	97	98	110	122	125	126	129	138

Source: Ref. 103.

[42]. Increased yields of dienes and short chain esters distinguished
the unsaturated esters, although the geometrical isomers differed
little. The diglyceride showed extensive fragmentation and the presence of trimethylamine was the distinguishing feature of the lecithin
pyrogram. This has also been observed in the Py-MS analysis of phospholipids which contain the ion series m/z 59, 71, and 89 derived
from fragmentation of choline [130].

Py-GC of vegetable oils have been reported [131,132] as has work
on the composition of surfactants [133] and the chain length of soaps.
Here, copyrolysis with acetate to give methyl ketones, which were
readily estimated by GC, was the technique adopted [134]. Mechanistic studies of these reactions have appeared [135] and lower chain
acids have been studied by laser techniques [136]. Quantitative determinations include lecithin, corn oil, and cholesteryl oleate estimations by pyrolysis and analysis with an uncoated GC column [137],
and the measurement of residual hexane in vegetable oils [138]. Py-
MS data on lipopolysaccharides has also been briefly mentioned (Table
7.12) [103].

Pyrolysis has been useful in the structural elucidation of various classes of lipid and reflects the importance of a knowledge of
pyrolysis pathways in analytical pyrolysis. Mycolic acids, for example, are α-branched β-hydroxy acids produced by several genera of
bacteria such as mycobacteria. The pyrolysis of these compounds
yields an aldehyde and a carboxylic acid which are characteristic
of the mycolic acid. Esterification and pyrolysis yields the carboxylic acid ester which is more effectively analyzed by GC (Fig.
7.10) [139]. Identification of the products allows the structure
of the mycolic acid to be deduced. Thus, palmitic aldehyde and
methyl palmitate are derived from methyl corynomycolate isolated

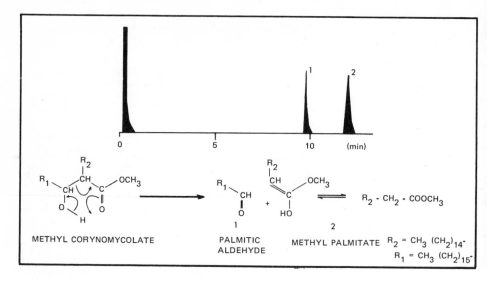

FIGURE 7.10 Pyrolysis of mycolic acid esters. (From Ref. 139, from the *Journal of Gas Chromatography*, by permission of Preston Publications, Inc.)

from *Cornyebacterium diphtheriae*. It was also possible to identify mixtures by this technique, but under the conditions used (on-column pyrolysis) some esters were too volatile for pyrolysis (e.g., Fig. 7.10; $R_1 = C_6H_{13}$; $R_2 = C_2H_5$) whereas others (e.g., Fig. 7.10; $R_1 \sim C_{52}H_{100}$; $R_2 = C_{22}H_{45}$) yielded a volatile ester only. Py-MS gave similar results and the MS fragmentation of the volatile samples paralleled the pyrolytic scheme, but aliphatic aldehyde fragments were very weak. The structure of mycolic acids varies with the strain, and pyrolysis studies readily reveal these differences [140,141]. The position of the double bond in unsaturated fatty acids may be determined by pyrolysis of the ozonides or by other oxidative methods [142-145] and the structure determination of cyclopropane and cyclopropene fatty acids has also involved a pyrolysis step [146].

A further class of lipophilic biomolecule is the steroids. Early use was made of the pyrolysis of these compounds in the structure elucidation of steroid side chains [147]. Successive pyrolysis, catalyzed by benzenesulfonyl chloride, and GC analysis gave 2-methylheptane from cholesterol and other steroids with a C_8 side-chain,

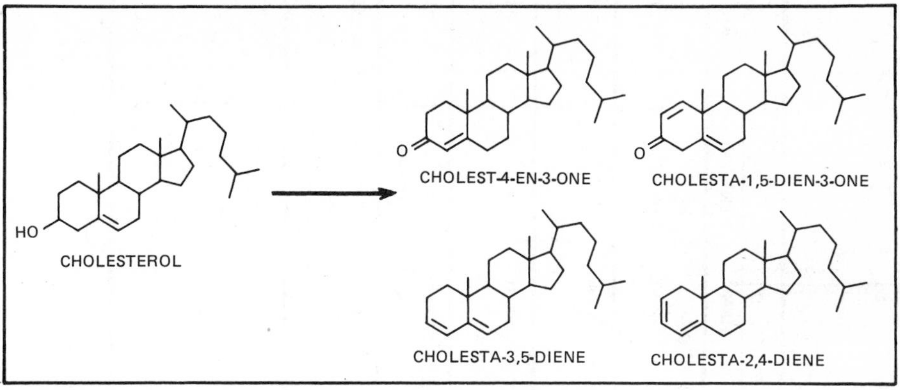

FIGURE 7.11 Pyrolysis of cholesterol.

whereas ergost-8-en-14-ol and analogues yielded 2,3-dimethylheptane.
Curie-point pyrolysis of cholesterol has shown that three main components are produced [148,149]. These are cholesta-3,5-diene (together with unresolved 2,4-isomer), cholest-4-en-3-one and cholesta-1,
5-dien-3-one (Fig. 7.11), although polycyclic hydrocarbons may be observed under furnace conditions [150]. Diene formation via pyrolysis
of the acetate or trimethylsilyl derivative of cholesterol; an elimination reaction noted during the GC analysis of these compounds [151]
was also observed [149]. In addition to the steroid fraction, volatiles including benzene, xylene, and styrene were identified. Fingerprint pyrograms have also been reported for bile acids [152] and the
stereochemistry of other steroids is reflected in the relative intensity of pyrolysis fragments [153]. Pyrograms, but again without fragment identification, have also been reported for the digitalis agly-
cones digitoxigenin and digoxigenin [154]. Other studies on the pyrolysis of steroids have been initiated to determine the origin of
artifacts observed during GC analysis. The 5,8-peroxides of ergos-
terol and of 7-dehydro-cholesterol were rapidly identified by pyrolysis and were readily distinguished from nonperoxide steroids [155].

Porphyrins have been shown to yield a characteristic volatile
fraction [156] and pyrrole derivatives [157]. The presence of the
$CH_3-CH(OH)-$ side chain enabled haemotoporphyrin to be distinguished

TABLE 7.13 Py-GC Conditions for Some Biological Molecules

SAMPLE	TEMP	COLUMN	STATIONARY PHASE	SUPPORT	TEMPERATURE			FLOW RATE
					INITIAL	PROGRAM	FINAL	
Amino acids and Proteins	770°	1.5m x 4mm	Carbowax 20M 8% KOH 2%	Chromosorb W AW (80-100)	100° 5 min	5°C/min	245° 10 min	50 ml/min (N_2)
Carbohydrates	460°	3m x 3mm	Carbowax 20M 5%	Chromosorb G AW (60-80)	80°	7.5°C/min	230°	40 ml/min (He)
Nucleic Acids[a]	850°	1m x 3mm	Poropak Q (100-120)		40°	5°C/min	230°	60 ml/min (He)
Lipids	770°	2m x 3mm	OV 17 3%	Chromosorb W AW	40° 5 min	5°C/min	250° 10 min	50 ml/min (N_2)
Steroids	770°	3.7m x 2mm	Carbowax 20M-TPA 5%	Anakrom ABS (110-120)	55° 6 min	6°C/min	165° 55 min	20 ml/min (N_2)
Acetyl choline	500°	1.2m x 3mm	OV 101 5% Dodecyldi methylene triamine succinamide 5%	Gas Chrom Q (80-100)	85°			25 ml/min (N_2)

[a]Conditions for carbohydrates may be more appropriate.

from protoporphyrin by the appearance of acetaldehyde in the pyrogram. This enabled quantification of the mixture to be achieved.
No pyrroles were detected in this work (probably due to GC conditions),
but these have since been identified [157]. At high pyrolysis temperatures polymethylpyrroles are abundant, but at lower temperatures
(~400°C) characteristic fragments, such as 2,3,5-trimethyl-4-ethyl-
pyrrole, produced by cleavage of the porphyrin ring at the =CH-
methene bridges are favored.

The analytical pyrolysis of cholinergic neurotransmitter substances has been discussed previously (Sec. 3.2.3). Table 7.13 holds
some typical Py-GC conditions for the analysis of biological molecules.

REFERENCES

1. Medley, E.E., Simmonds, P.G. and Manatt, S.L., A Py-GC-MS Study
    of the Actinomycete Streptomyces longisporoflavus, *Biomed. Mass
    Spectrom.*, *2*(1975)261-265.

2. Simmonds, P.G., Whole Microorganisms Studied by Py-GC-MS: Sig-
    nificance for Extra-terrestrial Life Detection Experiments,
    Appl. Microbiol., *20*(1970)567-572.

3. Simmonds, P.G., Shulman, G.P. and Stembridge, C.H., Organic
    Analysis by Py-GC-MS. A Candidate Experiment for the Biologi-
    cal Exploration of Mars, *J. Chromatogr. Sci.*, *7*(1969)36-41.

4. Shafizadeh, F., Combustion, Combustibility, and Heat Release of
    Forest Fuels, *AIChE Symposium Series (No. 177)*, *74*(1978)76-82.

5. Shafizadeh, F. and Bradbury, A.G.W., Smoldering Combustion of
    Cellulosic Materials, *J. Thermal Insul.*, *2*(1979)141-152.

6. Shafizadeh, F. and Chin, P.P.S., Thermal Deterioration of Wood,
    in I.S. Goldstein (Ed.), Wood Technology: Chemical Aspects,
    ACS Symposium Series No. 43, ACS, Washington, 1977, pp. 57-81.

7. Susott, R.A., Shafizadeh, F. and Aanerud, T.W., A Quantitative
    Thermal Analysis Technique for Combustible Gas Detection, *J.
    Fire Flamm.*, *10*(1979)94-104.

8. Cabradilla, K.E. and Zeronian, S.H. in R.M. Rowell and R.A.
    Young (Eds.), Modified Cellulosics, Academic Press, New York,
    1978, p. 321.

9. Susott, R.A., DeGroot, W.F. and Shafizadeh, F., Heat Content of
    Natural Fuels, *J. Fire Flamm.*, *6*(1975)311-325.

10. Shafizadeh, F., Chin, P-S. and DeGroot, W., Mechanistic Evalua-
    tion of Flame Retardants, *Fire Retard. Chem.*, *2*(1975)195-203.

11. Urbas, E. and Kullik, E., Py-GC Analyses of Untreated and Flame-
    proofed Wools, *Fire Mater.*, *2*(1978)25-26.

12. Shafizadeh, F., McIntyre, C., Lundstrom, M. and Fu, Y-L.,
    Chemical Conversion of Wood and Cellulosic Wastes, *Proc. Mon-
    tana Acad. Sci.*, *33*(1973)65-96.

13. Shafizadeh, F., Industrial Pyrolysis of Cellulosic Materials,
    J. Appl. Polym. Sci. (Symp.), *28*(1975)153-174.

14. Shafizadeh, F., Cochran, T.G. and Sakai, Y., Application of
    Pyrolytic Methods for the Saccharification of Cellulose, *AIChE
    Symposium Series (No. 184)*, *75*(1979)24-34.

15. Higman, E.B., Severson, R.F., Arrendale, R.F. and Chortyk, O.T.,
    Simulation of Smoking Conditions by Pyrolysis, *J. Agric. Food
    Chem.*, *25*(1977)1201-1207.

16. Severson, R.F., Schlotzhauer, W.S., Arrendale, R.F., Snook, M.E.
    and Higman, H.C., Correlation of Polynuclear Aromatic Hydrocar-
    bon Formation Between Pyrolysis and Smoking, *Beitrag. Tabak.*, *9*
    (1977)23-37.

17. Chopra, N.M., Campbell, B.S. and Hurley, J.C., Systematic Studies
    on the Breakdown of Endosulfan in Tobacco Smokes: Isolation and
    Identification of the Degradation Products from the Pyrolysis of
    Endosulfan I in a Nitrogen Atmosphere, *J. Agric. Food Chem.*, *26*
    (1978)255-258.

18. Spronck, H.J.W. and Salemink, C.A., Cannabis XVII: Pyrolysis
    of Cannabidiol. Structure of Two Pyrolytic Conversion Products,
    J. Royal Netherlands Chem. Soc., *97*(1978)185-186.

19. Perkins, E.G., Formation of Non-Volatile Decomposition Products
    in Heated Fats and Oils, *Food Technol.*, *21*(1967)125-130.

20. Perkins, E.G., Nutritional and Chemical Changes Occurring in
    Heated Fats: A Review, *Food Technol.*, *14*(1960)508-514.

21. Lien, Y.C. and Nawar, W.W., Thermal Decomposition of Some Amino
    Acids, *J. Food Sci.*, *39*(1974)911-913, 914-916.

22. Lien, Y.C. and Nawar, W.W., Thermal Interaction of Amino Acids
    and Triglycerides, *J. Food Sci.*, *39*(1974)917-919.

23. Takayama, S., Hirakawa, T., Tanaka, M., Kawachi, T. and Sugimura,
    T., In Vitro Transformation of Hamster Embryo Cells with a Gluta-
    mic Acid Pyrolysis Product, *Toxicol. Lett.*, *4*(1979)281-284.

24. Yasuda, T., Yamaizumi, Z., Nishimura, S., Nagao, M., Takahashi,
    Y., Fujiki, H., Sugimura, T. and Tsuji, K., Detection of Comu-
    tagenic Compounds Harman and Norharman, in Pyrolysis Products
    of Proteins and Food by GC-MS, *Chem. Abs.*, *92*(1980)127076j.

25. Ohnishi, A., Kato, K. and Takagi, E., Curie-point Pyrolysis of
    Cellulose, *Polym. J.*, *7*(1975)431-437.

26. Heyns, K., Stute, R. and Paulsen, H., Braunungsreaktionen und
    Fragmentierungen von Kohlenhydraten, T.I. Die Fluchtigen

Abbauprodukte der Pyrolyse von D-Glucose, *Carbohyd. Res.*, *2* (1966)132-149.

27. Heyns, K. and Klier, M., Braunungsreaktionen und Fragmentierungen von Kohlenhydraten. T.IV. Vergleich der Fluchtigen Abbau Produkte bei der Pyrolyse von Mono, Oligo- und Polysacchariden, *Carbohyd. Res.*, *6*(1968)436-448.

28. Stack, M.V., A Review of Py-GC of Biological Macromolecules, in C.L.A. Harbourn (Ed.), *Gas Chromatography 1968*, Institute of Petroleum, London, 1969, pp. 109-118.

29. Meuzelaar, H.L.C., Kistemaker, P.G. and Posthumus, M.A., Recent Advances in Py-MS of Complex Biological Molecules, *Biomed. Mass Spectrom.*, *1*(1974)312-319.

30. Irwin, W.J. and Slack, J.A., Analytical Pyrolysis in Biomedical Studies, *Analyst*, *103*(1978)673-704.

31. Irwin, W.J., Analytical Pyrolysis--An Overview, *J. Anal. Appl. Pyrol.*, *1*(1979)3-25, 89-122.

32. Simon, W. and Giacobbo, H., Thermische Fragmentierung und Strukturbestimmung organischen Verbindungen, *Chem.-Ing. Techn.*, *37* (1965)709-714.

33. Vollmin, J., Kriemler, P., Omura, I., Seibl, J. and Simon, W., Structural Elucidation with a Thermal Fragmentation-GC-MS Combination, *Microchem. J.*, *11*(1966)73-86.

34. Shulman, G.P. and Simmonds, P.G., Thermal Decomposition of Aromatic and Heteroaromatic Amino Acids, *Chem. Communs.*, *1968:* 1040-1042.

35. Simmonds, P.G., Medley, E.E., Ratcliff, Jr., M.A. and Shulman, G.P., Thermal Decomposition of Aliphatic Monoamino Mono-Carboxylic Acids, *Analyt. Chem.*, *44*(1972)2060-2066.

36. Ratcliff, Jr., M.A., Medley, E.E. and Simmonds, P.G., Pyrolysis of Amino Acids. Mechanistic Considerations, *J. Org. Chem.*, *39* (1974)1481-1490.

37. Patterson, J.M., Haidar, N.F., Papadopoulos, E.P. and Smith, Jr., W.T., Pyrolysis of Phenylalanine, 3,6-Dibenzyl-2,5-piperazinedione and Phenylethylamine, *J. Org. Chem.*, *38*(1973)663-666.

38. Posthumus, M.A. and Nibbering, N.M.M., Py-MS of Methionine, *Org. Mass Spectrom.*, *12*(1977)334-337.

39. Smith, R.M., Solabi, G.A., Hayes, W.P. and Stretton, R.J., Py-GC of Histidine and 3-Methylhistidine, *J. Anal. Appl. Pyrol.*, *1* (1980)197-201.

40. Slack, J.A., Some Applications of Py-GC-MS to the Identification of Drugs and Micro-organisms, Ph.D. Thesis, University of Aston in Birmingham, 1977.

41. Kojima, T. and Morishita, F., Application of Laser to Py-GC I, *J. Chromatogr. Sci.*, *8*(1970)471-473.

42. Means, J.C. and Perkins, E.G., Laser Py-GC-MS of Membrane Components, in C.E.R. Jones and C.A. Cramers (Eds.), *Analytical Pyrolysis,* Elsevier, Amsterdam, 1977, pp. 249-260.

43. Merritt, Jr. C., DiPietro, C. and Robertson, D.H., Characterisation of Amino Acids by Py-GC-MS of their Phenylhydantoin Derivatives, *J. Chromatogr. Sci., 12*(1974)668-671.

44. Merritt, Jr. C. and Robertson, D.H., The Analysis of Proteins, Peptides and Amino Acids by Py-GC-MS, *J. Gas Chromatogr., 5* (1967)96-98.

45. Mauger, A.B., Degradation of Peptides to Diketopiperazines. Application of Py-GC to Sequence Determinations in Actinomycins, *Chem. Communs., 1971*:39-40.

46. Schmid, J.P., Schmid, P.P. and Simon, W., Application of Curie-point Py-GC Using High-resolution Glass Open-tubular Columns, in C.E.R. Jones and C.A. Cramers (Eds.), *Analytical Pyrolysis,* Elsevier, Amsterdam, 1977, pp. 99-105.

47. Zemany, P.D., Identification of Complex Organic Materials, *Analyt. Chem., 24*(1952)1709-1713.

48. Meuzelaar, H.L.C., Kistemaker, P.G., Eshuis, W. and Boerboom, A.J.H., Automated Py-MS: Application to the Differentiation of Micro-organisms, *Adv. Mass Spectrom., 7B*(1978)1452-1457.

49. Windig, W., Kistemaker, P.G., Haverkamp, J. and Meuzelaar, H.L. C., The Effects of Sample Preparation, Pyrolysis and Pyrolysate Transfer Conditions on Py-MS, *J. Anal. Appl. Pyrol., 1*(1979)39-52.

50. Danielson, N.D., Glajch, J.L. and Rogers, L.B., Py-GC of Enzymes, *J. Chromatogr. Sci., 16*(1978)455-461.

51. Danielson, N.D. and Rogers, L.B., Determination of Tryptophan in Proteins by Py-GC, *Analyt. Chem., 50*(1978)1680-1683.

52. Stack, M.V., Applications of GC in Dental Research, *J. Chromatogr., 165*(1979)103-116.

53. Stack, M.V., Quantitative Resolution of Protein Pyrolysates by GC, *J. Gas Chromatogr., 5*(1967)22-24.

54. Stack, M.V., Py-GC of Amino Acids and Proteins, *Biochem. J., 96* (1965)56P.

55. Meuzelaar, H.L.C., Kistemaker, P.G., Posthumus, M.A. and Kistemaker, J., Curie-point Pyrolysis in Direct Combination with Low Voltage EIMS. New Method for the Analysis of Non-volatile Organic Compounds, *Analyt. Chem., 45*(1973)1546-1549.

56. Posthumus, M.A., Boerboom, A.J.H. and Meuzelaar, H.L.C., Analysis of Biopolymers by Curie-point Pyrolysis in Direct Combination with Low Voltage EIMS, *Adv. Mass Spectrom., 6*(1974)397-402.

57. Myers, A. and Smith, R.N.L., Application of Py-GC to Biological Materials, *Chromatographia, 5*(1972)521-524.

58. Flunker, G., Seidel, W. and Dohner, L., Moglichkeiten der Differenzierung von Virusstrukturproteinen mit der Py-GC, *Arch. Exper. Vet. Med., 32*(1978)501-509.

59. Bayer, F.L., Hopkins, J.J. and Menger, F.M., Py-GC of Biomedically Interesting Molecules, in C.E.R. Jones and C.A. Cramers (Eds.), *Analytical Pyrolysis*, Elsevier, Amsterdam, 1977, pp. 217-223.

60. Kirk, P.L., Identification by Means of Pyrolysis Products, *J. Gas Chromatogr., 5*(1967)11-14.

61. DeForest, P.R. and Kirk, P.L., Forensic Individualism of Hair, *Criminologist, 8*(1973)35-45.

62. Fujimaki, M., Kato, S. and Kurata, T., Pyrolysis of Sulphur-containing Amino Acids, *Agr. Biol. Chem., 33*(1969)1144-1151.

63. Higman, E.B., Schmeltz, I. and Schlotzhauer, W.S., Products from the Thermal Degradation of Some Naturally Occurring Materials, *J. Agric. Food Chem., 18*(1970)636-639.

64. Kato, S., Kurata, T., Ishitsuka, R. and Fujimaki, M., Pyrolysis of β-Hydroxy Amino Acids, Especially L-Serine, *Agr. Biol. Chem., 34*(1970)1826-1832.

65. Patterson, J.M., Haidar, N.F., Smith, Jr. W.T., Benner, J.F., Burton, H.R. and Burdick, D., Benzo[a]pyrene Formation in the Pyrolysis of Selected Amino Acids, Amines and Maleic Hydrazide, *J. Agric. Food Chem., 26*(1978)268-270.

66. Fagerson, I.S., Thermal Degradation of Carbohydrates: A Review, *J. Agr. Food Chem., 17*(1969)747-750.

67. Walter, R.H. and Fagerson, I.S., Volatile Compounds from Heated Glucose, *J. Food Sci., 33*(1968)294-297.

68. Sugisawa, H. and Edo, H., Thermal Polymerisation of Glucose, *Chemy. Ind., 1964*:892-893.

69. Prey, V., Eichberger, W. and Gruber, M., Die Thermische Zersetzung von D-Glucose, Teil 1. Die Trockene Pyrolyse von D-Glucose in Stickstoffstrom, *Starke, 29*(1977)60-65.

70. Prey, V. and Gruber, H., Die Thermische Zersetzung von D-Glucose unter Stickstoff und Wasserstoff, *Starke, 29*(1977)96-98.

71. Prey, V. and Gruber, H., Die Thermische Zersetzung von D-Glucose. Teil 3. Die Ausbeuten an Fluchtigen Produkten und die Quantitative Zusammensetzung der Pyrolyseprodukte, *Starke, 29*(1977)135-138.

72. Shafizadeh, F., Furneaux, R.H., Stevenson, T.T. and Cochran, T.G., Acid-catalysed Pyrolytic Synthesis and Decomposition of 1,4-3,6-Dianhydro-α-D-glucopyranose, *Carbohyd. Res., 61*(1978)519-528.

73. Baltes, W. and Schmahl, H.J., Hoch frequenzpyrolyse ausgewahlter Kohlenhydrate, *Z. Lebensm. Unters. Forsch., 167*(1978)69-77.

74. Johnson, R.R., Alford, E.D. and Kinzer, G.W., Formation of Sucrose Pyrolysis Products, *J. Agric. Food Chem., 17*(1969)22-24.

75. Shafizadeh, F., Furneaux, R.H., Cochran, T.G., Scholl, J.P. and Sakai, Y., Production of Levoglucosan and Glucose from Pyrolysis of Cellulosic Materials, *J. Appl. Polym. Sci., 23*(1979)3525-3539.

76. Shafizadeh, F. and Fu, Y.L., Pyrolysis of Cellulose, *Carbohyd. Res., 29*(1973)113-122.

77. Kato, K., Pyrolysis of Cellulose. Part III. Comparative Studies of the Volatile Compounds from Pyrolysates of Cellulose and its Related Compounds, *Agr. Biol. Chem., 31*(1967)657-663.

78. Schwenker, Jr. R.F. and Beck, Jr. L.R., Study of the Pyrolytic Decomposition of Cellulose by GC, *J. Polym. Sci. (Part C), 2* (1963)331-340.

79. Lai, Y-Z. and Shafizadeh, F., Thermolysis of Phenyl β-D-Glucopyranoside Catalysed by Zinc Chloride, *Carbohyd. Res., 38*(1974) 177-187.

80. Shafizadeh, F., Lai, Y-Z. and McIntyre, C.R., Thermal Degradation of 6-Chlorocellulose and Cellulose-Zinc Chloride Mixture, *J. Appl. Polym. Sci., 22*(1978)1183-1193.

81. Shafizadeh, F. and Bradbury, A.G.W., Thermal Degradation of Cellulose in Air and Nitrogen at Low Temperatures, *J. Appl. Polym. Sci., 23*(1979)1431-1442.

82. Bradbury, A.G.W., Sakai, Y. and Shafizadeh, F., A Kinetic Model for Pyrolysis of Cellulose, *J. Appl. Polym. Sci., 23*(1979)3271-3280.

83. Bradbury, A.G.W. and Shafizadeh, F., Role of Oxygen Chemisorption in Low-Temperature Ignition of Cellulose, *Combust. Flame, 37*(1980)85-89.

84. Franklin, W.E. and Rowland, S.P., Effects of Phosphorus-containing Flame Retardants on Pyrolysis of Cotton Cellulose, *J. Appl. Polym. Sci., 24*(1979)1281-1294.

85. Franklin, W.E., Direct Pyrolysis of Cellulose and Cellulose Derivatives in a MS with a Data System, *Analyt. Chem., 51*(1979) 992-996.

86. Weijman, A.C.M., Cell-wall Composition and Taxonomy of Cephaloascus fragrans and some Ophiostomataceae, *J. Microbiol. Serol., 42*(1976)315-324.

87. Schulten, H-R. and Gortz, W., Curie-point Py-FIMS of Polysaccharides, *Analyt. Chem., 50*(1978)428-433.

88. Cabradilla, K.E. and Zeronian, S.H., Effect of Changes in Supramolecular Structure on the Thermal Properties and Pyrolysis of Cellulose, in R.M. Rowell and R.A. Young (Eds.), *Modified Cellulosics,* Academic Press, New York, 1978, pp. 321-339.

89. Min, K., Vapour-phase Thermal Analysis of Pyrolysis Products from Cellulosic Materials, *Combust. Flame, 30*(1977)285-294.

90. Brown, W.P. and Tipper, C.F.H., The Pyrolysis of Cellulose Derivatives, *J. Appl. Polym. Sci., 22*(1978)1459-1468.

91. Greenwood, C.T., The Thermal Degradation of Starch, *Adv. Carbohyd. Chem., 22*(1967)483-515.

92. Sawardecker, J.S., Slonekker, J.H. and Dimler, R.J., Detection and Quantitative Determination of Anhydroglucose by GC, *J. Chromatogr., 20*(1965)260-265.

93. Bryce, D.J. and Greenwood, C.T., Aspects of the Thermal Degradation of Starch, *Starke, 15*(1963)166-170.

94. Schulten, H-R., FDMS and its Application in Biochemical Analysis, in D. Glick (Ed.), *Methods in Biochemical Analysis,* Vol. 24, John Wiley and Sons, New York, 1977, pp. 313-448.

95. Schulten, H-R., Py-FI- and Py-FD-MS of Biomacromolecules, Micro-organisms and Tissue Materials, in C.E.R. Jones and C.A. Cramers (Eds.), *Analytical Pyrolysis,* Elsevier, Amsterdam, 1977, pp. 17-28.

96. Shafizadeh, F., McGinnis, G.D. and Philpot, C.W., Thermal Degradation of Xylan and Related Model Compounds, *Carbohyd. Res., 25*(1972)23-33.

97. Ohnishi, A., Takagi, E. and Kato, K., Curie-point Py-GC of Xylan, *Carbohyd. Res., 50*(1976)275-278.

98. Ohnishi, A., Kato, K. and Takagi, E., Pyrolytic Formation of 3-Hydroxy-2-penteno-1,5-lactone from Xylan, Xylo-oligo-saccharides and Methyl Xylopyranosides, *Carbohyd. Res., 58*(1977)387-395.

99. Zamarani, A., Roda, G. and Lanzarini, G., Richerche sulla Pirolisi delle Sostanze Naturali, Nota I- Pirolisi delle Pectine, *Industrie Agric., 9*(1971)35-41.

100. Herlitz, E., Lofroth, G. and Widmark, G., Analysis of Vapourisable Organic Compounds formed at γ-Irradiation or Pyrolysis of Some Crystal Modifications of Glucose, *Acta Chem. Scand., 19* (1965)595-600.

101. Shafizadeh, F., McGinnis, G.D., Susott, R.A. and Meshreki, M.H., Thermolysis of Derivatives of Amino Sugars, *Carbohyd. Res., 33*(1974)191-202.

102. Schlotzhauer, W.S., Chortyk, O.T. and Austin, P.R., Pyrolysis of Chitin, a Potential Tobacco Extender, *J. Agric. Food Chem., 24*(1976)177-180.

103. Haverkamp, J., Meuzelaar, H.L.C., Beuvery, E.C., Boonekamp, P.M. and Tiesjema, R.H., Characterisation of Neisseria meningitidis Capsular Polysaccharides Containing Sialic Acid by Py-MS, *Analyt. Biochem., 104*(1980)407-418.

104. Huis in't Veld, J.H.J., Meuzelaar, H.L.C. and Tom, A., Analysis of Streptococcal Cell Wall Fractions by Curie-point Py-GC, *Appl. Microbiol., 26*(1973)92-97.

105. Kistemaker, P.G., Meuzelaar, H.L.C. and Posthumus, M.A., Rapid and Automated Identification of Micro-organisms by Curie-point Pyrolysis Techniques. II. Fast Identification of Microbiological Samples by Curie-point Py-MS, in G. Heden and T. Illeni (Eds.), *New Approaches to the Identification of Micro-organisms,* John Wiley, New York, 1975, pp. 179-191.

106. Maters, W.L., Meent, V.d.D., Schuyl, P.J.W., deLeeuw, J.W., Schenk, P.A. and Meuzelaar, H.L.C., Curie-point Pyrolysis in Organic Geochemistry, in C.E.R. Jones and C.A. Cramers (Eds.), *Analytical Pyrolysis,* Elsevier, Amsterdam, 1977, pp. 203-216.

107. Metzger, J., CIMS of Lignins, *Z. Anal. Chem., 295*(1979)45-46.

108. Jennings, Jr., E.C. and Dimick, K.P., GC of Pyrolytic Products of Purines and Pyrimidines, *Analyt. Chem., 34*(1962)1543-1547.

109. Turner, L.P., Characterisation of Nucleotides and Nucleosides by Py-GC, *Analyt. Biochem., 28*(1969)288-294.

110. Turner, L.P. and Barr, W.R., Py-GC of Ribonucleosides, Ribonucleotides and Dinucleotides, *J. Chromatogr. Sci., 9*(1971) 176-181.

111. Posthumus, M.A., Nibbering, N.M.M., Boerboom, A.J.H. and Schulten, H-R., Py-MS Studies on Nucleic Acids, *Biomed. Mass Spectrom., 1*(1974)352-357.

112. Levsen, K. and Schulten, H-R., Analysis of Mixtures by CAMS: Pyrolysis Products of DNA, *Biomed. Mass Spectrom., 3*(1976) 137-139.

113. Schulten, H-R., Beckey, H.D., Boerboom, A.J.H. and Meuzelaar, H.L.C., Py-FDMS of DNA, *Analyt. Chem., 45*(1973)2358-2362.

114. Charnock, G.A. and Loo, J.L., MS Studies of DNA, *Analyt. Biochem., 37*(1970)81-84.

115. Wiebers, J.L., Sequence Analysis of Oligodeoxyribonucleotides by MS, *Analyt. Biochem., 51*(1973)542-556.

116. Wiebers, J.L., Detection and Identification of Minor Nucleotides in Intact DNA by MS, *Nucleic Acids Res., 3*(1976)2959-2970.

117. Wiebers, J.L. and Shapiro, J.A., Sequence Analysis of Oligodeoxyribonucleotides by MS. I. Dinucleoside Monophosphate, *Biochemistry, 16*(1977)1044-1050.

118. Gauding, D. and Jankowski, K., Exact Mass Analysis of DNA Fragments, *Org. Mass Spectrom., 15*(1980)78-79.

119. Jankowski, K. and Soler, F., MS of DNA: Part 2. Quantitative Estimation of Base Composition, *Eur. J. Mass Spectrom., 1* (1980)45-52.

120. Burgard, D.R., Perone, S.P. and Wiebers, J.L., Sequence Analysis of Oligodeoxyribonucleotides by MS. 2. Application of Computerised Pattern Recognition to Sequence Determination of Di-, Tri- and Tetra- nucleotides, *Biochemistry, 16*(1977)1051-1057.

121. Gross, M.L., Lyon, P.A., Dasgupta, A. and Gupta, N.K., MS Studies of Probe Pyrolysis Products of Intact Oligoribonucleotides, *Nucleic Acids Res., 5*(1978)2695-2704.

122. Perkins, E.G., Nutritional and Chemical Changes Occurring in Heated Fats: A Review, *Food Technol., 14*(1960)508-514.

123. Perkins, E.G., Formation of Non-Volatile Decomposition Products in Heated Fats and Oils, *Food Technol., 21*(1967)125-130.

124. Nawar, W.W., Thermal Degradation of Lipids. A Review, *J. Agric. Food Chem., 17*(1969)18-21.

125. Higman, E.B., Schmeltz, I., Higman, H.C. and Chortyk, O.T., Studies on the Thermal Degradation of Naturally Occurring Materials. II. Products from the Pyrolysis of Triglycerides at 400°, *J. Agric. Food Chem., 21*(1973)202-204.

126. Levy, E.J. and Paul, D.G., The Application of Controlled Partial Gas Phase Thermolytic Dissociation to the Identification of GC Effluents, *J. Gas Chromatogr., 5*(1967)136-145.

127. Holman, R.T., Deubig, M. and Hayes, H., Py-GC of Lipids. I. MS Identification of Pyrolysis Products of Hydrocarbons, *Lipids, 1*(1966)247-253.

128. Perkins, E.G. and Johnston, P.V., Py-GC of Phosphoglycerides: An MS Study of the Products, *Lipids, 4*(1969)301-303.

129. Horning, M.G., Casparinni, G. and Horning, E.C., The Use of Gas Phase Analytical Methods for the Analysis of Phospholipids, *J. Chromatogr. Sci., 7*(1969)267-275.

130. Weijman, A.C.M., The Application of Curie-point Py-MS in Fungal Taxonomy, in C.E.R. Jones and C.A. Cramers (Eds.), *Analytical Pyrolysis*, Elsevier, Amsterdam, 1977, pp. 225-233.

131. Janak, J., Identification of the Structure of Non-volatile Organic Substances by GC of Pyrolytic Products, *Nature, 185* (1960)684-686.

132. Groten, B., Application of Py-GC to Polymer Characterization, *Analyt. Chem., 36*(1964)1206-1212.

133. Liddicoet, T.H. and Smithson, L.H., Analysis of Surfactants Using Py-GC, *J. Amer. Oil Chem. Soc., 42*(1965)1097-1102.

134. Nakagawa, T., Miyajima, K. and Uno, T., Py-GC of Long Chain Fatty Acid Salts, *J. Chromatogr. Sci., 8*(1970)261-265.

135. Hites, R.A. and Biemann, K., On the Mechanism of Ketonic Decarboxylation. Pyrolysis of Calcium Decanoate, *J. Amer. Chem. Soc., 94*(1972)5772-5777.

136. Kojima, T. and Morishita, F., Application of Laser to Py-GC I, *J. Chromatogr. Sci.,* 8(1970)471-473.

137. Karmen, A., Walker, T. and Bowman, R.C., A Sensitive Method for Quantitative Microdetermination of Lipids, *J. Lipid Res.,* 4(1963)103-106.

138. Hirayama, S. and Imai, C., Rapid Determination of Residual Hexane in Oils by GC Using Pyrolyser, *J. Amer. Oil Chem. Soc.,* 54(1977)190-192.

139. Etemadi, A.H., The Use of Py-GC and MS in the Study of the Structure of Mycolic Acids, *J. Gas Chromatogr.,* 5(1967)447-456.

140. Lechevalier, M.P., Horan, A.C. and Lechevalier, H., Lipid Composition in the Classification of Nocardiae and Mycobacteria, *J. Bacteriol.,* 105(1971)313-318.

141. Lechevalier, M.P., Lechevalier, H. and Horan, A.C., Chemical Characteristics and Classification of Nocardiae, *Can J. Microbiol.,* 19(1973)965-972.

142. Davison, V.L. and Dutton, H.J., Microreactor Chromatography. Quantitative Determination of Double Bond Positions by Ozonization-Pyrolysis, *Analyt. Chem.,* 38(1966)1302-1305.

143. Privett, O.S., Determination of the Structure of Unsaturated Fatty Acids via Degradative Methods, *Prog. Chem. Fats,* 9(1966) 91-117.

144. Downing, D.T. and Green, R.S., Rapid Determination of Double-bond Positions in Mono-enoic Fatty Acids by Periodate-Permanganate Oxidation, *Lipids,* 3(1968)96-100.

145. Nickell, E.C. and Privett, O.S., A Simple, Rapid Micromethod for the Determination of Unsaturated Fatty Acids via Ozonolysis, *Lipids,* 3(1968)166-170.

146. Gellerman, J.L. and Schlenk, H., Pyrolysis for Structure Determination of Cyclopropane and Cyclopropene Fatty Acids, *Analyt. Chem.,* 38(1966)72-76.

147. Cox, J.S.G., High, L.B. and Jones, E.R.H., The Determination of Steroid Side-chains, *Proc. Chem. Soc., 1958*:234-235.

148. Gassiot-Matas, M. and Julia-Danes, E., Py-GC of some Sterols, *Chromatographia,* 5(1972)493-501.

149. Gassiot-Matas, M. and Julia-Danes, E., Py-GC of some Sterols (II), *Chromatographia,* 9(1976)151-156.

150. Falk, H.L., Goldfein, S. and Steiner, P.E., The Products of Pyrolysis of Cholesterol at 360°C and Their Relation to Carcinogens, *Cancer Res.,* 9(1949)438-447.

151. Poole, C.F. and Morgan, E.D., Anomalies in the GC of Cholesterolheptafluorobutyrate, *J. Chromatogr.,* 90(1974)380-381.

152. Bayer, F.L., Hopkins, J.J. and Menger, F.M., Py-GC of Biomedically Interesting Molecules, in C.E.R. Jones and C.A. Cramers (Eds.), *Analytical Pyrolysis,* Elsevier, Amsterdam, 1977, pp. 217-223.

153. Menger, F.M., Hopkins, J.J., Cox, G.S., Maloney, M.J. and Bayer, F.L., Py-GC of Structurally Related Steroids, *Analyt. Chem., 50*(1978)1135-1137.

154. Reiner, E., The Role of Py-GC in Biomedical Studies, in C.E.R. Jones and C.A. Cramers (Eds.), *Analytical Pyrolysis*, Elsevier, Amsterdam, 1977, pp. 49-56.

155. Blondin, G.A., Kulkarni, B.D., John, J.P., van Aller, R.T., Russell, P.T. and Nes, W.R., Identification of Steroidal 5,8-Peroxides by GC, *Analyt. Chem., 39*(1967)36-40.

156. Levy, R.L., Gesser, H., Halevi, E.A. and Saidman, S., Py-GC of Porphyrins, *J. Gas Chromatogr., 2*(1964)254-255.

157. Whitter, D.G., Bentley, K.E. and Kuwada, D., Pyrolysis Studies. Controlled Thermal Degradation of Mesoporphyrin, *J. Org. Chem., 31*(1966)322-324.

Chapter 8

Taxonomy

8.1. INTRODUCTION

The classification and identification of microorganisms or, indeed, of pathological conditions is often a tedious task. Such work is important in clinical medicine, environmental health, and many industrial processes and traditionally involves a wide range of morphological, serological, and biochemical analyses. These tests are time consuming to perform, are liable to error without appropriate controls or precautions and may be subject to cross-reactions. Problems may be severe with pathogenic organisms, particularly when subjective interpretation of serological or biochemical tests is required. Further, the current schemes used as the basis for classification may not be appropriate, due to, for example, variability in conventional taxonomic characteristics [1]. The often arbitrary nature of classification by traditional techniques has thus encouraged the application of numerical methods of taxonomy [2-4]. These methods are valid in the study of microorganisms [5], especially in view of the interest in rapid, universal, and automated identification techniques [6,7], which provide a range of numerical data describing the sample.

The ideal taxonomic technique should be

Universal: Standard analytical conditions must be applicable to a wide range of samples.

Specific: Different clases of sample must be readily differentiated, while samples of the same class are grouped together.

Reproducible: Conclusions must be reliable and have universal
   applicability. They are not system or operator dependent.

Rapid: Minimal sample preparation is required and the system
   must be capable of analyzing a large number of samples in
   a short period of time.

Sensitive: Small amounts of sample must be demanded to minimize
   subculturing.

Interpretive: Analytical information should enable chemical or
   biochemical inferences about sample composition to be made.

Automated: The instrument analysis and data handling should be
   capable of unsupervised operation to reduce the expertise
   demanded by the analysis.

The methods which have been used for taxonomic purposes include
several chromatographic techniques such as gel electrophoresis and
gas chromatography. The GC of cell wall fractions has been rewarding
[8], particularly when the GC-MS combination is used, but derivitization after isolation is necessary. The long elution time required
for many biological molecules reduces the through-put significantly.
Of all the available methods for rapid, instrumental taxonomic classification, those based upon analytical pyrolysis appear to combine
most of the above-named features. Indeed, one set of analytical conditions may be appropriate for the analysis of a range of samples
such as bacteria, fungi, proteins, pathological tissue, and body
fluids. The Py-MS approach, in particular, offers a rapid turn over
of samples (ca. 1 min per analysis) and gives direct chemical information on the nature of the sample. This is also available with Py-
GC-MS, although longer analysis times--necessary for the elution of
more polar products--are necessary. Both systems may be fully automated for unattended operation (Secs. 3.2.4 and 4.2). No comprehensive taxonomic scheme based upon pyrograms or mass pyrograms has been
proposed, but the techniques allow distinctions to be made between
very closely allied organisms, such as those differing in one antigenic component [9,10]. The analyses are relatively easy to undertake and although not yet as popular as GC identification of isolated
microbiological components, the advantages offered by ease of sample
preparation, the wide applicability, the rapidity of the analysis and
automation are undeniable attributes. It has been suggested that

pyrolysis techniques offer little advantage over more conventional approaches [8]. This may have been true in the early 1970s (when the statement, mainly referring to fingerprint pyrograms determined under furnace conditions, was made), but it is doubtful if a similar comment could be made today. The advances in system design, the appreciation of the variables involved in analytical pyrolysis, the modern emphasis upon structural elucidation and chemical interpretation of the data, and the refined methods of data handling currently available all combine to make the pyrolysis approach a powerful analytical technique.

8.2. SOURCES OF VARIABILITY

8.2.1. Pyrolysis-Analysis

The control of the pyrolysis and analysis conditions to ensure reproducibility is of paramount importance in taxonomic studies. The principles of small sample size and rapid temperature rise times should be followed and effective transfer of the pyrolysate to the analytical device should be ensured by the avoidance of cold spots. Over 200 products may originate from the pyrolysis of microorganisms [11]. With Py-GC, capillary columns, although not essential, are advocated due to their ability to separate many more of these components. This not only gives more data points for comparison, but also removes the variability caused by peaks which represent coeluting components. A precolumn-back-flushing technique is available to reduce long-term column degeneration which causes progressive loss of resolution and poor long term performance from capillary columns (Fig. 3.6). Discussions of Py-GC data are also inhibited in many cases by the lack of identification of the diagnostic peaks. This means that empirical comparisons of the data, which are dependent upon the analytical conditions and are probably valid for one laboratory only, are common. Py-MS, too, may lose information due to the concurrence of isomeric or isobaric ions. Peaks due to such combinations may well be expected to show greater variation than single-component peaks.

To ensure adequate control of pyrolysis and analysis, a standard
system should be adopted. The variables have been discussed in Chaps.
3 and 4 and a summary, in the form of checklists, is provided in Appendix 2.

Data handling is a further essential consideration. In very few
cases are unique fragments observed or expected. It has been found
that the pyrograms of some organisms did indeed contain a component,
not necessarily fully resolved, which was not present in the pyrograms from other species [12]. Thus, pyrazine characterized *Micrococcus lutea, Bacillus subtilis* var. *niger* yielded naphthalene and
Streptomyces longisporoflavus gave methylnaphthalene. This probably
reflects the small number of organisms which have been studied in
this detail. When peak identification becomes more routine it would
be surprising if this trend was extended, or even maintained. Certainly, the similarity of the pyrolysis products from the Gram negative *Pseudomonas putida* [13] and the Gram positive *Bacillus s.* [12]
does not encourage optimism. Pyrograms may thus be expected to differ in the relative intensities of various fragments only, and identification must be based upon pattern recognition techniques.

Visual comparison of pyrograms has been reported on many occa-
sions to be an adequate means of identification. Such assertions,
however, are frequently unsupported by statistical data on the reproducibility of replicate determinations, or the deviations expected
between different samples (i.e., inner and outer variances). Care
must also be taken to ensure that library data is authentic and can
be reproduced by the current analytical system. This condition is
less important for Py-MS, for the rapid through-put of samples enables standards to be incorporated with the unknowns. In view of the
success of numerical methods, the calculation of a similarity coefficient (Table 5.5) or a nearest-neighbor matrix (Tables 5.9, 5.13,
and 5.15), weighted for inner and outer variance (Table 5.10) would
clearly be more appropriate than visual comparisons. These approaches
demand intensity measurements, peak selection and normalization
(Chap. 5). Care must be taken with analog Py-GC signals to ensure
that overlapping peaks, tailing, or a drifting base line do not

invalidate the numerical transformations. Each pyrogram should be
normalized to the same total response, and it is advisable to auto-
scale each peak to a zero mean and unit standard deviation to reduce
the disproportionate significance of very intense peaks.

8.2.2. Sample Preparation

In addition to these general problems, associated to some extent with
many applications of analytical pyrolysis, there are others which are
unique to taxonomic applications. These are concerned with sample
preparation and although little quantitative or mechanistic information is available, sufficient evidence is to hand to suggest that
rigorous control should be applied wherever possible.

Culture Media

The growth of microorganisms may be critically dependent upon the
growth medium. It is possible to subculture colonies of the same
organism to provide very different biochemical profiles. This may
be achieved by using growth media which have been depleted of one
essential factor. Glucose-limited, magnesium-limited and phosphate-
limited organisms are examples of those which may be cultured in this
way [14]. In the case of bacteria, such strains have a considerable
variation in their susceptibility towards antibiotics. This reflects
biochemical differences in the nature of the cell wall and it would
be surprising if such variations were not exposed during analytical
pyrolysis of such samples. Indeed, the analytical pyrolysis of
growth-limited cultures may be a rapid indicator of biochemical differences, although no such work has appeared to date.

   Media may also control the formation of extracellular components.
This is a well-recognized factor in antibiotic production, but may be
of wider significance especially as many pathogenic bacteria produce
exotoxins, e.g., Clostridia. The production of an extracellular
staphylocoagulase enzyme from strains of *Staphylococcus aureus,* a
potential pathogen, has been shown to be dependent upon culture conditions [15]. The level of magnesium was critical, with an optimum

of 50 to 100 µM with higher concentrations causing inhibition of enzyme production. Inhibition was also induced by glucose levels
higher than 0.2%. Other features which were of importance were temperature, oxygen tension, pH, incubation time, size of inoculum, and
whether the culture was static or agitated. Trace additives such as
glycerol, thiamine, nicotinic acid, and biotin were also of vital
importance.

The extent of the effect of media and cultivation conditions
upon the pyrograms of microorganisms cannot be assessed at present.
However, pyrograms are clearly dependent upon growth conditions.
Pyrograms from bacteria grown in a nitrogen-rich medium showed many
peaks in common with protein pyrograms, but significant differences
were observed when the medium was changed [16,17]. Thus, fungi grown
in a nitrogen-rich medium differed significantly from those grown in
a carbohydrate-rich medium. Such differences were also revealed in
the elemental analysis of the organisms: C, 37.4%; N, 7.1% (cf. C,
43.6%; N, 4.0%). It was concluded that organisms grown in different
media should not be compared, but that the variation induced by such
changes may well be of significance in establishing identity. More
detailed study with *Bacillus* spp. has shown that only small differences between batches of the same growth medium are evident, but
large deviations may result if different media are used. Figure 8.1
illustrates the effect [18]. Peak 1 was characteristic of pigmentation and was particularly evident with Eugon Agar. Other reports
show similar findings [19,20] and multivariate statistical techniques
readily expose culture differences in both bacterial [21] and fungal
[22] samples.

In view of this work it would seem appropriate to investigate
these variables in detail. Chemically-defined media in which all
components of the growth medium are specified, would be an approach
to this [23]; indeed, a beginning has been made [22]. The use of
continuous culture (chemostat) would enable significant amounts of
organism to be collected [23]. These conditions would enable a detailed study into the effects of controlled growth parameters on pyrogram reproducibility to be undertaken. If combined with structure

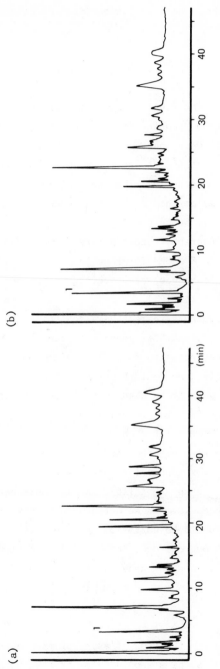

FIGURE 8.1 Variation of pyrograms of *Bacillus globigii* dependent upon culture medium: (a) trypticase soy agar and (b) eugon agar. (Adapted from Ref. 18.)

elucidation of the products and correlation with the biochemical
properties of the organisms under test, important results would
surely emerge.

Culture Age

The effect of culture age was found to be a smaller source of the
variation in pyrograms from Mycobacteria than culture medium.
Nevertheless, differences were observed and it was possible to vary
the culture age so that certain features were emphasized to accentu-
ate characteristic differences [19]. Due to the slow growth of Mycobacteria, cultures were maintained for 2 to 6 weeks prior to analysis
with identification usually being possible after 2 to 3 weeks of
growth. This compares favorably with the more conventional approach
which usually demands 4 to 8 weeks for results to be available.

     More significant differences were observed in the pyrograms of
Clostridia [20] and of Bacilli [24]. Here, sporulation was evident
and pronounced changes in the pyrograms heralded this event. Figure
8.2 records typical data. Peak 2 is characteristic of the changes
and clearly indicates the necessity for standardized culture periods.
Certainly, no valid conclusions could be drawn if variable sporing
and nonsporing cultures were to be used without regard. To avoid
this, a culture period of 24 h was recommended.

     No data appears to be available on the variability of pyrolysis
associated with inoculum size or other stages of growth such as log,
lag or stationary phases. Evidence is available to indicate that
biochemical differences are apparent [15], and it would be unusual
that under no circumstances were these detectable by pyrolysis.
Perhaps this area, too, would be worthy of more definitive work.
The technique of synchronous culture, in which all organisms mature
at the same rate, would give a homogeneous sample and so validate
the comparisons [25].

Sampling

Opinions differ as to the best method of sampling from a culture.
It is generally held that samples should be as free as possible from

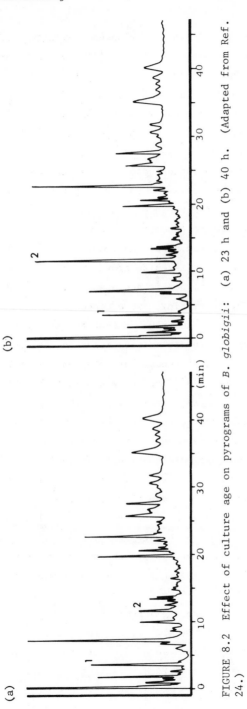

FIGURE 8.2 Effect of culture age on pyrograms of *B. globigii*: (a) 23 h and (b) 40 h. (Adapted from Ref. 24.)

medium contamination. Harvested colonies from liquid or gel culture
are thus washed by resuspending in distilled water several times before further treatment. A technique which facilitates this approach
ensures that no residual nutrient medium, particularly lumps or insoluble material from agar cultures, contaminates the colony. Here,
membrane filters (pore size 0.45 μm) are placed on the surface of
nutrient agar plates and a resuspended culture is spread over the
filter. The plates are incubated under standard conditions, after
which the cells may be harvested by scraping from the filter. Little
difference was observed between washed and unwashed cells and no contamination by membrane components was observed [26].

Direct harvesting and pyrolysis from agar plates is also possible. Providing care is taken to prevent removal of the solid
medium, excellent performance is achieved [27,28]. Indeed, reproducibility and characteristicity were improved when compared to other
sampling procedures. It is probable that the increased characteristicity of pyrograms is due to extracellular metabolites which are removed with most sampling methods, but not when direct application to
the pyrolysis wire is used. Washing was said to have a particularly
deleterious effect upon reproducibility, perhaps due to lysis of bacterial cells. Although encouraging, the complexity of media necessary
for the growth of some pathogenic organisms (blood agar, egg additives
or even meat broths) suggests caution in extending this approach.
Other studies using mycobacteria have grown the organisms in a fil-
tered medium (0.45 μm). Cells were harvested by centrifugation at
20,000 g for 30 min and were washed three times with an inorganic
phosphate-saline buffer which was prefiltered (0.22 μm) and which
gave no background when pyrolyzed. The buffer was chosen to minimize
osmotic lysis and washing with distilled water was not attempted [11].

The pyrograms of growth media have also been reported. These
have similarities to those of microorganisms [29], but may be distinguished, particularly if high-resolution columns are used.

Sampling of fungi has also been undertaken in different ways.
Liquid cultures may be harvested by filtration [22], by centrifugation

[30], or by sedimentation [31]. Colonies may also be grown on cel-
lophane membranes situated on top of a nutrient agar medium [32].
As with the analogous bacterial system [26], the colony is essentially
free from medium contamination. Care must also be taken to ensure
that samples are comparable for differences between pyrograms from
young and mature mycelia and from spores have been demonstrated [33,
34]. Spores may be isolated by collecting and blending mycelia grown
in liquid culture. The mycelial remains are removed by filtration
through cotton wool and the spores were collected by centrifugation
following ultrasonic vibration of the filtrate [33]. Alternatively,
spores may be obtained directly by washing the mycelial mats with
methanol [35].

Clearly, as individual samples become larger and more hetero-
geneous, the sampling protocol should become more rigid to ensure
comparability.

Subculturing

Progressive subculturing of bacterial colonies may result in attenuation of the strain or mutation with resultant loss of characteristicity. Cultured animal cells may revert to an undifferentiated
type. Such problems must be appreciated to ensure reliable sample
control [36,37].

Lyophilization

Except when colonies have been sampled directly from a culture plate,
freeze drying has generally been undertaken. This enables more efficient suspension and loading onto the pyrolysis wire. Reports have,
however, appeared where suspensions of washed organisms have been
loaded onto the pyrolysis wire and air-dried before analysis [38,39].
Little detailed work has appeared on the effect of these various
treatments, and it would appear essential to prepare the sample in an
identical manner each time, and from this viewpoint freeze-drying has
certain advantages.

Sterilization

When highly virulent pathogenic organisms are studied, safety considerations have usually demanded that sterilization of the colonies
prior to analysis has been undertaken. Autoclaving, which involves
wet heat at 121°C for 15 min, has proven to be satisfactory [19], although spores may require considerably longer for complete death.
Surprisingly, these vigorous conditions do not appear to reduce the
characteristicity of the pyrograms from organisms treated in this
way, despite the complexity of the denaturation and hydrolytic reactions which must be initiated. Indeed, it has been shown that pyrolysis wires, loaded directly from a culture plate, may be sterilized
prior to posting to the analytical center [27]. Such samples, when
stored at 4°C, maintain their integrity for several days and yield
stable pyrograms. These results are important in that central laboratories may be provided with samples from national and international sources without hazard or loss of performance. Despite this
promise, it must be recognized that little detailed information is
available on the effects of sterilization on pyrogram reproducibility
(e.g., time-temperature profiles) and perhaps further work is required to confirm the generality of this approach.

A further method of sterilization is through the incorporation
of a bacterial agent. This substantially decreases the temperature
requirements, but contaminates the sample with the additive, and
steps must be taken to ensure its removal. Formalin (0.25%) [40] and
phenol (0.5%) [41] have been used to wash cultures directly from culture plates. Sterilization was effected by standing overnight [40],
and this was followed by water washes to remove the bactericide [40,
41]. Irradiation is a further means of sterilization which, in addition to cultures and suspensions, could be used directly on freeze-
dried samples [42]. No reports of the effect on pyrograms have appeared, but the technique may have profound effects upon the integrity of biological molecules [43-46].

In addition to fingerprint identification of unknown samples,
pyrolysis has also been used to establish taxonomic relationships

between groups of organisms. The major features and differences between pyrograms may not, however, parallel the gross morphological
or even biochemical differences which are the basis of traditional
classifications. Thus, clinical properties, such as haemolytic capabilities or morphological features such as whether bacteria are
bacilli of cocci, or the occurrence and appearance of flagella are
not generally indicated by state-of-the-art studies. This is an
area for future concern. Nevertheless, closely related organisms do
have similar pyrograms, and studies on isolated cell components such
as cell-walls, flagella and DNA add to the information available [47].
Thus, although analytical pyrolysis may not replace the microscope or
test tube, in this area it is certainly a powerful complementation to
more standard techniques.

8.3. BACTERIA

The pyrolysis products from bacteria are essentially those which are
expected from a knowledge of the pyrolysis of biomacromolecules. A
Py-GC-MS run on *B. subtilis* var. *niger* illustrates a typical product
distribution [12,48]. The pyrogram is displayed in Fig. 8.3 and
Table 8.1 records the component identification and proposed origins.

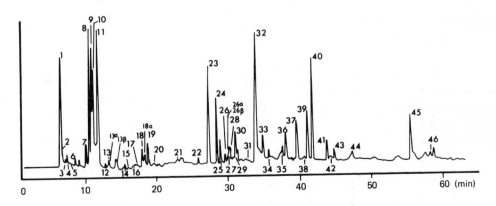

FIGURE 8.3   Capillary column pyrogram from *B. subtilis* var. *niger*.
(See Table 8.1 for component identification.)   (From Ref. 12, by
permission of Heyden & Son Ltd.)

TABLE 8.1 Bacterial Pyrolysis Products (Py-GC)

PEAK	COMPONENT		PEAK	COMPONENT		PEAK	COMPONENT	
1	Carbon dioxide		9	Acrylonitrile	(P,N)	33	C_4-Alkylpyrroles	(P,Po)
1	Water		9	Acetonitrile	(P,N)	34	C_4-Alkylpyrroles	(P,Po)
1	Carbonyl sulphide		10	Propionitrile	(P,N)	35	Propionamide	(P-C)
1	Ammonia		11	Toluene	(P)	36	Phenylacetonitrile	(P)
1	Hydrogen sulphide		12	Butyronitrile	(P,N)	37	Tolunitrile	(P)
1	Methane	(P)	13	isoButyronitrile	(P)	38	o-Cresol	(P)
1	Ethylene	(P,L)	13	Et-benzene	(P)	39	Phenol	(P)
1	Ethane	(P)	13	isoValeronitrile	(P)	40	p-Cresol	(P)
1	Propane	(P,L)	15	Pr-benzene	(P)	41	Naphthalene	
1	Propene	(P)	16	o-Xylene	(P)	42	Et-phenol	(P)
1	Butene	(P,L)	17	Pyridine	(P,N)	43	Xylenol	(P)
1	isoButene	(P)	18	C_3-Alkylbenzenes	(P)	44	Xylenol	(P)
1	isoPentene	(P)	18	isoCapronitrile	(P)	45	Indole	(P)
1	Me-butadiene	(C)	19	Styrene	(P)	46	Me-indole	(P)
1	cycloPentadiene	(C)	20	Picoline	(P,N)	*		
1	Me-cyclopentadiene	(C)	21	Di-Me-pyridines	(P)	1	isoButane	
1	Methanethiol	(P)	22	Furfural	(C)	1	Butane	
1	Ethylene oxide	(P)	23	Pyrrole	(P,Po)	1	Butadiene	
1	Acrolein	(C,L)	24	Me-pyrrole	(P,Po)	1	Trimethylamine	
1	Propionaldehyde	(C)	25	Me-pyrrole	(P,Po)	1	Pentene	
1	Acetone	(C)	26	5-Me-furfural	(C)	4	Dimethyl sulphide	
2	isoButyraldehyde	(C)	26	Di-Me-pyrroles	(P,Po)	5a	isoHexene	
3	Furan	(C)	26	Di-Me-pyrroles	(P,Po)	13a	Me-pentanone	
4	Butan-2-one	(C)	27	Benzonitrile	(P)	13b	Xylenes	
5	2-Me-furan	(C)	28	Di-Me-pyrroles	(P,Po)	18a	Pyrazine	
6	isoValeraldehyde	(C)	29	Acetophenone		19a	Me-styrene	
7	Di-Me-furan	(C)	30	Furfuryl alcohol	(C)	23a	Furyl Me ketone	
7	Benzene	(P)	31	C_3-Alkylpyrroles	(P,Po)	41a	Me-naphthalene	
8	Pentan-2-one	(C)	32	Acetamide	(P-C)			

*Components not present in all organisms studied. Origin of frag-
ments: P, protein; C, carbohydrate; N, nucleic acid; L, lipid; Pp,
porphyrin.

Source: Ref. 12, by permission of Heyden & Son Ltd.

Major products resulting from proteins (11, toluene; 23, pyrrole; 40,
p-cresole; 45, indole) may be observed, whereas carbohydrate residues
are mainly detected as small fragments (8, pentan-2-one) with furan
compounds being found as low intensity peaks (3, furan; 22, furfural;
26, 5-methylfurfural). Acetamide is a further major pyrolysis product. This probably arises from the bacterial cell wall which constitutes 20-35% of the dry weight of the cell. The wall comprises
of polysaccharide chains (N-acetylmuramic acid) which are cross-linked

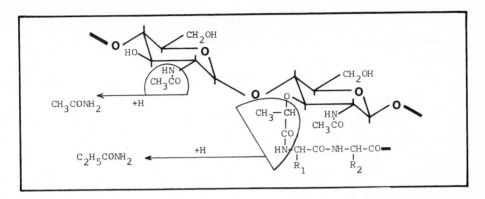

FIGURE 8.4 Pyrolysis of the glycosaminopeptide polymer of the bacterial cell wall.

TABLE 8.2 Bacterial Pyrolysis Products Not Common to All Organisms Studied

B. subtilis	M. luteus	S. longisporoflavus
		Isobutane
Butane		Butane
Butadiene		Butadiene
Trimethylamine	Trimethylamine	
	Pentene	Pentene
Dimethylsulfide		Dimethylsulfide
Isohexene		
	Methylpentanone	Methylpentanone
	m- and p-Xylene	m- and p-Xylene
	Pyrazine	
	Methylstyrene	Methylstyrene
Furylmethyl ketone		
Naphthalene		
		Methylnaphthalene

Source: Ref. 12, by permission of Heyden & Son Ltd.

by oligo-peptide chains, giving a rigid glycosaminopeptide (peptido-
glycan) matrix. Pyrolysis of such polymers should lead to large
amounts of acetamide (Fig. 8.4).

Three organisms have been studied (*B. subtilis* var. *niger, Micrococcus luteus* and *Streptomyces longisporoflavus*), and a high similarity between the product distributions was observed. The 15 most
intense peaks for *B. subtilis* were also those of *S. longisporoflavus*
and 13 of the largest 15 peaks from *M. luteus* were also common to
this group. Unique fragments, too, were observed and these are recorded in Table 8.2

The identification of bacterial pyrolysis has also been undertaken using Py-MS techniques [49]. Field ionization was used to ensure a high proportion of molecular ions and high-resolution mass
spectrometry enabled component identification to be achieved via accurate mass measurements. *Pseudomonas putida* was the organism
studied and, despite the differences in organisms and in technique,
a surprising degree of similarity between these and the Py-GC-MS results was apparent. The mass pyrogram is displayed in Fig. 8.5 and
the Table 8.3 records the component identification. Over 200 fragments were detected and of the components listed in Table 8.1 (Py-GC-
MS), 60 also appear in Table 8.3 (Py-FIMS).

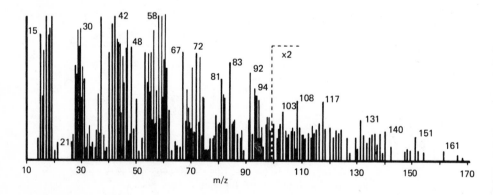

FIGURE 8.5 Mass pyrogram from *Pseudomonas putida* (high-resolution
FIMS) (see Table 8.3 for component identification). (From Ref. 49,
adapted with permission from H-R. Schulten et al., High resolution
FIMS of bacterial pyrolysis products, *Anal. Chem.* *45*:191-195. Copyright 1973 American Chemical Society.)

TABLE 8.3 Bacterial Pyrolysis Products from *P. putida* (Py-FIMS)

m/z	COMPONENT	m/z	COMPONENT	m/z	COMPONENT
16	Methane	67	Pyrrole	97	Furfurylamine
17	Ammonia	67	Me-propenenitrile	98	Furfuyl alcohol
18	Water	68	Furan	98	Heptene
26	Acetylene	68	Me-butadiene	100	Me-pentanone
27	Hydrogen cyanide	69	C_3H_3NO	101	$C_5H_{11}NO$
28	Carbon monoxide	69	Butanenitrile	103	Benzonitrile
28	Ethylene	69	Me-propanenitrile	104	Styrene
30	Formaldehyde	70	Propynoic acid	105	C_7H_7N
30	Ethane	70	Me-butene	106	Et-benzenes
31	Methylamine	70	Pentene	106	Xylenes
32	Oxygen	71	HO-propanenitrile	107	Di-Me-pyridines
32	Methanol	72	Butanone	108	Cresols
33	Hydroxylamine	72	Me-propanal	109	C_3-Alkylpyrroles
34	Hydrogen sulphide	73	Propionamide	110	$C_7H_{10}O$
35	NH_4OH	74	Thiapropane	111	C_6H_9NO
36	HCl	74	Propionic acid	112	$C_7H_{12}O$
41	Acetonitrile	76	Propanethiol	112	C_8H_{16}
42	Propene	78	HO-ethanethiol	113	$C_4H_3NO_3$
44	Carbon dioxide	78	Benzene	113	$C_5H_{11}NO$
44	Ethylene oxide	79	Pyridine	114	$C_7H_{14}O$
44	Propane	80	Pyrazine	115	$C_5H_9NO_2$
45	Formamide	81	Me-pyrroles	116	Indene
46	Formic acid	82	Me-furan	117	Indole
46	Ethanol	83	Me-butanenitrile	117	Tolunitrile
47	Me-NHOH	84	Me-pentene	117	Phenylacetonitrile
48	Methanethiol	84	Hexene	118	Me-styrene
52	Cyanogen	85	HO-butanenitrile	120	C_3-Alkylbenzenes
53	Acrylonitrile	86	Pentanone	122	Et-phenols
54	Propynal	86	Me-butanal	122	Xylenols
54	Butadiene	87	Butyramide	123	C_4-Alkylpyrroles
55	Propanenitrile	88	Butanoic acid	126	C_9-Alkenes
56	Propenal	89	Aminobutanal	129	C_9H_7N
56	Butene	92	Glycerol	130	$C_7H_{14}O_2$
56	Me-propene	92	Toluene	131	Me-indole
58	Acetone	93	Picolines	133	$C_9H_{11}N$
58	Propanal	94	Di-Me-sulphide	134	$C_8H_{10}N_2$
58	Butane	94	Phenol	135	$C_4H_9NO_4$
59	Acetamide	94	Me-pyrazine	136	$C_9H_{12}O$
60	Acetic acid	95	Hydroxypyridines	138	$C_8H_{10}O_2$
62	Dimethyl sulphide	95	Di-Me-pyrroles	140	$C_9H_{16}O$
63	Nitric acid	96	Furfural	142	$C_8H_{14}O_2$
64	Sulphur dioxide	96	Di-Me-furan		

Source: Ref. 49, adapted with permission from H-R. Schulten et al., High Resolution FIMS of Bacterial Pyrolysis Products, *Analyt. Chem.* 45:191-195. Copyright 1973 American Chemical Society.

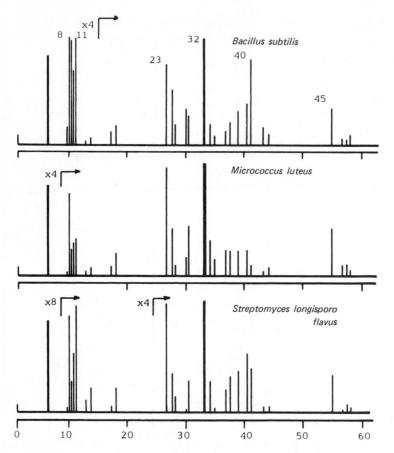

FIGURE 8.6   Bar-graph pyrograms showing variation with microorganism. (Components are identified in Table 8.1.)   (Adapted from Ref. 12, by permission of Heyden & Son Ltd.)

Although much similarity between the pyrolysis products of microorganisms is observed, the intensities of the various components differs, in many cases substantially, and comparisons of pyrograms enables various studies to be undertaken.   Figure 8.6 compares in bar graph form the major peaks in the pyrograms of *B. subtilis*, *M. luteus*, and *S. longisporoflavus*.

Significant differences between these organisms may be noted with peaks 11 (toluene), 23 (pyrrole), and 40 (*p*-cresol) (all from proteins) showing sufficient variation to distinguish between the species.

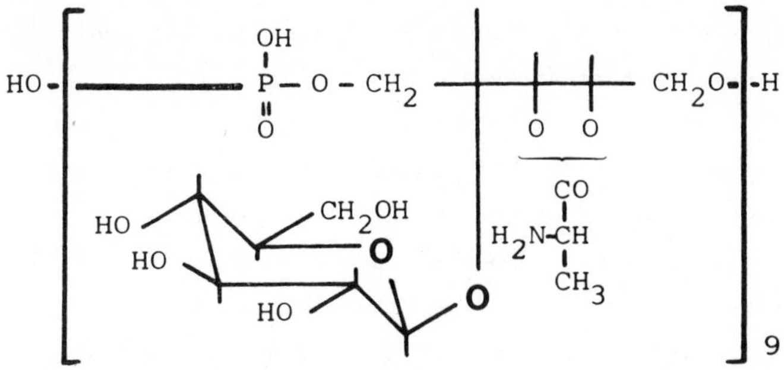

FIGURE 8.7 Cell wall teichoic acid: *B. subtilis.*

Component identification also offers a unique and rapid means of profiling microorganisms. Techniques for isolating discrete constituents of the bacterial cell are now well advanced. Thus, cell walls or their polysaccharide components, flagella, and biomacromolecules such as DNA may be readily obtained. Variations in whole-cell and cell-constituent pyrolysis, in conjunction with product identification and the current state of knowledge concerning the origin of such fragments yields a powerful technique awaiting exploitation. For example, the cell wall of *M. luteus* is composed mainly of the glycosaminopeptide polymer (85% of the dry weight cell wall) with minor amounts of amino acids, amino sugars, and glucose. This gives an intense acetamide peak in the pyrogram. In contrast, the cell wall of *B. subtilis* contains teichoic acid derivatives (Fig. 8.7) which account for 50% of the cell-wall dry weight in addition to other polymers and the glycosaminopeptide. This results in reduced amounts of acetamide in the pyrogram of *B. subtilis.*

Isolated cell fractions have also been used to improve the differentiation of Salmonella species. No component identification was undertaken, but pyrograms from flagella or from DNA were more characteristic than those from whole cells or from cell walls (Table 5.4) [47]. The power of this approach has been revealed by a Py-GC study of streptococci [9]. The strain Z_3III and a mutant strain Z_3 which lacked the type III polysaccharide antigen were studied. The antigenic component was known to be associated with the outer layer of

the cell wall, thus whole cell, cell-wall and isolated antigen pyrograms were obtained. Partial pyrograms are depicted in Fig. 8.8.
These show significant variation in the intensities of the later
peaks in the pyrgram of whole cells. These differences were largely
paralleled by the pyrograms from the isolatec cell walls. The pyrolysis of the isolated antigenic component gave a much simpler pyrogram. This differed significantly from other antigen types showing
the analytical method to be quite specific. Moreover, subtraction
of the Z_3 pyrogram from that of the Z_3III sample gave a residual pyrogram showing a high degree of similarity with that from the purified antigen. Although no peak identification was undertaken in this
work and hence no firm structural conclusions may be drawn, the Py-MS
approach to the same problem enables direct chemical interpretation
[10]. Figure 8.9 displays typical mass pyrograms and again indicates
that the major differences between the two strains are effectively
due to the antigenic component. The major components (m/z 84, 98,
110, 128) may readily be assigned to carbohydrate residues (Table
7.6). The analysis of other isolated components such as polysaccharides (Fig. 7.6) and mycolic acids (Fig. 7.10) are signposts to this
new, molecular phase of analytical pyrolysis.

      At present, most pyrolysis studies of bacteria are concerned
with the identification of samples through a comparison of pyrograms.
First studies, using Py-GC, were concerned with the possibilities for
the detection of extraterrestrial life [16]--an area since developed
to an amazingly sophisticated degree--when it was shown that micro-
organisms give well-defined and characteristic pyrograms. The taxonomic potentialities were recognized soon after, and applications
were described by Reiner in 1965 [50] and these have now been realized
in many important applications [51]. Emphasis has centered upon
pathogenic organisms in attempts to develop rapid identification
schemes. The early work demonstrated that differences between the
pyrograms of closely related organisms were apparent, and that these
differences were reproducible and characteristic. Thus, cultures
could be identified through a comparison of their pyrograms with those

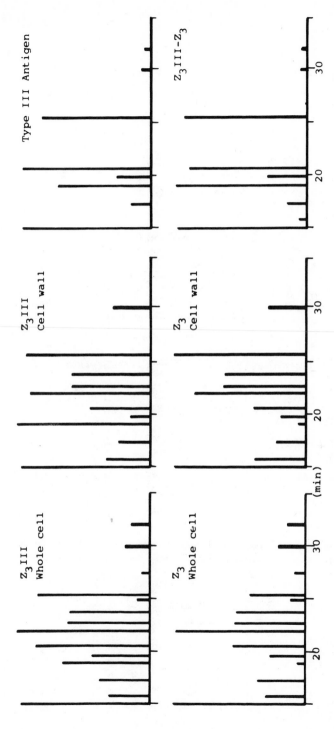

FIGURE 8.8 Partial bar-graph pyrograms of whole cells, cell walls and type III antigen from Z_3III and Z_3 Streptococcal strains. (Adapted from Ref. 9.)

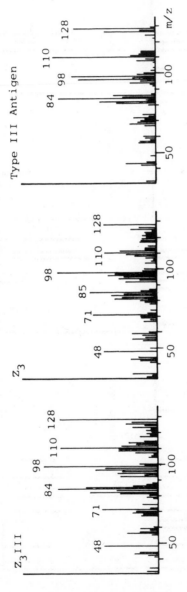

FIGURE 8.9 Mass pyrograms from whole cells and type III antigen from Z_3III and Z_3 Streptococcal strains. (Adapted from Ref. 10.)

of a library collection. Specific areas of the pyrogram were highlighted as especially diagnostic and in many cases only three or four
peaks were necessary to ensure correct identification. As these
techniques broadened in application and were in use in several laboratories, it became necessary to establish more rigorous control [11].
Py-MS offered advantages over Py-GC methods, particularly in the
speed of analysis which enabled controls to be analyzed in with the
batch of unknown samples, and this technique, too, was under active
development [52]. This was accompanied by the recognition that a
numerical comparison of pyrograms and mass pyrograms was more appropriate than a visual check for establishing similarity (Chap. 5).

A typically important group of organisms which have been studied
in depth are the Mycobacteria which include the tuberculis bacillus.
These are slow growing bacteria which may be quite tedious to identify
by biochemical means. Py-GC, however, has proved an effective and
rapid alternative. Figure 8.10 illustrates the difference between
the pyrograms of various Mycobacteria. Such dissimilarities were

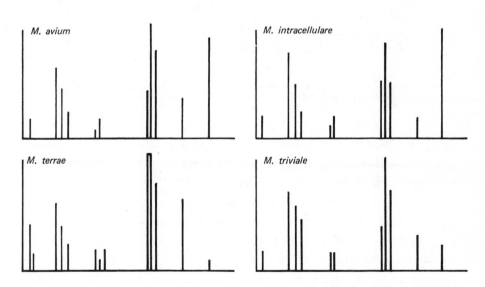

FIGURE 8.10 **Partial bar** graph pyrograms from Mycobacteria. (Adapted
from Ref. 54.)

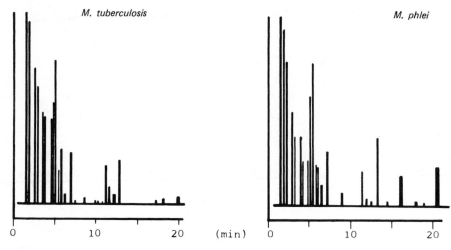

FIGURE 8.11 Bar pyrograms from Mycobacteria. (Adapted from Ref. 55.)

TABLE 8.4 Retention Time[a] Markers to Ensure Comparability of Bacterial Pyrograms

Pentan-2-one	Toluene	Pyrrole	Acetamide	Phenol	Indole
4	10	26	32	40	50

[a]Typical retention times given in minutes.

consistant over a larger number of analyses from various strains of
the organisms and allowed positive recognition to be achieved [19,53,
54]. Other work has confirmed this approach [55]. However, the pyrograms reported (Fig. 8.11) [55] showed little similarity to those
reported earlier [19,53,54]. For this reason, it is advisable to
adopt standard pyrolysis and analysis conditions to ensure that published data has significance for more than one laboratory. The identification of all or some of the components would also aid this objective and it has been suggested that furan or acetone (early), pro-
pionamide (medium) and indole (late) would be suitable internal standards to allow tuning of the pyrogram (cf. Fig. 8.3) [11]. Recommended retention time targets, using typical conditions (Table 8.6)
are recorded in Table 8.4.

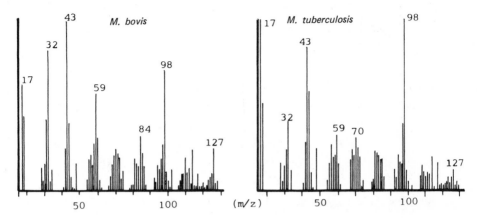

FIGURE 8.12   Mass pyrograms from Mycobacteria.   (Adapted from Ref. 27.)

Py-MS prospects for the identification of Mycobacteria have also been examined [27].   Comparisons are aided by virtue of the reproducible mass scale which enables chemical interpretation of pyrogram differences to be made.   Thus, the mass pyrograms from *Mycobacterium tuberculosis* and *M. bovis* (Fig. 8.12) reveal that M. bovis us characterized by lower protein content (m/z 17, 34, 48, 92, 94, 108, 117; Table 7.1) and a higher carbohydrate content (m/z 32, 43, 60, 84, 102, 110, 126; Table 7.6).   The visual appearance of the mass pyrograms was often sufficient to allow identification to be made and nonlinear maps of the autoscaled distances obtained using the 40 most significant peaks (i.e., small inner variance, high outer variance) gave excellent recognition of 15 strains from 8 species.   Unknown strains were also classified well, although some association of closely related strains was evident [27].   The complete reassembly of the MS system, including a new ionization filament, retained the major characteristic features of each organism without correction.

Other attemps to characterize bacteria have involved probe pyrolysis techniques.   Mass pyrograms were characterized by products derived from phospholipids and ubiquinones [56].   Reproducibility over a 3-month period was established and significant ions ranged up to m/z 796.   Considerable differences were observed between

Staphylococcus epidermis (m/z *204*, 215 272, 314, 368) and *Staph.
aureus* (m/z 200, *214*, 282, 430, 538, 638, 697) while other species
examined (*Neisseria, Proteus, Pseudomonas, Salmonella*), all showed a
base peak at m/z 313. All but Pseudomonas additionally yielded an
ion at m/z 728 ($C_{49}H_{76}O_4$), believed to be the M + 2 ion of ubiquinone-8.
Mass pyrograms were nevertheless characteristic and species identification was readily achieved. Linear-programmed thermal degradation
MS has also shown promise [57,58]. Here specific fragments, again
mainly derived from lipids, produced at specific temperatures, were
detected by chemical ionization. Ion profiles enabled 10 organisms,
including *Pseudomonas, Escherichia,* and *Bacillus* species, to be identified using the presented algorithms.

One further technique for the study of organisms by pyrolysis
involves silylation of the more polar products. The chromatogram
from these derivatives is also characteristic of the organism [59].

Many reports of the successful differentiation of bacterial
strains have appeared. Table 8.5 records a selection of these
sources. Most organisms have been selected because of their importance in human illness and the identification of many closely related
strains of pathogenic bacteria is possible. Reports have also included comments upon the reproducibility of the technique and on the
instrumental, mathematical or biochemical parameters necessary for
useful results. These include a comprehensive study of the variables
which may influence the analysis [11], numerical methods [21,64], the
effect of column performance on long-term reproducibility [39] and
comparisons with literature data to establish the prospects for
interlaboratory comparisons [68]. Comment has also been made comparing the effectiveness of Py-GC and Py-MS in the differentiation of
micro-organisms [28]. It was held that Py-GC had a somewhat better
performance than Py-MS (~10%), but at a cost of longer analysis times.
No recent comparisons of the two techniques has appeared and until
such definitive work is reported, it would seem that either technique,
with appropriate controls, is adequate for taxonomic applications.
The choice of method is determined by the availability of apparatus
and projected through-put of samples.

TABLE 8.5 Pyrolysis Studies of Bacteria

ORGANISM	R E F E R E N C E	ORGANISM	R E F E R E N C E
Acholeplasma	11	Micrococcus	11,12,21,48,78
Arthrobacter	57	Microbacterium	21
Azotobacter	17	Moraxella	21
Bacillus	11,12,17,18,24,26,48,57	Mycobacterium	19,27,50,53,54,55,63,65,80,81
Bacteroides	60	Mycoplasma	11
Bordetella	78	Neisseria	52,56,66,67
Cellulomonas	17	Proteus	38,56,62,70
Citrobacter	11,38,62	Pseudomonas	17,21,38,49,56,57,62,73
Clostridium	17,20,61,63,70,72,73,78	Rhizobium	76
Enterobacter	38,62	Salmonella	41,47,56,77,78,79
Erivinia	57	Serratia	70
Escherichia	38,50,57,59,62,63,70,71	Shigella	50,63,71,78
Klebsiella	10,38,62,80	Staphylococcus	56,63
Lactobacillus	21	Streptococcus	9,10,28,39,50,63,68,69,70,75
Leptaspira	40,52	Streptomyces	12
Listeria	64,65	Vibrio	10,74

TABLE 8.6 Py-GC Conditions for Taxonomic Applications

| PYROLYSIS | COLUMN | STATIONARY PHASE | SUPPORT | TEMPERATURE | | | FLOW RATE |
				INITIAL	PROGRAM	FINAL	
610°C	50-150 m x 0.5 mm	Carbowax 20M	WCOT	5 °C 5 min	5 °C/min or 10 °C/min to 175 C then 2 C/min to 200 C	200°C 30 min	5-15 ml/min (He)
	2.1m x 4mm	8% Carbowax 20M 2% KOH	Chromosorb W AW (80-100)	50°C 5 min	2.5°C/min	200°C 10 min	50 ml/min (N$_2$)

Recommended conditions for the Py-GC of microorganisms are re-corded in Table 8.6.

8.4. FUNGI

Although fungi are important in ecology and as pathogens, as food spoilage organisms, as food and as sources of chemicals and drugs, the study of these organisms by pyrolysis methods has not as yet in-volved the range of techniques which have been applied to bacteria. In particular, although variation in pyrograms dependent upon the nitrogen content of mycelia has been observed [17], no full identif-ication of the components has appeared. Preliminary evidence sug-gests that significant differences between the pyrolysis products from fungi and those from bacteria may be evident [12]. Thus, of the 15 major products from *Aspergillus niger*, only five (acrylonitrile, acetonitrile, toluene, isovaleronitrile, pyrrole, and acetamide) are among the 15 major bacterial products. Furthermore, at least three intense components were absent from bacterial pyrograms. In view of the ability of fungi to produce specific metabolites, e.g., penicil-lin and griseofulvin from *Penicillium* and aflatoxin and aspergillic acid from *Aspergillus*, specific fragments may be expected. At present, no studies in this area have been initiated, although it has been reported that various unique peaks were obtained from strains of *Aspergillus flavus* [35]. Figure 8.13 compares the pyrograms derived from a bacterium, a yeast and a fungus. Considerable similarity is apparent and it would seem that the internal markers suggested for bacteria to ensure that pyrograms are suitable for interlaboratory comparisons (Table 8.4 and Table 8.6) would also be appropriate for fungal analysis.

Py-MS studies confirm this view. Thus, proteins (m/z 17, 34, 48, 67, 92, 94, 108, 117) and polysaccharides (m/z 32, 43, 60, 82, 96, 98, 110, 112, 126) are strongly represented. In addition, peaks characteristic of phospholipids (m/z 59, 71, 89) and amino sugars (m/z 95, 97, 109, 125, 137, 151) were evident [30]. Biochemical differences between whole cells and cell walls may be readily revealed.

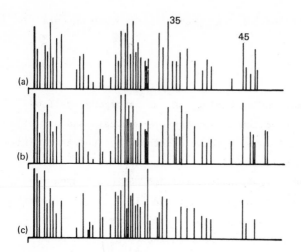

FIGURE 8.13 Bar-graphy pyrograms from a bacterium, a yeast, and a
fungus: (a) *Micrococcus luteus*, bacterium, (b) *Saccharomyces cerevisiae*, yeast, and (c) *Rhizopus nigricans*, fungus. (Adapted from
Ref. 11, from the *Journal of Chromatographic Science*, by permission
of Preston Publications, Inc.

Figure 8.14 compares the mass pyrograms from whole cells of *Endomyces
decipiens* with those from cell walls, defatted cells and the lipid
extract. The cell walls show a marked reduction in the intensity of
the phospholipid and protein derived components indicating the removal of cytoplasmic proteins and membrane phospholipid. In particular, the complete absence of peaks at m/z 48, 89, and 92 (protein
derived) is an indicator which may be used to monitor the purity of
the preparation. It should be noted that low-resolution MS cannot
resolve isomers and certain peaks may be associated with several
fragments. Acetamide ($M^{+\cdot}59$), derived from N-acetylglucosamine
(Fig. 8.4) is thus coincident with trimethylamine ($M^{+\cdot}59$), a product
from lecithin-type phospholipids. Such problems may be resolved by
Py-GC or by high-resolution MS, but chemical differences, too, may
be exploited. The defatted cells reveal a marked reduction in the
intensity of the m/z 59 peak, which becomes the major component of
the lipid residue. Peaks at m/z 58, 59, 71, 89, 147, 161, 163, and
177 are indicative of the presence of phosphatidylcholine and N,N-
dimethylphosphatidylethanolamine in this fraction.

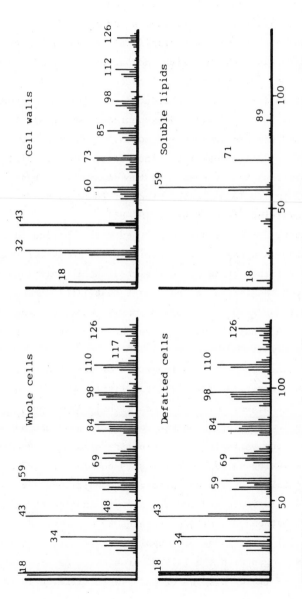

FIGURE 8.14 Mass pyrograms derived from the fungus *Endomyces decipiens*. (From Ref. 30.)

The direct chemical information available from Py-MS studies
has a further immediate use. Gross classification of fungi may be
achieved by comparisons of cell-wall carbohydrate composition. In
particular, the presence of chitin is important in the taxonomy of
yeasts and yeastlike fungi. This may be readily exposed by the presence of ions at m/z 95, 97, 109, 125 and at m/z 137, 151, 167.
Figure 8.15 shows that these are significant peaks with *Petriella*
setifera (10% chitin), whereas with *Saccharomyces cerevisiae* (1%
chitin) these ions are significantly reduced in intensity. The presence of other sugars may also be deduced from the mass pyrograms.
Thus, m/z 96 and 98 characterize deoxypentoses, while m/z 126 arises
from hexoses and m/z 128 from deoxyhexoses. Thus when deoxyhexoses
are present, the 128:126 peak ratio increases. In *Petriella*, this
ratio is significant and is caused by the deoxyhexose sugar rhamnose.
Although still in its infancy, the study of fungi by Py-MS will
surely be a rewarding field of study [82-85].

In contrast, Py-GC has been concerned with empirical differences
between pyrograms and has been used principally to differentiate between various strains, or to probe their taxonomic relationships.
The lack of peak identification has meant that little comparison
between the various reported pyrograms may be made, although the incorporation of internal standards represents substantial progress
[86].

The pyrograms reported from fungi are of variable quality (cf.
two studies on *Aspergillus* [34,35]), and little application of high-
resolution GC techniques have appeared [11,86]. The peak overlap
resulting from low-resolution packed columns is no doubt a significant
contribution to the variable reproducibility of some components [33].
Acceptable resolution on packed columns is possible [22,87,88] and
the use of controlled growth and pyrolysis conditions, combined with
multivariate statistical techniques (Chap. 5) enabled satisfactory
discrimination between *Penicillium* strains to be achieved.

Mathematical comparisons of fungal pyrograms has been a dominant
feature. These began with the initial studies of *Aspergillus* [35],
when four species (three strains of each) were differentiated correctly

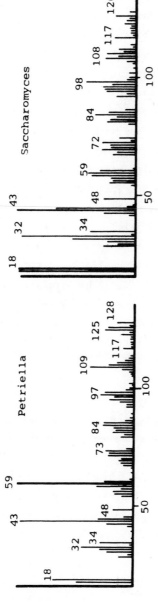

FIGURE 8.15 Mass pyrograms from *Petriella setifera* and *S. cerevisiae*. (From Ref. 30.)

and were extended to *Penicillium* species [89], when 27 pyrograms were reproduced and used to assess and quantify the variability of the strains examined.

In parallel with this work, the taxometric map procedure, based upon nearest-neighbor classifications, was under development. Dermatophytes were the organisms studied, and although culture media and colony age were controlled classifications were poor [32]. With improvements in GC, the resolution and reproducibility were enhanced and distances using all pyrogram peaks--but in contrast to bacteria not the 19 largest--enabled differentiation of organisms such as *Arthroderma, Microsporium* and *Nannizzia* to be achieved [90,91]. Further studies developed these techniques [86,92], but showed that chemotaxonomic relationships may not parallel those determined from traditional techniques. However, intensities were graded (0,1,2) so that much information may well be lost. Certainly, the use of normalized, autoscaled intensities gives excellent differentiation of Penicillum strains [22,87,88]. Work has also been reported on fungi attacking wheat [93,94] and eucalyptus roots [95]. Table 8.7 displays sources of studies of fungi by pyrolysis techniques.

Occasional reports have also appeared on the Py-GC of other organisms, but no comprehensive information is available. Thus, although satisfactory individual applications have been reported, insufficient data is available to assess their merits as standard techniques. Algae [59,97-99], plankton [59], plant seeds [99], and cockroaches [101] have been examined to date. Large samples and stopped flow techniques have perhaps aggravated the problems of comparability and reproducibility and it should be noted that sophisticated multivariate analysis of older data [102] cannot improve the reliability of the analysis. Biochemical interpretation of results have also appeared and a ligninlike polymer present in the cell walls of Staurastrum was postulated to be responsible for resistance to microbial attack [100].

TABLE 8.7 Pyrolysis Studies of Fungi

Organism	Reference	Organism	Reference
Arachniotus	92	Penicillium	22,23,87,88,89
Arthroderma	86,90	Petriella	30
Aspergillus	34,35	Phymatotrichum	93
Candida	17	Pisolithus	95
Chrysosporium	92	Ramichloridium	83
Endomyces	30	Rhizopus	11
Europhium	30,82	Saccharomyces	11,30,39
Gliocladium	92	Saprolegnia	82
Gymnoascus	92	Sporothrix	96
Leptodontium	83	Tilletia	94
Microsporum	90	Trichophyton	90,92
Nannizzia	86,90,91	Trichosporon	30
Ophiostoma	30,96		

8.5. PATHOLOGY

Pathological conditions are frequently accompanied by biochemical changes. These may range from the loss of control of cell growth, resulting in the tissue proliferation which characterizes cancers, to the presence of unusual metabolites due to malfunction of enzyme systems. The classical example here is phenylketonuria, an inborn metabolic disease, where phenylalanine cannot be converted into tyrosine. This condition is accompanied by urinary excretion of phenylalanine metabolites such as phenylpyruvic, phenylpropionic, phenyllactic, and phenylglycollic acids. The early diagnosis of such conditions is essential so that appropriate therapy may be begun before irreversible damage has been sustained.

The changing biochemical profile suggests that pyrolysis methods may be applicable in this area in two ways: the screening of samples

for abnormality using fingerprint comparisons with normal samples,
or to provide biochemical evidence on the nature of abnormalities
through peak identification techniques. The potential of analytical
pyrolysis for the recognition of pathological conditions has, indeed,
been demonstrated. However, although such results hold promise,
systematic study has only recently been undertaken and a rich field
of possible applications is still awaiting exploitation. In particular, the potential for early diagnosis must be established and reproducibility and specificity aspects should be studied in depth.
This is particularly important as the potential variability of samples
is immense. In addition, the identification of characteristic fragments has been described in only few instances. Thus, comparisons of
literature pyrograms are difficult to undertake and their biochemical
significance cannot be assessed. As with bacteria and fungi, the
identification of the internal marker peaks is recommended to impart
some measure of standardization to future reports (Tables 8.4 and
8.8).

The application of Py–GC to the detection of plant infections
was reported shortly after the initial studies on bacterial differentiation [103]. A comparison of the pyrograms from normal and infected
samples of barley, oats, wheat and tobacco was undertaken. The presence and distribution of barley yellow dwarf virus (oats), potato
virus Y (tobacco), yellow rust (barley and wheat), and mildew (wheat)
were readily established. It was suggested that the differences
caused by the fungal infections were due directly to the presence of
the fungus, rather than any biochemical modification of the host. It
would be appropriate, therefore, that extensions of this work should
involve the determination of the latent period, the period following
infection when gross fungal presence is not observed, to establish
whether early detection of the infection is possible. The presence
of fungi on Eucalyptus roots may also be demonstrated by Py–GC
techniques [95].

The application of Py–GC to mammalian cells has been described.
Normal cells gave characteristic and reproducible pyrograms and differences were apparent between normal and pathological cells.

TABLE 8.8 Proposed Structures for the Pyrolysis Products of Human Tissues (+, Present in: B, Brain; L, Liver; K, Kidney; S, Spleen)

PEAK	COMPONENT	B	L	K	S
1	Hex-1-ene	+	+	+	
2	Hept-1-ene	+	+	+	+
3	Prop-2-en-1-ol				+
4	Octane	+	+	+	
5	Oct-1-ene	+	+	+	
6	Nonane			+	
7	Non-1-ene	+	+	+	
8	Propenenitrile	+	+	+	
9	Decane...				
	2-Methylnonane			+	
10	Acetonitrile	+	+	+	
11	Toluene...	+	+	+	+
12	Cycloheptatriene	+	+	+	
	Dec-1-ene	+	+	+	
13	Undecane...	+		+	
	2-Methyldecane			+	+
14	Ethylbenzene			+	
15	isoValeronitrile	+	+	+	+
16	Undec-1-ene	+	+	+	+
17	Methyldecenes	+	+	+	
18	Pyridine		+		
19	Dodecane				
20	n-Propylbenzene				
21	Dodec-1-ene	+	+	+	
22	Undeca-1,10-diene	+	+	+	
23	4-Methylpentanenitrile				+
24	Cyclo-octatetraene		+	+	+
25	Tridecane	+	+	+	
26	Tridec-1-ene	+		+	
27	Dodeca-1,11(or 5,7)diene	+			
28	Tetradecane...	+		+	
29	2-Methyltridecane		+	+	
	Tetradec-1-ene	+		+	
30	2-Methyltetradecane...	+		+	
	Pentadecane				
31	Pyrrole				
32	Pentadecenes	+	+	+	
33	Trideca-1,12-diene	+		+	
34	2,6,11-Trimethyldodecane		+	+	
35	Hexadec-1-ene			+	
36	2-Hydroxymethylfuran		+	+	
37	Tetradeca-1,13-diene	+		+	
38	Heptadecane...		+	+	
	2-Methylhexadecane				
39	Heptadecene	+	+	+	
40	Hexanoic acid			+	+

Source: Ref. 107.

Pyrograms were presented which characterized normal and leukaemic
white blood cells and normal, cancer, and transformed cells [104].
Certain tissues (e.g., kidney) did not show any interspecies variation (mouse and rabbit). A possible rapid method for the prenatal
detection of genetic disorders using lyophilized amniotic fluid, collected by amniocentesis, has also been described [105]. Profiles
from normal samples were reproducible and comparable. Skin fibro-
blasts grown from patients with hereditable disorders were used to
model the amniotic fluid components expected from a genetic disease
carrier. Characteristic pyrograms were obtained for five genetic
disorders, including Huntington's Chorea, a progressively debilitat-
ing disease which is not usually manifested until the fourth decade
of life. This work was continued with the demonstration of differences between normal mouse cells, and those from lymphoblastic leuk-
aemia and from tumors [73]. Human cell work has shown that brain,
kidney, liver, and spleen may be differentiated on the basis of their
pyrograms, but that cystic fibrosis produces pyrograms very similar
to normal samples [106]. The presence of various other malignant
states, however, may be detected readily.

     The pyrolysis products obtained from human brain, liver, kidney
and spleen tissues are recorded in Table 8.8 [107]. The extent of
the lipid content is remarkable and contrasts markedly with bacterial
pyrograms. Mass fragmentography and a novel multivariate classification (linear warping function) were also described. Linear-programmed
thermal degradation MS of normal and malignant human white blood cells
has also been described [108]. Here, typical ion profiles (e.g., m/z
259, 367) are used and enable the clustering of malignant cells from
the same disease state and the partitioning of cells from different
disease states to be achieved.

     A study of haemoglobin has shown that normal adult haemoglobin
may be distinguished from that of the foetus (replacement of β-chains .
by γ-chains) and from that of patients suffering from sickle-cell
anemia, in which a glutamyl residue replaces that of valine at position 6 in the β-chains [109]. A Py-GC-MS study of hand epithelial
tissue has shown that the pyrogram is characterized by the products

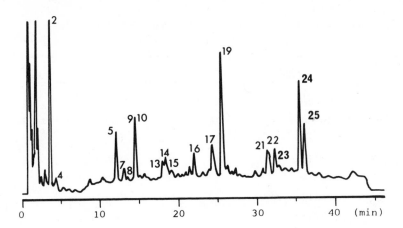

FIGURE 8.16 Pyrogram from lyophilized urine. (See Table 8.9 for component identification.) (From Ref. 120, by permission of Heyden & Son Ltd.)

from aromatic amino acids, with carbon dioxide, acetonitrile, butadiene, a C_5 nitrile, toluene, ethyl-benzene, styrene, phenol, and p-cresol being identified [110]. When combined with numerical comparisons of pyrograms and further component identification, these methods will surely be a powerful aid to rapid diagnosis.

The profiling of urine and other body fluids to reveal biochemical disorders is now highly advanced. GC-MS applications have been highly rewarding, with many new metabolites and disease states being recognized [111-118]. Potential exists for the application of analytical pyrolysis in this area, too, despite the power and ascension of the current methodology. Pyrograms from normal urine solids have proven to be characteristic and reproducible [119-121]. The low resolution pyrogram is displayed in Fig. 8.16 and the components are identified in Table 8.9. The major fragments are derived from amino acid pyrolysis, but are mediated considerably by the presence of large amounts of urea. Hippuric acid (benzoyl glycine) is the next most common component (~700 mgm excreted in 24 h). The pyrogram is characterized by expected fission processes, including decarboxylation to yield N-methylbenzamide, and an elimination reaction yields benzonitrile (Fig. 8.17). When excess urea is present, the profile

TABLE 8.9 Major Pyrolysis Products from Lyophilized Urine and Components

Peak	$M^{+ \cdot}$	Component	Peak	$M^{+ \cdot}$	Component
1	78	Benzene	15	59	Acetamide
2	92	Toluene	16	117	Phenylacetonitrile
3	104	Styrene	17	94	Phenol
4	79	Pyridine	18	182	Dibenzyl
5	67	Pyrrole	19	108	p-Cresol
6	106	Benzaldehyde	20	122	Ethylphenol
7	81	Methylpyrrole	21	117	Indole
8	81	Methylpyrrole	22	135	N-Methylbenzamide
9	103	Benzonitrile	23	131	3-Methylindole
10	95	Ethylpyrrole	24	121	Benzamide
11	121	Phenylethylamine	25	135	Phenylacetamide
12	120	Acetophenone	26	145	Ethylindole
13	93	Aniline	27	137	p-Aminoethylphenol
14	104	Nicotinonitrile	28	160	3-Aminoethylindole

Source: Ref. 120.

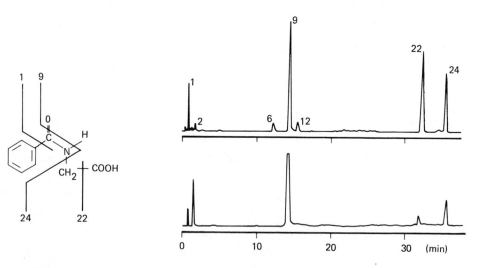

FIGURE 8.17 Pyrolysis of hippuric acid with and without added urea. (For component identification, see Table 8.8.) (From Ref. 120, by permission of Heyden & Son Ltd.)

changes dramatically and the elimination, no doubt base catalyzed by
the urea, is now the predominant mode of decomposition. The appearance of pyrograms from phenylalanine, tyrosine and tryptophan are
shown in Fig. 8.18, together with one from tryptophan in the presence
of urea to illustrate the profound change. Clearly, such variations
and their origin, e.g., effect of diet on the profile of excreted
nitrogenous compounds, must be appreciated and controlled for viable
applications to materialize. One preliminary study has appeared
[122] and it has been shown that Py-GC-MS profiles from urine may reveal the presence of the genetic disease methylmalonic aciduria.

Although such promise is rewarding, Py-GC probably has little
advantage over GC-MS techniques for easily handled samples such as
urine, although there are obvious advantages when whole cells or
macromolecular species are monitored. Py-MS, however, has the added
advantage of a very short analysis time. A throughput of one sample
per minute is possible, which compares very favorably indeed with a
typical GC-MS profile run which may take 1 h to complete. Such a
rate would enable a far more effective screening program to be operated than may presently be accomplished. Furthermore, the presence
of abnormal metabolites in many genetic disorders should be readily
detected by Py-MS, so that direct chemical interpretation, in addition to showing abnormality when compared to normal profiles, should
be possible. To date, no formal reports of this aspect of Py-MS have
appeared, but in view of the many biological tissues and fluids
(blood, urine, saliva, cerebrospinal fluid, lymph, bile, lachrymal
and sebaceous secretions, and biopsy material, particularly urine
and saliva [118,123] which are available through noninvasive sampling
techniques) such work will surely be a potent force in the early
diagnosis of genetic disorders and other pathological conditions.

Py-MS reports have been restricted to a study of blood (Py-FDMS)
which may be used to monitor inorganic cations (Table 8.10) [123],
and to the detection of polymer particles produced by the mechanical
wear of artificial joints [124]. In this last example, dry tissue
sections were subjected to probe pyrolysis. At low temperatures
(300°C) the mass pyrogram was characterized by fragments from amino

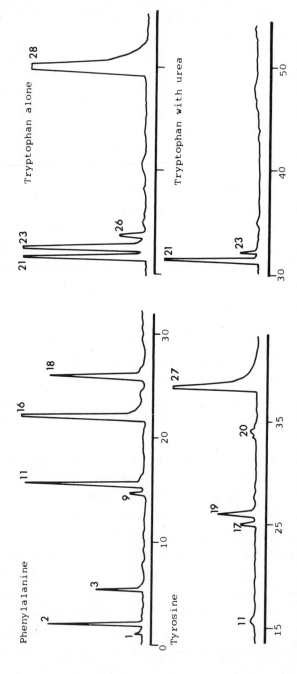

FIGURE 8.18 Pyrolysis of phenylalanine, tyrosine, and tryptophan. (For component identification, see Table 8.8.)

TABLE 8.10 Py-FDMS Products of Human Blood

m/z	Component	m/z	Component	m/z	Component
17	Ammonia	29	C_2H_5	43	C_3H_7
18	Water	30	Formaldehyde	44	Ethylene oxide
18	Ammonium	30	Ethane	44	Propane
19	Hydronium	39	^{39}K	84	C_4H_6NO
23	Sodium	40	^{40}Ca	85	^{85}Rb
27	C_2H_3	41	^{41}K	87	^{87}Rb
28	CO	43	CHNO	132	Cs
28	Ethylene	43	C_2H_5N		

Source: Ref. 123.

acids, but above 400°C ions from the polymer, released into the tissue by wear of the artificial joint, were significant (Figs. 4.7 to 4.9). A quantitative assessment of the tissue levels of the polymer was also undertaken. This was achieved by producing a calibration curve from known ratios of the joint polymer [polyethyleneterephthalate (PET)] and an internal standard [polypropyleneterephthalate (PPT)] with the ratio of the intensities of the peaks at m/z 193 (PET) and m/z 207 (PPT) being quantified. Addition of the internal standard to the joint tissue and subsequent pyrolysis enabled the amount of the polymer contaminant (17 μg of polymer in 400 μg of tissue) to be estimated. These results are important in view of the possibility of side reactions induced by wear particles, e.g., necrosis, granuloma, joint-loosening, and the sensitivity and specificity of the Py-MS approach are compelling advantages. The quantitative application of Py-MS techniques is also in its infancy, but with appropriate care and calibration, the areas of useful application in the biomedical field could appear prodigious.

REFERENCES

1. Hardie, J.M. and Bowden, G.H., Physiological Classification of Oral Viridans Streptococci, *J. Dent. Res.*, *55*(1976)166-176.

2. Sneath, P.H.A. and Sokal, R.R., *Numerical Taxonomy*, Freeman,
    San Francisco, 1973.

3. Clifford, H.T. and Stephenson, W., *An Introduction to Numerical
    Classification*, Academic Press, New York, 1975.

4. Pankhurst, R.J. (Ed.), *Biological Identification with Computers*,
    Academic Press, New York, 1975.

5. Wayne, L.G., Andrade, L., Froman, S., Kapler, W., Kubala, E.,
    Meissner, G. and Tsukamura, M., A Co-operative Numerical Analy-
    sis of Mycobacterium gastri, M. kansasii and M. marinum, *J. Gen.
    Microbiol.*, *109*(1978)319-327.

6. Johnston, H.H. and Newson, S.W.B. (Eds.), *Rapid Methods and
    Automation in Microbiology*, Learned Information (Europe) Ltd.,
    Oxford, 1976.

7. Heden, C-G. and Illeni, T. (Eds.), *New Approaches to the Iden-
    tification of Micro-organisms*, John Wiley, New York, 1975.

8. Mitruka, B.M., *GC Applications in Microbiology and Medicine*,
    John Wiley, New York, 1975.

9. Huis in't Veld, J.H.J., Meuzelaar, H.L.C. and Tom, A., Analysis
    of Streptococcal Cell Wall Fractions by Curie-point Py-GC, *Appl.
    Microbiol.*, *26*(1973)92-97.

10. Kistemaker, P.G., Meuzelaar, H.L.C. and Posthumus, M.A., Rapid
    and Automated Identification of Micro-organisms by Curie-point
    Pyrolysis Techniques. II. Fast Identification of Microbiologi-
    cal Samples by Curie-point Py-MS, in C-G. Heden and T. Illeni
    (Eds.), *New Approaches to the Identification of Micro-organisms*,
    John Wiley, New York, 1975, pp. 179-191.

11. Quinn, P.A., Development of High Resolution Py-GC for the Iden-
    tification of Micro-organisms, *J. Chromatogr. Sci.*, *12*(1974)
    796-806.

12. Medley, E.E., Simmonds, P.G. and Manatt, S.L., A Py-GC-MS Study
    of the Actinomycete Streptomyces longisporoflavus, *Biomed. Mass
    Spectrom.*, *2*(1975)261-265.

13. Schulten, H-R., Beckey, H.D., Meuzelaar, H.L.C. and Boerboom,
    A.J.H., High Resolution FIMS of Bacterial Pyrolysis Products,
    Analyt. Chem., *45*(1973)191-195.

14. Brown, M.R.W. and Hodges, N.A., Growth and Sporulation Charac-
    teristics of *Bacillus megaterium* under Different Conditions of
    Nutrient Limitation, *J. Pharm. Pharmacol.*, *26*(1974)217-227.

15. Engels, W., Kamps, M. and van Boven, C.P.A., Influence of Cul-
    tivation Conditions on the Production of Staphylocoagulase by
    Staphylococcus aureus 104, *J. Gen. Microbiol.*, *109*(1978)237-243.

16. Oyama, V.I., Use of GC for the Detection of Life on Mars,
    Nature, *200*(1963)1058-1059.

17. Oyama, V.I. and Carle, G.C., Py-GC Application to Life Detection and Chemotaxonomy, *J. Gas Chromatogr.*, 5(1967)151-154.

18. Oxborrow, G.S., Fields, N.D. and Puleo, J.R., Py-GC of the Genus Bacillus: Effect of Growth Media on Pyrochromatogram Reproducibility, *Appl. Environ. Microbiol.*, 33(1977)865-870.

19. Reiner, E., Beam, R.E. and Kubica, G.P., Py-GC Studies for the Classification of Mycobacteria, *Amer. Rev. Resp. Dis.*, 99(1969) 750-759.

20. Cone, R.D. and Lechowich, R.V., Differentiation of Clostridium botulinum types A, B and E by Py-GC, *Appl. Microbiol.*, 19(1970) 138-145.

21. MacFie, H.J.H., Gutteridge, C.S. and Norris, J.R., Use of Canonical Variates Analysis in Differentiation of Bacteria by Py-GC, *J. Gen. Microbiol.*, 104(1978)67-74.

22. Blomquist, G., Johansson, E. and Wold, S., Data Analysis of Py-GC by Means of SIMCA Pattern Recognition, *J. Anal. Appl. Pyrol.*, 1(1979)53-65.

23. Evans, C.G.T., Herbert, D. and Tempest, D.W., The Continuous Cultivation of Micro-organisms. II. Construction of a Chemostat, *Meth. Microbiol.*, 2(1970)277-328.

24. Oxborrow, G.S., Fields, N.D. and Puleo, J.R., Py-GC Studies of the Genus Bacillus. Effect of Growth Time on Pyrochromatogram Reproducibility, in C.E.R. Jones and C.A. Cramers (Eds.), *Analytical Pyrolysis*, Elsevier, Amsterdam, 1977, pp. 69-76.

25. Mitchison, J.M., *The Biology of the Cell Cycle*, Cambridge University Press, Cambridge, 1971, p. 25.

26. Oxborrow, G.S., Fields, N.D. and Puleo, J.R., Preparation of Pure Microbiological Samples for Py-GC Studies, *Appl. Environ. Microbiol.*, 32(1976)306-309.

27. Meuzelaar, H.L.C., Kistemaker, P.G., Eshuis, W. and Engel, H.W. B., Progress in Automated and Computerised Characterisation of Micro-organisms by Py-MS, in H.H. Johnston and S.W.B. Newsom (Eds.), *Rapid Methods and Automation in Microbiology*, Learned Information (Europe) Ltd., Oxford, 1976, pp. 225-230.

28. Meuzelaar, H.L.C., Kistemaker, P.G. and Tom, A., Rapid and Automated Identification of Micro-organisms by Curie-point Pyrolysis Techniques. I. Differentiation of Bacterial Strains by Fully Automated Curie-point Py-GC, in G.-G. Heden and T. Illeni (Eds.), *New Approaches to the Identification of Micro-organisms*, John Wiley, New York, 1975, pp. 165-178.

29. Fontanges, R., Blandenet, G. and Queignec, R., Difficultes d'Application de la Chromatogrie en phase Gazeuse a l'Identification des Bacteries, *Ann. Inst. Pasteur*, 112(1967)10-23.

30. Weijman, A.C.M., The Application of Curie-point Py-MS in Fungal
     Taxonomy, in C.E.R. Jones and C.A. Cramers (Eds.), *Analytical
     Pyrolysis*, Elsevier, Amsterdam, 1977, pp. 225-233.

31. Taylor, J.J., Ex vivo Determination of Potentially Virulent
     Sporothrix schenctii, *Mycopathologia, 58*(1976)107-114.

32. Sekhon, A.S. and Carmichael, J.W., Py-GC of Some Dermatophytes,
     Can. J. Microbiol., 18(1972)1593-1601.

33. Burns, D.T., Stretton, R.J. and Jayatilake, S.D.A.K., Py-GC as
     an Aid to the Identification of Penicillium Species, *J. Chroma-
     togr., 116*(1976)107-115.

34. Stretton, R.J., Campbell, M. and Burns, D.T., Py-GC as an Aid
     to the Identification of Aspergillus Species, *J. Chromatogr.,
     129*(1976)321-328.

35. Vincent, P.G. and Kulik, M.M., Py-GC of Fungi: Differentiation
     of Species and Strains of Several Members of the Aspergillus
     flavus Group, *Appl. Microbiol., 20*(1970)957-963.

36. Nichols, W.W. and Murphy, D.G. (Eds.), Differentiated Cells in
     Ageing Research, *Int. Rev. Cytol. (Suppl. 10),* 1979.

37. Nierlich, D.P., Regulation of Bacterial Growth, RNA and Protein
     Synthesis, *Ann. Rev. Microbiol., 32*(1978)393-432.

38. Needleman, M. and Stuchbery, P., The Identification of Micro-
     organisms by Py-GC, in C.E.R. Jones and C.A. Cramers (Eds.),
     Analytical Pyrolysis, Elsevier, Amsterdam, 1977, pp. 77-88.

39. Stack, M.V., Donoghue, H.D. and Tyler, J.E., Discrimination
     between Oral Streptococci by Py-GC, *Appl. Environ. Microbiol.,
     35*(1978)45-50.

40. Reiner, E., Hicks, J.J. and Sulzer, C.R., Leptospiral Taxonomy
     by Py-GC, *Can. J. Microbiol., 19*(1973)1203-1206.

41. Reiner, E., Hicks, J.J., Ball, M.M. and Martin, W.J., Rapid
     Characterisation of Salmonella Organisms by Means of Py-GC,
     Analyt. Chem., 44(1972)1058-1061.

42. Jefferson, S., Manual on Radiation Sterilisation of Medical and
     Biological Materials, Int. Atomic Energy Author., Vienna, 1973.

43. Simic, M.G., Radiation Chemistry of Amino Acids and Peptides in
     Aqueous Solutions, *J. Agric. Food Chem., 26*(1978)6-14.

44. Diehl, J.F., Adam, S., Delincee, H. and Jakubick, V., Radiolysis
     of Carbohydrate-containing Foodstuffs, *J. Agric. Food Chem., 26*
     (1978)15-20.

45. Ward, J.F., Chemical Consequences of Irradiating Nucleic Acids,
     J. Agric. Food Chem., 26(1978)25-28.

46. Nawar, W.W., Reaction Mechanisms in the Radiolysis of Fats: A
     Review, *J. Agric. Food Chem., 26*(1978)21-25.

47. Emswiler, B.S. and Kotula, A.W., Differentiation of Salmonella serotypes by Py-GC, *Appl. Environ. Microbiol.*, *35*(1978)97-104.

48. Simmonds, P.G., Whole Organisms Studied by Py-GC-MS: Significance for Extraterrestrial Life Detection Experiments, *Appl. Microbiol.*, *20*(1970)567-572.

49. Schulten, H-R., Beckey, H.D., Meuzelaar, H.L.C. and Boerboom, A.J.H., High Resolution FIMS of Bacterial Pyrolysis Products, *Analyt. Chem.*, *45*(1973)191-195.

50. Reiner, E., Identification of Bacterial Strains by Py-GC, *Nature*, *206*(1965)1272-1274.

51. Drucker, D.B., GC Chemotaxonomy, in J.R. Norris (Ed.), Methods in Microbiology, Academic Press, London, *9*(1976)51-125.

52. Meuzelaar, H.L.C. and Kistemaker, P.G., A Technique for Fast and Reproducible Fingerprinting of Bacteria by Py-MS, *Analyt. Chem.*, *45*(1973)587-590.

53. Reiner, E. and Kubica, G.P., Predictive Value of Py-GC in the Differentiation of Mycobacteria, *Amer. Rev. Resp. Dis.*, *99*(1969) 42-49.

54. Reiner, E., Hicks, J.J., Beam, R.E. and David, H.L., Recent Studies on Mycobacterial Differentiation by Means of Py-GC, *Amer. Rev. Resp. Dis.*, *104*(1971)656-660.

55. Wickman, K., Py-GC of Mycobacteria, *Acta. path. microbiol. Scand. (Section B Suppl.)*, *259*(1977)49-53.

56. Anhalt, J.P. and Fenselau, C., Identification of Bacteria using MS, *Analyt. Chem.*, *47*(1975)219-225.

57. Risby, T.H. and Yergey, A.L., Identification of Bacteria using Linear-programmed Thermal Degradation MS. The Preliminary Investigation, *J. Phys. Chem.*, *80*(1976)2839-2845.

58. Risby, T.H. and Yergey, A.L., Linear-programmed Thermal-degradation MS, *Analyt. Chem.*, *50*(1978)327A-334A.

59. Derenbach, J.B. and Ehrhardt, M., Polar Pyrolysis Products for a Sensitive Fingerprint Characterisation of Organisms by GC, *J. Chromatogr.*, *105*(1975)339-343.

60. Ericsson, I., Larsson, L. and Mardh, P-A., Py-GC of Microorganisms. Influence of Various Pyrolysis Parameters on the Yield of Volatile Organic Products, *Acta. path. microbiol. scand. (Sect. B. Suppl.)*, *259*(1977)43-47.

61. Larsson, L. and Mardh, P-A., Application of GC to Diagnosis of Micro-organisms and Infectious Diseases, *Acta. path. microbiol. scand. (Sect. B. Suppl.)*, *259*(1977)5-15.

62. Mitchell, A., Needleman, M. and Strafford, B., Identification of Micro-organisms by Py-GC and Py-MS, *Australian J. Pharm. Sci.*, *7*(1978)25-28.

63. Reiner, E., Studies on Differentiation of Micro-organisms by Py-GC, *J. Gas Chromatogr.*, *5*(1967)65–67.

64. Eshuis, W., Kistemaker, P.G. and Meuzelaar, H.L.C., Some Numerical Aspects of Reproducibility and Specificity, in C.E.R. Jones and C.A. Cramers (Eds.), *Analytical Pyrolysis*, Elsevier, Amsterdam, 1977, pp. 151–166.

65. Meuzelaar, H.L.C., Kistemaker, P.G., Eshuis, W. and Boerboom, A.J.H., Automated Py-MS: Application to the Differentiation of Micro-organisms, *Adv. Mass Spectrom.*, *7B*(1978)1452–1457.

66. Meuzelaar, H.L.C. and in't Veld, R.A., A Technique for Curie-point Py-GC of Complex Biological Samples, *J. Chromatogr. Sci.*, *10*(1972)213–216.

67. Haverkamp, J., Meuzelaar, H.L.C., Beuvery, E.C., Boonekamp, P.M. and Tiesjema, R.H., Characterisation of Neisseria meningitidis Capsular Polysaccharides Containing Sialic Acid by Py-MS, *Analyt. Biochem.*, *104*(1980)407–418.

68. Stack, M.V., Donoghue, H.D., Tyler, J.E. and Marshall, M., Comparison of Oral Streptococci by Py-GC, in C.E.R. Jones and C.A. Cramers (Eds.), *Analytical Pyrolysis*, Elsevier, Amsterdam, 1977, pp. 57–68.

69. Meuzelaar, H.L.C., Ficke, H.G. and den Harinck, H.C., Fully Automated Curie-point Py-GC, *J. Chromatogr. Sci.*, *13*(1975)12–17.

70. Gutteridge, C.S., MacFie, H.J.H. and Norris, J.R., Use of Principal Components Analysis for Displaying Variation between Pyrograms of Micro-organisms, *J. Anal. Appl. Pyrol.*, *1*(1979)67–76.

71. Reiner, E. and Ewing, W.H., Chemotaxonomic Studies of Some Gram Negative Bacteria by means of Py-GC, *Nature*, *217*(1968)191–194.

72. Reiner, E. and Bayer, F.L., Botulism. A Py-GC Study, *J. Chromatogr. Sci.*, *16*(1978)623–629.

73. Reiner, E., The Role of Py-GC in Biomedical Studies, in C.E.R. Jones and C.A. Cramers (Eds.), *Analytical Pyrolysis*, Elsevier, Amsterdam, 1977, pp. 49–56.

74. Haddadin, J.M., Stirland, R.M., Preston, N.W. and Collard, P., Identification of Vibrio Cholerae by Py-GC, *Appl. Microbiol.*, *45*(1973)40–43.

75. Drucker, D.B., Holmes, C., Stack, M.V., Donoghue, H.D., Tyler, J.E. and Marshal, M., Comparison of Two GC Techniques for Identifying Streptococci of Several Lancefield Groups, *J. Appl. Bacteriol.*, *39*(1975)viii.

76. Oleinikov, R.R., Murzakov, B.G., Shemakhanova, N.M. and Oreshkin, A.E., Competitive Analysis of the Effectiveness of Rhizobuim Strains from Root Nodules and the Roots of Leguminous Plants by Means of Py-GC, *Izvest. Akad. Nauk, SSSR. (Ser. Biolog.)*, *6* (1971)873–878.

77. Lambert, N.G., Petrow, S., Kasatiya, S.S. and Boulaine, M., Variation des Produits de Pyrolyse de Salmonella typhimurium au cours des Subcultures, *Ann. Microbiol.*, *127B*(1976)309-315.

78. Bzdega, J., Chromatograms of Pyrolysis Products as applied for Differentiation of Clostridium perfringens type A Strains, *Med. Doswiad. Mikrobiol.*, *29*(1977)79-83.

79. Menger, F.M., Epstein, G.A., Goldberg, D.A. and Reiner, E., Computer Matching of Pyrolysis Chromatograms of Pathogenic Micro-organisms, *Analyt. Chem.*, *44*(1972)423-424.

80. Quinn, P.A., Swanson, J. Meuzelaar, H.L.C. and Kistemaker, P.G., The Classification and Identification of Micro-organisms by Py-MS, in C.E.R. Jones and C.A. Cramers (Eds.), *Analytical Pyrolysis*, Elsevier, Amsterdam, 1977, pp. 408-409.

81. Etemadi, A.H., The Use of Py-GC and MS in the Study of the Structure of Mycolic Acids, *J. Gas Chromatogr.*, *5*(1967)447-456.

82. Weijman, A.C.M., Cell-wall Composition and Taxonomy of Cephaloascus fragrans and some Ophiostomataceae, *J. Microbiol. Serol.*, *42*(1976)315-324.

83. De Hoog, G.S., Rhinocladiella and Allied Genera, *Stud. Mycol.*, *15*(1977)1-140.

84. Weijman, A.C.M., Carbohydrate Composition and Taxonomy of the Genus Dipodascus, *J. Microbiol. Serol.*, *43*(1977)323-331.

85. Weijman, A.C.M. and Meuzelaar, H.L.C., Biochemical Contributions to the Taxonomic Status of Endogonaceae, *Can. J. Botany*, *57*(1979) 284-291.

86. Brosseau, J.D. and Carmichael, J.W., Py-GC Applied to a Study of Variation in Arthroderma tuberculatum, *Mycopathalogia*, *63* (1978)67-79.

87. Blomquist, G., Johansson, E., Soderstrom, B. and Wold, S., Reproducibility of Py-GC Analyses of the Mould Penicillium brevi-compactum, *J. Chromatogr.*, *173*(1979)7-17.

88. Blomquist, G., Johansson, E., Soderstrom, B. and Wold, S., Classification of Fungi by Means of Py-GC Pattern Recognition, *J. Chromatogr.*, *173*(1979)19-32.

89. Kulik, M.M. and Vincent, P.G., Py-GC of Fungi: Observations on Variability Among Nine Penicillium Species of the Section Asymmetrica, Subsection Fasciculata, *Mycopath. Mycol. Applicata*, *51* (1973)1-18.

90. Carmichael, J.W., Sekhon, A.S. and Sigler, A., Classification of Some Dermatophytes by Py-GC, *Can. J. Microbiol.*, *19*(1973) 403-407.

91. Sekhon, A.S. and Carmichael, J.W., Column Variation Affecting a Py-GC Study of Strain Variation in Two Species of Nannizzia, *Can. J. Microbiol.*, *19*(1973)409-411.

92. Sekhon, A.S. and Carmichael, J.W., Classification of Some Gym-
     noascaceae by Py-GC Using Added Marker Compounds, *Sabouraudia,*
     13(1975)83-88.

93. Gunasekeran, M. and Pifer, L., Physiological Studies on Phyma-
     totrichum omnivorum. VIII. Chemotaxonomic Studies, *Mycologia,*
     70(1978)1164-1172.

94. Gunasekaran, M., Weber, D.J. and Hess, W.M., Differentiation
     of Races of Tilletia Caries and T. foetida by Py-GC, *Mycologia,*
     71(1979)1066-1071.

95. Seviour, R.J., Chilvers, G.A. and Crow, W.D., Characterisation
     of Eucalypt Mycorrhizas by Py-GC, *New Phytol., 73*(1974)321-332.

96. Taylor, J.J., Ex Vivo Determination of Potentially Virulent
     Sporothrix schenckii, *Mycopathologia, 58*(1976)107-114.

97. Nichols, H.W., Anderson, D.J., Shaw, J.I. and Sommerfeld, M.R.,
     Py-GC Analysis of Chlorophycean and Rhodophycean Algae, *J.*
     Phycol., 4(1968)362-368.

98. Sprung, D.C. and Wujek, D.E., Chemotaxonomic Studies of Pleur-
     astrum Chodat by Means of Py-GC, *Phycologia, 10*(1971)251-254.

99. Hethelyi, I. and Tetenyi, P., Application of Py-GC in Medicinal
     Plant Research, *Herba Hungarica, 18*(1979)87-96.

100. Gunnison, D. and Alexander, M., Basis for the Resistance of
     Several Algae to Microbial Decomposition, *Appl. Microbiol., 29*
     (1975)729-738.

101. Hall, R.C. and Bennett, G.W., Py-GC of Several Cockroach
     Species, *J. Chromatogr. Sci., 11*(1973)439-443.

102. Garbary, D. and Mortimer, M., Use and Analysis of Py-GC in
     Algal Taxonomy, *Phycologia, 17*(1978)105-107.

103. Myers, A. and Watson, L., Rapid Diagnosis of Viral and Fungal
     Diseases in Plants by Py-GC, *Nature, 223*(1969)964-965.

104. Reiner, E. and Hicks, J.J., Differentiation of Normal and
     Pathological Cells by Py-GC, *Chromatographia, 5*(1972)525-528.

105. Reiner, E. and Hicks, J.J., Py-GC Studies on Amniotic Fluid:
     A Prospective Method for Confirming Inborn Errors in Foetuses,
     Chromatographia, 5(1972)529-531.

106. Reiner, E., Abbey, L.E. and Moran, T.F., Py-GC of Normal Human
     Cells and Amniotic Fluid, *J. Anal. Appl. Pyrol., 1*(1979)123-132.

107. Reiner, E., Abbey, L.E., Moran, T.F., Papamichalis, P. and
     Schafer, R.W., Characterisation of Normal Human Cells by Py-
     GC-MS, *Biomed. Mass Spectrom., 6*(1979)491-498.

108. Yergey, A.L., Risby, T.H. and Golomb, H.M., Monitoring Normal
     and Malignant Human White Blood Cells by the Use of LPTD-MS,
     Biomed. Mass Spectrom., 5(1978)47-51.

109. Bayer, F.L., Hopkins, J.J. and Menger, F.M., Py-GC of Biomedically Interesting Molecules, in C.E.R. Jones and C.A. Cramers (Eds.), *Analytical Pyrolysis*, Elsevier, Amsterdam, 1977, pp. 217-223.

110. Wuepper, J.L., Py-GC-MS of Intractable Materials, *Analyt. Chem.*, *51*(1979)997-1000.

111. Jellum, E., Profiling of Human Body Fluids in Healthy and Diseased States Using GC-MS with Special Reference to Organic Acids, *J. Chromatogr.*, *143*(1977)427-462.

112. Pierce, S.K., Gearhart, H.L. and Payne-Bose, D., The Analysis of Human Breath and Urine for Organic Components with Chromatographic and MS Techniques. A Review, *Talanta*, *24*(1977)473-481.

113. Issachar, D. and Yinon, J., Screening of Organic Acids in Urine by CIMS, *Biomed. Mass Spectrom.*, *6*(1979)47-56.

114. Krotoszynski, B., Gabriel, G. and O'Neill, H., Characterisation of Human Expired Air: A Promising Investigative and Diagnostic Technique, *J. Chromatogr. Sci.*, *15*(1977)239-244.

115. Jellum, E., GC-MS in the Study of Inborn Errors of Metabolism. An Overview, *Proc. Symp. Mass Spectrom. (Tubingen)*, *1*(1977) 146-164.

116. Spiteller, G., Profile, Ein Abbild des Stoffwechselgeschehens, *Proc. Symp. Mass Spectrom. (Tubingen)*, *1*(1977)165-179.

117. Lovett, A.M., Reid, N.M., Buckley, J.A., French, J.B. and Cameron, D.M., Real Time Analysis of Breath Using an Atmospheric Pressure Ionisation MS, *Biomed. Mass Spectrom.*, *6*(1979)91-97.

118. Horning, M.G., Brown, L., Nowlin, J., Lertrtanangkoon, K., Kellaway, P. and Zion, T.E., Use of Saliva in Therapeutic Drug Monitoring, *Clin, Chem.*, *23*(1977)157-164.

119. Irwin, W.J. and Slack, J.A., Identification of Formulated Drugs by Py-GC-MS, in C.E.R. Jones and C.A. Cramers (Eds.), *Analytical Pyrolysis*, Elsevier, Amsterdam, 1977, pp. 107-116.

120. Irwin, W.J. and Slack, J.A., The Identification of Ibuprofen and Analogues in Urine by Py-GC-MS, *Biomed. Mass Spectrom.*, *5* (1978)654-657.

121. Slack, J.A., Some Applications of Py-GC-MS to the Identification of Drugs and Micro-organisms, Ph.D. Thesis, University of Aston in Birmingham, 1977.

122. Roy, T.A., Application of Py-GC-MS to the Study of Metabolic Disorders, *Analyt. Lett.*, *B11*(1978)175-182.

123. Schulten, H-R., *FDMS and its Application in Biochemical Analysis. Methods in Biochemical Analysis,* D. Glick (Ed.), John Wiley, New York, 1977, pp. 313-448.

124. Buchhorn, G., Luderwald, I., Ringsdorf, H. and H-G. Willert, MS Detection of Polymer Particles in Capsule Tissue Surrounding Joint Endoprostheses of Polyethyleneterephthalate, *Proc. Symp. Mass Spectrom. (Tubingen)*, *1*(1977)319-329.

Chapter 9

Drugs and Forensic Science

9.1. DRUGS

The analytical pyrolysis of drugs has largely been limited to Py-GC
techniques which have proven to be suitable for structure elucidation, fingerprint identification, and quantitative analysis of various medicinal and natural products. Although many of the studies
reported previously dealt with compounds which were difficult to
analyze by GC and would now perhaps be more appropriately undertaken
by alternative methods such as HPLC, analytical pyrolysis may still
have a role to play in drug analysis. In particular, the technique
involves little sample preparation and may be used for pure compound,
formulated drug or to detect the active principle and metabolites in
body fluids. Further, it is possible to analyze a wide variety of
drugs using similar analytical systems. The simple molecular structure of many drug substances also means that pyrolysis pathways may
be readily elucidated and that chemical interpretation of the pyrogram is possible.

9.1.1. Sulfonamides

The antibacterial sulfonamides are derivatives of 4-aminobenzenesul-
fonamide. Usually a heterocyclic substituent is present on the sulfonamide N atom. The pyrolysis of these compounds proceeds in a very
simple manner with fission about the sulfonamido group resulting in
the extrusion of sulfur dioxide and the formation of two aromatic

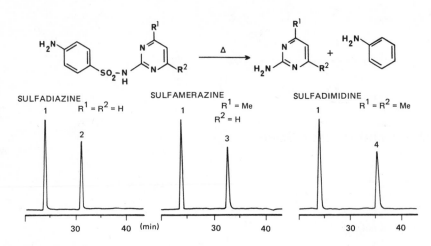

FIGURE 9.1 Pyrolysis of pyrimidine sulfonamides: (1) aniline; (2) 2-aminopyrimidine; (3) 2-amino-4-methylpyrimidine; (4) 2-amino-4,6-dimethylpyrimidine. (From Ref. 2.)

amines. One of these is always aniline, which therefore serves as an internal standard, and the other is derived from the heterocyclic fragment and characterizes the sulfonamide [1,2]. This process is shown for the pyrimidine sulfonamides sulfadiazine, sulfamerazine, and sulfadimidine in Fig. 9.1. Products such as diarylamines ($M-SO_2$) produced by extrusion of sulfur dioxide, which are common in electron-impact fragmentation reactions, are not evident in the pyrogram, although biphenyl, carbazole, and diphenylamine were present in the pyrolysis residues from benzenesulfonanilide. Each sulfonamide produces a unique pyrogram and the heterocyclic amino component is sufficient for identification to be made (Fig. 9.2).

Heterocyclic sulfonamides with methoxy substituents gave somewhat more complex pyrograms, due to the occurrence of transmethylation reactions. Thus, sulfametopyrazine and sulfamethoxypyridazine (no parent amine detected) gave N-methylalinine in addition to the usual products. The dimethoxy compound, sulfadimethoxine, also yielded N,N-dimethylaniline (Fig. 9.3). In contrast, sulfamethoxy-diazine shows the expected products only. Here, the methoxy substituent is in position 5 and only those α- or γ- to the ring N-atom

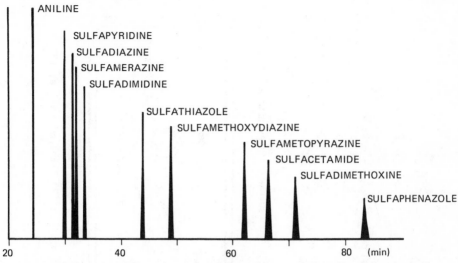

FIGURE 9.2 Identification of sulfonamides. Retention behavior of heterocyclic amine pyrolysis product. (From Ref. 1.)

are activated (Fig. 9.4). Sulfacetamide, a compound without a heterocyclic residue, also displayed a range of pyrolysis products, including acetanilide, a transacetylated product (Fig. 9.5).

The techniques developed for the identification of the pure drugs may also be applied to formulated systems, such as tablets. Many excipients, e.g., filler, binder, disintegrant, lubricant, wetting agent, are used so that pyrograms from the direct analysis of tablet drugs may contain peaks characterizing additives such as lactose, starch, and magnesium stearate. Sulfonamides are high-dosage medicaments and the excipients contribute but little to the pyrogram which thus is almost identical to that of the pure drug. Moreover, mixed sulfonamide preparations may be readily identified, and Fig. 9.6 shows that the pyrogram from sulfatriad tablets, which contain sulfadiazine, sulfamerazine, and sulfathiazole, is essentially due to the component sulfonamides.

Sulfonamides may also be detected in urine. Although the pyrogram from urine solids has many peaks, the appearance is highly reproducible (Fig. 8.16). Thus, drugs excreted in urine modify the

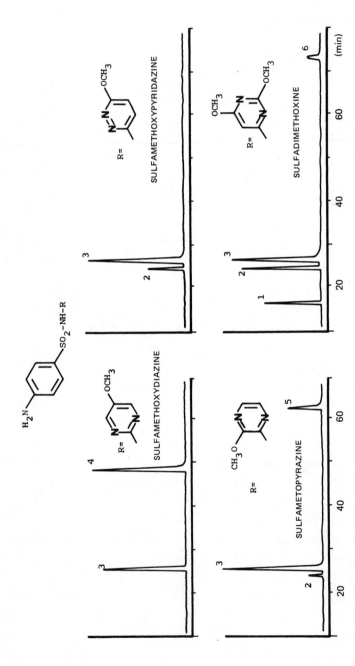

FIGURE 9.3 Pyrolysis of methoxy sulfonamides: (1) N,N-dimethylaniline; (2) N-methylaniline; (3) aniline; (4) to (6) parent amines.

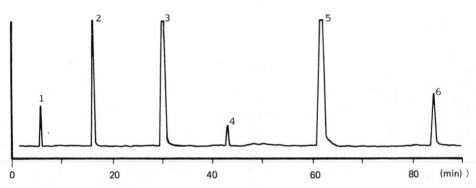

FIGURE 9.4   Transmethylation in pyrolysis of methoxy sulfonamides.

FIGURE 9.5   Pyrolysis of sulfacetamide: (1) CO_2, (2) SO_2, (3) acetonitrile, (4) acetic acid, (5) aniline, (6) acetanilide.

pyrogram and may be readily identified [3,4].   This is illustrated
in Fig. 9.7 for sulfatriad and the presence of the three sulfonamides
is clearly indicated.   One further feature is evident.   Sulfonamides
undergo N-acetylation as one metabolic conversion.   The presence of
the N-acetyl metabolite is revealed by the presence of acetanilide
in the pyrogram.   Other sulfonamides behave similarly.   The pyrolysis
of saliva as a noninvasive sampling technique has also been proposed,
but not reported in full [5].   Although these results were undertaken
by Py-GC techniques, the presence of unique fragments should enable
Py-MS to be readily applied to this area.

A study has also appeared on the Py-MS and Py-FDMS of polymeric
sulfonamides.   These polymers were synthesized as potential antibacterial agents and carriers for antitumor agents, and the structures
of monomeric and dimeric subunits were elucidated [6].   The hydrolysis and pyrolysis of pyrimidine sulfonamides have also been compared
[7] and a Py-GC study of saccharin and sodium saccharin has shown

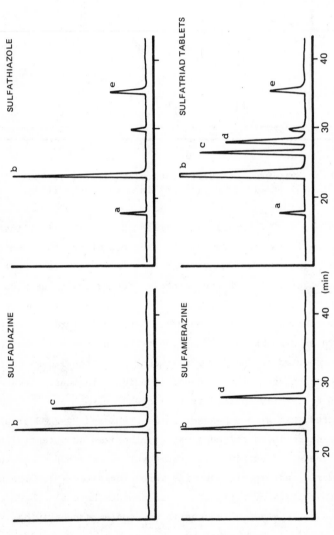

FIGURE 9.6 Identification of sulfatriad tablets (a, benzonitrile; b, aniline; c, 2-aminopyrimidine; d, 2-amino-4-methylpyrimidine; e, 2-aminothiazole). (Adapted from Ref. 2.)

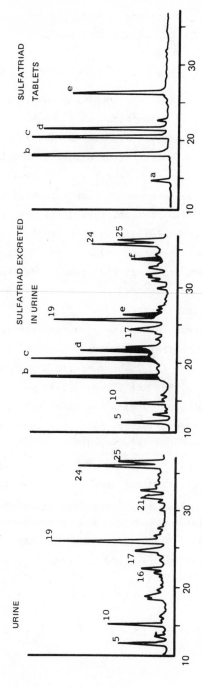

FIGURE 9.7 Identification of sulfatriad in urine (f, acetanilide; other components as in Figs. 8.16 and 9.9). (Adapted from Ref. 4.)

that characteristic pyrograms are produced, which enable the detection of the sweetening agent in soft drinks [8].

9.1.2. β-Lactam Antibiotics

Pyrolysis of penicillins and cephalosporins involves fragmentation
at the amido group and in most instances a characteristic fragment
is obtained [9]. These are illustrated in Fig. 9.8. When no characteristic fragment was apparent (for example, ampicillin and benzylpenicillin which both produce benzonitrile, phenylacetonitrile and
dibenzyl) significant differences in peak intensity were obtained.
In only one instance was differentiation difficult; the pyrograms of
benzylpenicillin and carbenicillin were quite similar, although
there was a significant increase in the yield of dibenzyl from the
latter drug. As carbenicillin is the carboxyl derivative of benzylpenicillin and decarboxylation will be a preferred pyrolysis pathway,
such behavior is not unexpected.

Quantitative analysis using the pyrograms was also possible.
Here, an intense, symmetrical peak formed by a unimolecular fragmentation process was chosen (i.e., benzonitrile, phenylacetonitrile,
or 1,3-dimethoxybenzene, but not dibenzyl); it was found that over a
range of 10 ng to 100 μg an excellent linearity of response was
achieved. The plots were, however, made on a log-log plot. It
would be interesting to see these results over a more conservative
analytical range with a rectilinear data plot to assess their potential for precise quantitative analysis. Under the conditions used
for the analysis, the pyrolysis of the drug deposited on the pyrolysis probe was essentially complete. This was determined as the
cracking severity (CS) [10], by measuring the intensity of the analytical peak for an initial (I_1) and a second (I_2) pyrolysis:

$$CS = \frac{100 I_1}{I_1 + I_2}$$

The cracking severity ranged from 97.9 to 100%, with a coefficient
of variation of 0.2 to 0.9%.

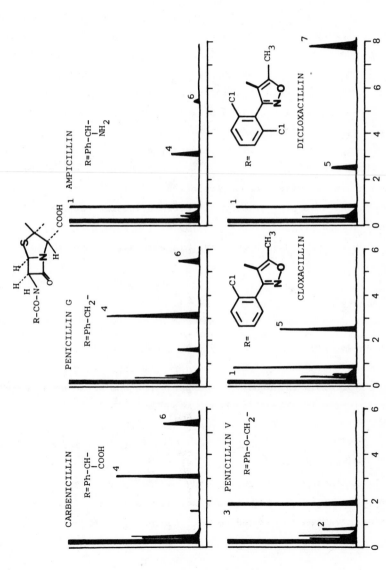

FIGURE 9.8 Py-GC of penicillins. Major products are (1) benzonitrile; (2) phenyl acetate; (3) phenol; (4) phenylacetonitrile; (5) 2-chlorobenzonitrile; (6) dibenzyl; and (7) 2,6-dichlorobenzonitrile. (Adapted from Ref. 9, from the *Journal of Chromatographic Science*, by permission of Preston Publications, Inc.)

Four penicillins and three cephalosporins were common to a further study which compared Py-GC and Py-MS of β-lactam antibiotics
[11]. Unique mass pyrograms were obtained for each of six penicillins and six cephalosporins and identification of fragments was
undertaken by Py-GC-MS. The differentiation of the penicillins using
the most intense peaks is displayed in Fig. 9.9. Py-MS appeared more
effective than Py-GC in characterizing these compounds. Thus, ampicillin and spectacillin, the 1,4-dihydro derivative, differ markedly
and benzylpenicillin and carbenicillin are now readily distinguishable. The 6-aminopenicillanic acid residue was characterized by
peaks at m/z 115 (4,5-dihydrodimethylthiazole), m/z 81 (1- and 2-
methyl pyrrole), m/z 125 (unidentified) and m/z 100, which was composed of several different ions in varying proportion. It is interesting to note the absence of dibenzyl in the Py-MS runs, due to the
suppression of bimolecular events by the very high vacuum pyrolysis
conditions designed into the analytical system [12].

Py-MS was also effective in differentiating the cephalosporins
and peaks at m/z 113 and either m/z 155 or 156 were characteristic
of the ring system. Mass pyrograms [11] were again superior to
pyrograms [9] in the identification of individual members.

9.1.3. Quaternary Ammonium Compounds

The predictable thermal fragmentation reactions of quaternary ammonium compounds which lead efficiently to products readily analyzed by
GC has meant that Py-GC of this group of compounds has been well explored. Typical is the study of the parasympathomimetic agent carpronium chloride [13,14]--3-methoxycarbonylpropyl-trimethylammonium
chloride. This drug has no UV chromophore so that the direct and
sensitive detection after HPLC analysis is difficult. However, analytical pyrolysis offers a rapid, sensitive and selective method for
the determination of this drug in body fluids. Mass fragmentography
was used to achieve the required sensitivity and specificity and
deuterio derivatives were used as the internal standards.

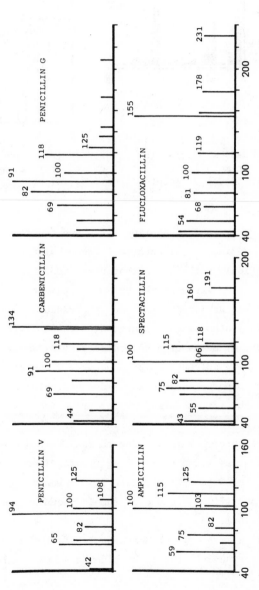

FIGURE 9.9 Py-MS of penicillins. Major ions (92, toluene; 94, phenol; 103, benzonitrile; 104, styrene; 106, benzaldehyde; 108, anisole; 117, phenylacetonitrile; 117, indole; 118, benzofuran; 121, 2-fluorobenzonitrile; 133, phenoxyacetonitrile; 134, 3H-benzo[b]furan-2-one; 155, 2-chloro-6-fluorobenzonitrile). (Adapted from Ref. 11.)

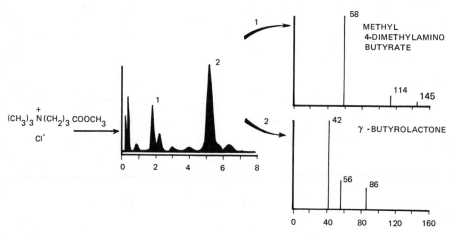

FIGURE 9.10 Py-GC-MS of carpronium chloride. (Other products include methyl chloride, trimethylamine, methyl chlorobutyrate, methyl butyrate and 1-methylpyrrolidone). (Adapted from Ref. 13, by permission of Heyden & Son Ltd.)

Pyrolysis of carpronium chloride was shown to yield two major products, methyl 4-dimethylaminobutyrate (demethylation) and butyro-lactone (elimination), with the lactone peak predominating (Fig. 9.10). Quantification was achieved by mass fragmentography using the butryolactone molecular ion ($M^{+\cdot}86$) and a tetradeutero derivative ($M^{+\cdot}90$) as internal standard. Extraction of the drug from blood and urine was achieved by means of a periodine complex which allowed ex-traction into organic solvents to be achieved. Ion-exchange treat-ment followed by evaporation gave a residue suitable for analysis. Although this method was satisfactory for urine samples, larger errors were encountered when pharmacokinetic profiles from blood levels were constructed. This was due to the limit of sensitivity of the assay which was found to be 20 ng (coefficient of variation = 5%) or 5 ng (CV = 15 to 20%). Figure 9.10 reveals the sensitivity of the assay is sacrificed because of the low abundance of the ana-lytical peak (m/z 86).

An alternative strategy depends upon the fact that the relative yields of butyrolactone and the dimethylamino compound are determined

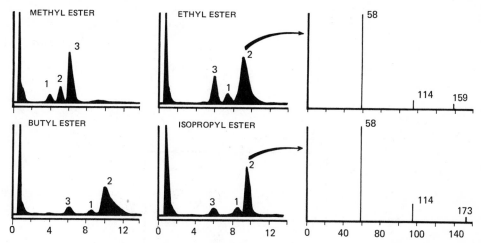

FIGURE 9.11 Py-GC-MS of carpronium chloride homologs (1, chlorobuty-rate; 2, dimethylaminobutyrate; 3, γ-butyrolactone). (Adapted from Ref. 14, by permission of Heyden & Son Ltd.)

by the nature of the ester function. This is illustrated in Fig. 9.11 which shows when the isopropyl ester is used, the amine is the predominant product. The presence of this species, rather than buty-rolactone, allows a fragment ion (m/z 114, $M-OC_3H_7$) to be chosen for the analytical peak [14]. As a greater proportion of the ion current is associated with this species, an enhanced sensitivity (5 ng ml^{-1}, CV = 10%), suitable for pharmocokinetic studies, is available. The $^2H_9$ derivative of carpronium chloride, used as the internal standard, was followed by means of the ion at m/z 120. The sample preparation now involves one further step with transesterification being achieved by treatment with isopropanol and HCl. The conversion rate was monitored by using the ethyl ester as internal standard, with measurements of the intensities of ions at m/z 145 (carpronium), 159 (ethyl homologue), and 173 (isopropyl derivative).

The detailed pyrolysis pathway was also examined by means of various deuterium labeled compounds, when it was found that an N to O transmethylation reaction was evident (Fig. 9.12) [15].

Other work on quaternary ammonium compounds has dealt with the identification of antiseptics [16], the estimation of cetrimide in

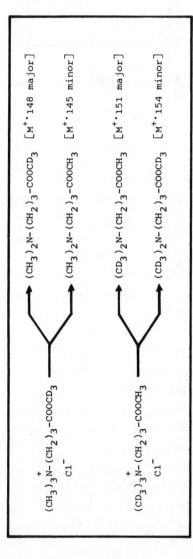

FIGURE 9.12 Pyrolytic transmethylation of carpronium chloride. (From Ref. 15.)

formulations [17], the detection and quantification of herbicides such as paraquat in urine [18], water [19], and in cannabis samples [20]. Tensioactive quaternary compounds have also been studied by Py-GC and characteristic products, expected from known thermal degradation pathways, are obtained [21].

9.1.4. Barbiturates

The pyrolysis of a series of sodium salts from barbiturates was one of the first biomedical applications of Py-GC [22,23]. Volatile hydrocarbons characterized the pyrogram which enabled individual drugs to be identified. The procedure was used to identify veronal (5,5-diethylbarbituric acid) extracted from the urine of a child who had consumed tablets of unknown drug content [23]. Complex rearrangements were evident for phenobarbitone yielded *o*-xylene. Further work enabled 27 medicinal barbiturates to be distinguished [24]. Comparisons of the pyrograms from the free acids, their sodium salts and mixtures with potassium carbonate was undertaken, and although the ionic state modified the pyrogram, similar product distributions were obtained in most cases. The use of isothermal conditions in these early studies meant that no single set of analytical conditions was ideal for all analyses. Analyses at column temperatures of 110°C and 150°C were reported and bar graphs enabled identification of the original drug to be rapidly undertaken (Fig. 9.13).

The major fragments were subsequently identified and, in contrast to earlier reports [22], nitriles were predominant peaks [25]. This discrepancy is probably due to different chromatographic conditions, enabling these components to be eluted.

A range of products was detected and pyrolysis of barbitone sodium yielded HCN, butane-, pentane-, hexane-, propene-, and 2-methylpropene nitriles, together with acetone, benzene, dimethylcarbonate, isobutyraldehyde, and water. These were of minor importance and the major component of the pyrogram was shown to be 2-ethylbutane nitrile. This fragment maintains the characteristic substituents at the 5 position of the barbituric acid ring system and thus is characteristic of the drug under test (Fig. 9.14). Amylobarbitone, yielding

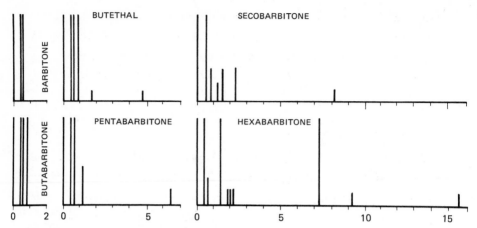

FIGURE 9.13 Pyrograms of some barbiturates. (Adapted from Ref. 24;
reprinted with permission from D. F. Nelson and P. L. Kirk, Identification of substituted barbituric acids by GC of their pyrolysis products, *Analyt. Chem. 34*:899-903. Copyright 1962 American Chemical
Society.)

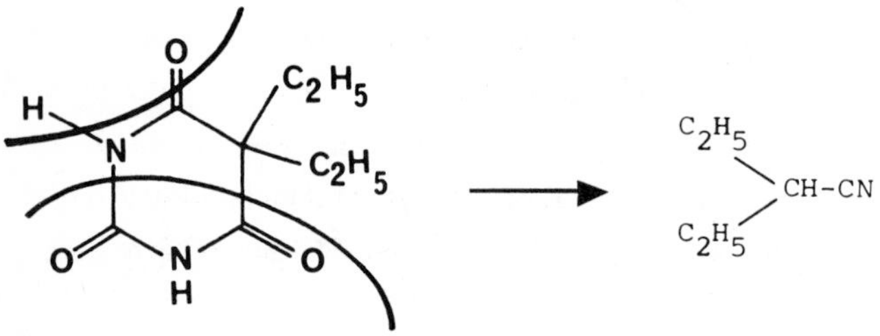

FIGURE 9.14 Pyrolysis of barbitone.

5-methyl-3-ethylhexane nitrile, and butobarbitone, yielding 2-ethyl-
hexanenitrile, were among other barbiturates successfully differentiated. The one exception to this behavior was the N-methylated derivative of barbitone. This drug yielded methyl isocyanate as a
major early fragment, together with a large unidentified component
of long retention time.

Quantitative analysis of barbiturates in pharmaceutical preparations has also been reported [26]. Vinbarbital and aprobarbital are
coformulated in tablets. Separation of the drugs from four inactive

substances by conversion to sodium salts and pyrolysis, enabled the proportions of the two compounds to be estimated. The analysis was complicated in that one peak from aprobarbital was coincident with the major peak from aprobarbital, and the intensity thus required correction for this contribution. Pyrograms from seconal sodium and phenobarbitone sodium have also been discussed [27].

9.1.5. Alkaloids

The identification of alkaloids for forensic purposes has received attention. Twenty-one morphine related alkaloids and other compounds (phenacetin, theobromine, caffeine) were differentiated using the intensities of the lower hydrocarbon fragments (C_1-C_4) and trimethylamine [28], although morphine and heroin were resolved by the production of acetic acid on pyrolysis of the latter drug. Comprehensive work has also appeared on the classification of indole alkaloids [29]. Products were identified by retention time comparisons and various classes of alkaloid were differentiated on the basis of their pyrograms (Fig. 9.15). Methoxy substituents on the indole ring also yielded characteristic fragments.

Quantitative estimation of atropine has been described, using two major pyrolysis fragments [30]; a scheme for the differentiation of narcotics and psychotropic agents (e.g., amphetamine) has also been proposed [31], and brief reports of several other alkaloids including cocaine [23], atropine [23], quinine [28], strychnine [28], colchicine [28], and ajmaline [29] have appeared. Most of these reports appeared in the developmental era of Py-GC and it is possible that with a more modern approach significant improvement on the earlier work could be made. Thus, although HPLC now dominates the analysis of many drug substances [32] and direct probe MS techniques have made significant contributions to the rapid identification of drugs of abuse [33], the application of Py-MS may yet have potential as a rapid screening method in this area.

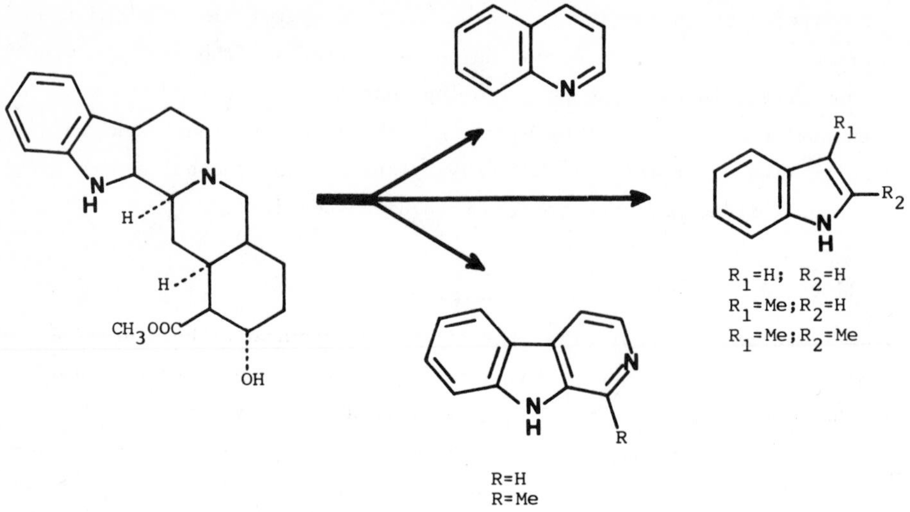

FIGURE 9.15 Pyrolysis of yohimbine. (From Ref. 29.)

9.1.6. Tobacco

The thermal reactions which accompany the smoking of tobacco and cannabis have been extensively studied [34-37]. Pyrolysis has proven to
be a useful process to model these transformations, and proposals for
the origin of many of the components of tobacco smoke have been made.
The formation of carcinogenic polynuclear aromatic hydrocarbons is a
particularly important aspect [38], and studies of polyenes have been
undertaken in attempts to reveal possible pathways [39]. Thus β-
carotene, when pyrolyzed at 300°C, yielded 1,2,3,4-tetrahydro-1,1,6-
trimethylnaphthalene (ionene), together with smaller amounts of toluene, *m*-xylene, and 2,6-dimethylnaphthalene. The effect of contaminants such as DDT [40,41], endosulfan [42], and paraquat [20] have
also been monitored.

The pyrolysis of tobacco alkaloids has also been described, and
products such as quinoline, isoquinoline, and nicotinonitrile are
derived from nornicotine and mysomine [43]. At lower temperatures,
alkaloids may be released from tobaccos without degradation [44].
Nicotine is quantitatively evolved at 100°C, but a temperature of
300°C proved optimum for the less volatile nornicotine. Quantification of the alkaloidal content, and that of neophytadiene from various tobaccos was readily achieved. Indeed, this Py-GC method was

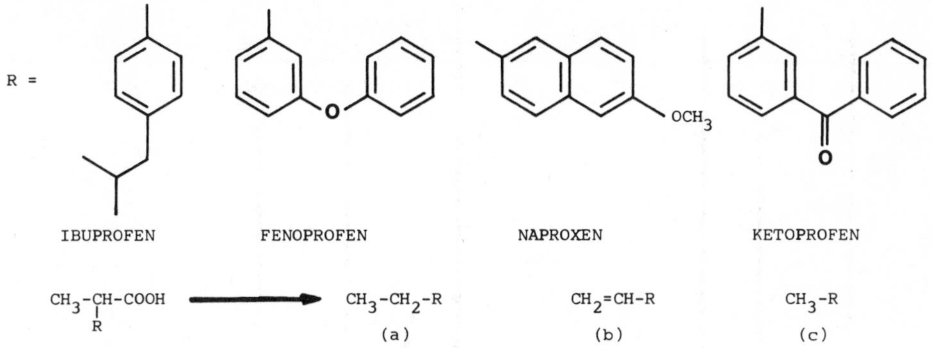

R =

IBUPROFEN   FENOPROFEN   NAPROXEN   KETOPROFEN

$CH_3-\underset{R}{CH}-COOH \longrightarrow CH_3-CH_2-R \qquad CH_2=CH-R \qquad CH_3-R$

(a)   (b)   (c)

FIGURE 9.16 Pyrolysis of anti-inflammatory propionic acids.

superior to the traditional GC assay of nornicotine, which involves considerable sample preparation.

9.1.7. Anti-inflammatory Propionic Acids

The pyrolysis of these compounds proceeds by decarboxylation to yield an ethyl derivative, or by notional decarboxylation and elimination to yield a vinyl compound (Fig. 9.16) [46,47]. In view of the pyrolytic elimination of formic acid from 4-phenylbutanoic acid to yield allylbenzene [45], it is probable that a similar one-step elimination process occurs here. Pyrograms are characterized by two principal peaks, and the retention time serves to identify the drug (Fig. 9.17). The ethyl derivative was the most intense fragment for all of the propionic acids, with the exception of Ibuprofen. Fenoprofen additionally yielded a methyl derivative.

The pyrograms were effectively unchanged when tablets were analyzed, and identification of the active principal could be rapidly achieved. Moreover, the presence of these drugs in the urine of patients undertaking therapy was readily revealed with the characteristic urine pyrogram (Fig. 8.16) being modified by that of the drug (Fig. 9.18). Ibuprofen undergoes substantial metabolic conversion, involving oxidation and conjugation. The pyrogram from the urine of patients undergoing treatment with this drug contains peaks due also to these metabolites. Although a more complex pyro-

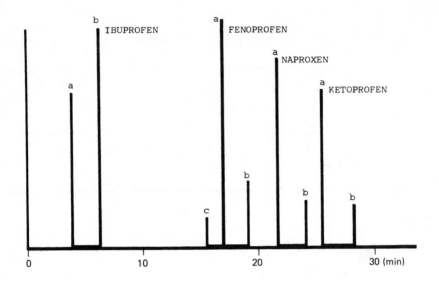

FIGURE 9.17 Identification of anti-inflammatory propionic acids:
(a) ethyl, (b) vinyl, and (c) methyl. (From Ref. 48, by permission
of Heyden & Son Ltd.)

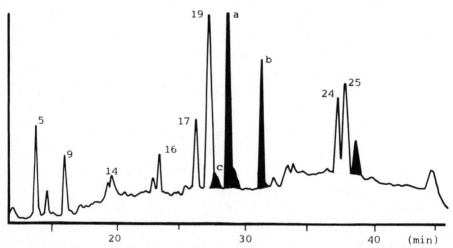

FIGURE 9.18 Detection of fenoprofen in urine. (Adapted from Ref.
48, by permission of Heyden & Son Ltd.)

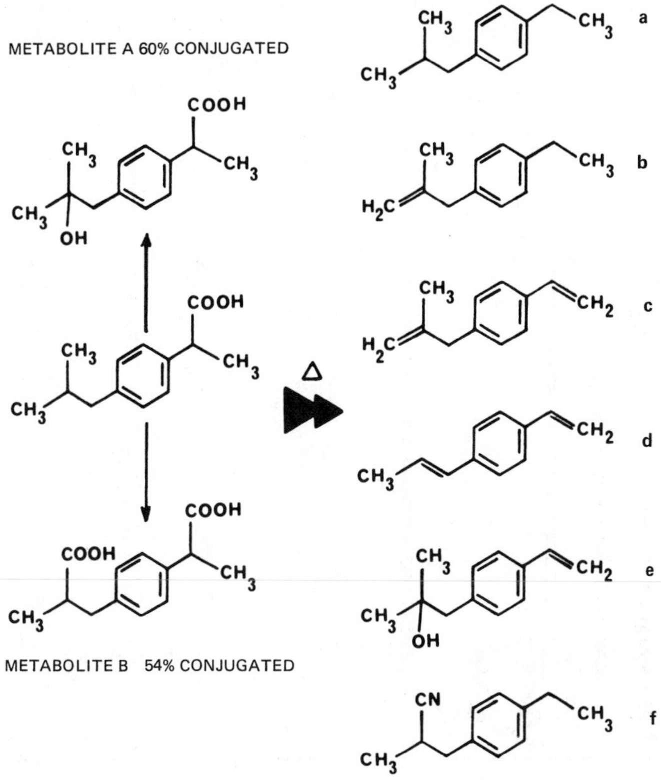

FIGURE 9.19 Pyrolysis of ibuprofen and metabolites in urine. (From Ref. 48, by permission of Heyden & Son Ltd.)

gram is obtained, the presence of Ibuprofen is shown clearly (Figs. 9.19 and 9.20). It should be noted that reaction with excess urea is observed, and the nitrile product is clear evidence for this [48].

9.1.8. Miscellaneous Drugs

Among other drugs studied by pyrolysis methods have been antibiotics such as lincomycin and analogues. Identification of pyrolysis products enabled the nature and position of substituents to be established [49]. Polyene antifungal compounds such as candicidin, levorin, and trichomycin have also been studied [50] by Py-GC, and the antibiotic lasalocid undergoes a retroaldol rearrangement to give quantitative yields of a volatile ketone [51].

Digitized pyrograms from a series of 19 phenothiazines (including chlorpromazine, fluphenazine, and trifluoperazine) have been

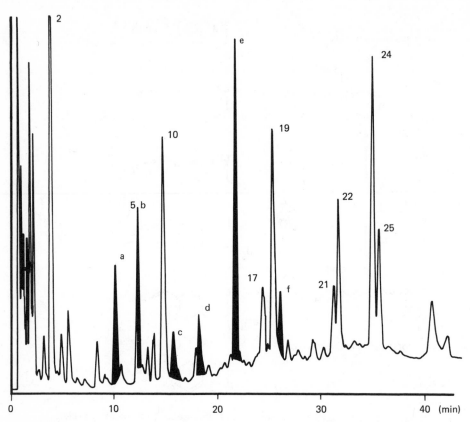

FIGURE 9.20 Pyrogram of ibuprofen and metabolites in urine. (Letters
refer to components in Fig. 9.19; numbers to those in Fig. 8.16.)
(Adapted from Ref. 48, by permission of Heyden & Son Ltd.)

reported and although unique fragments were generally not encountered,
the pyrograms were specific and allowed identification to be made [52].
Anthocyanins [53], dye substances [12,54], and the novel anticancer
drug cis-platin complexed with DNA [55] have also been studied, as
have cardenolides including digoxigenin and digitoxigenin [56].

In addition to this work, there are also many reports of drug
analysis which have involved thermal degradation, either in the GC
inlet port or in the MS source prior to detection [57,58]. These
range from derivatization procedures such as flash-methylation [59,
60], through cyclization [61] or decomposition [62–66] to produce
volatile fragments, to many unrecognized degradation reactions.

9.2. FORENSIC SCIENCE

Although the identification and quantification of drugs, e.g., drugs
of abuse and alcohol blood levels, are of forensic interest, the major
impact of analytical pyrolysis in forensic science has been in the
identification of polymeric substances. Individuality of samples is
all important and two situations may be encountered. The more common is a comparison of a sample found at the scene of the crime with
that found associated with a suspect. Such samples may consist of a
wide variety of materials, both natural and synthetic (e.g., blood,
skin, hair; wood, soil, paint, fibers, plastics, adhesives, waxes,
glass, metals), and the identicalness of the samples is used to establish the presence of the suspect at the scene of the crime. To
perform this analysis, it is essential that a full and reproducible
profile of the sample is available. It is not sufficient to show
that two samples are similar--they must be proven to be of identical
origin, if such is the case. Additionally, the amount of sample
available is usually restricted--perhaps a tiny paint flake--neces-
sitating a highly sensitive technique with efficient sample-handling
characteristics. Pyrolysis gas chromatography has been shown to be
a sensitive and discriminating tool in this respect [67-69], and with
appropriate care to ensure reproducibility [70-72] the technique has
proven to be of outstanding utility to the forensic scientist. Indeed, but few laboratories do not resort to this mode of analysis.

The applications of pyrolysis mass spectrometry are also under
investigation [73,74]. The substantial reduction in analysis time
is a major advantage of this technique. Custom-built mass spectrometers are available and have clear advantages [75], but as many
forensic science laboratories now have access to GC-MS systems for
drug analysis, the use of short empty columns to connect a pyrolyzer
to the mass spectrometer via the separator interface [76], or insertion probe pyrolyzers [77] are viable alternatives. Single mass pyrograms may be recorded at a precise time after pyrolysis [11], but
much better reproducibility is achieved if time-averaged mass pyrograms are collected. This requires data system facilities.

The second situation encountered is one in which a single sample--perhaps left at the scene of the crime--is available. Here, information on possible origins, to suggest certain fields of investigation, is sought. Typically, a paint flake from a hit-and-run accident is available and the make and probable year of manufacture of the vehicle is required. Analytical pyrolysis is a viable technique in this area, where a library of pyrograms is searched to enable fingerprint identification. The use of Py-GC-MS-DS is of significant value when more detailed information is required [78].

9.2.1. Paints

Automotive

Much work has been concerned with the differentiation and characterization of automobile finishes [79-82]. These frequently consist of several layers of different colors and composition. Visual classification using the layer structure and color is possible, and in combination with information on the chemical composition of the paint identification may be achieved. This is based typically upon the polymeric organic binders used in the paint, which are of various chemical types. Priming coats produced by dipping are frequently epoxy ester or short-chain alkyd-melamine formaldehyde resins. This is frequently covered by several surfacing layers composed of similar polymers, but also containing styrene. A high level of pigmentation (50 to 80%) is present. The top coat is highly variable and three major bases are used.

1. *Thermosetting alkyds.* These resins are polyesters which contain a variety of components such as pentaerythritol, glycerol, maleic acid, o-phthalic acid and unsaturated fatty acids. An amino resin (butylated melamine-formaldehyde) is incorporated to produce a thermosetting product which yields a hard, resistant surface.

2. *Thermoplastic acrylics.* These paints contain large amounts of poly(methyl methacrylate) together with smaller amounts of other acrylate polymers. Cellulose acetate butyrate and plasticizers such as phthalate esters or coconut oil alkyds are also incorporated into these finishes.

3. *Thermosetting acrylics*. These paints are p
    rived from co-polymers of styrene with acr)
    methylacrylates. A thermosetting product
    the incorporation of compounds such as hyd
    methacrylate. These cross-link with a but,
    formaldehyde resin on stoving, to provide a resistant
    finish.

Processes such as electropainting and the use of nonaqueous dis-
persions require different formulations so that the detailed composition of automotive finishes will also depend upon manufacturing
strategy.

The pyrograms from each type of finish is characteristic of the
monomeric composition and the type of resin is readily determined.
Furthermore, individual variation within each group is profound.
The result is that pyrograms from automotive finishes used by different manufacturers and by the same manufacturer on different models
may vary substantially. Changes which result from the availability
of different polymeric formulations or to process development make
the variability also depend upon time and may, therefore, provide
information on the age of the vehicle. Pyrograms illustrating
typical product distributions are depicted in Fig. 9.21 [79]. Pyrolysis of the thermosetting acrylic finish reveals methyl methacrylate
as the sole monomer, whereas the thermosetting acrylics contain
styrene and other acrylate and methacrylate esters. Acrolein and
butanol, derived from glycerol and butylated melamine-formaldehyde
resin respectively, were characteristic peaks from the alkyds, which
also yielded benzene, possibly from phthalate additives. The effect
of vehicle age and color is illustrated in Fig. 9.22 [71]. The pyrograms from a series of alkyd paints produced under standardized
conditions, together with the identification of the components, have
been reproduced. A scheme for identification, based upon the three
most intense peaks in the pyrogram was provided [72]. A quantitative
analysis of the components of thermosetting acrylic finishes has also
been undertaken [83].

Recent emphasis on the forensic analysis of paints has involved Py-MS applications, and both 70-eV EI [73] and CI [74] work
has been described. Both techniques show promise for the rapid

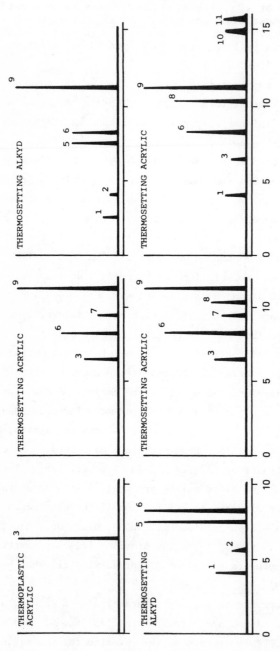

FIGURE 9.21 Major peaks in the pyrogram from automotive finishes (1, acrolein; 2, benzene; 3, methyl methacrylate; 4, ethyl methacrylate; 5, isobutanol; 6, butanol; 7, butyl acrylate; 8, butyl methacrylate; 9, styrene; 10, unidentified; 11, 2-ethylhexyl acrylate). (Adapted from Ref. 79.)

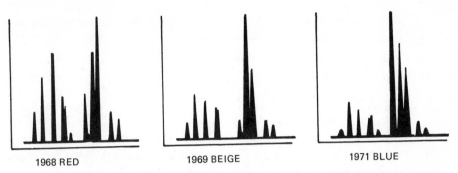

1968 RED   1969 BEIGE   1971 BLUE

FIGURE 9.22 Partial pyrograms showing variation of Volkswagen finish composition with age and color. (Adapted from Ref. 71.)

identification of samples, but in comparison with Py-GC, some information may be lost. This is particularly true of the CI analysis of
acrylates which essentially leads to very simple mass pyrograms consisting of the quasimolecular ions of the monomers. Thus, isomers
are not resolved. Nevertheless, an interlaboratory trial involving
seven different paint samples and an unknown was successfully accomplished, with each sample providing a characteristic mass pyrogram.
The different product transfer conditions which operate in Py-MS may
allow additional information to be extracted. Thus, benzene is the
sole indicator of benzoic and phthalic acid derivatives in the Py-GC
of alkyd paints. With Py-EIMS these are also revealed directly by
ions at m/z 105 and 104, respectively. The mass pyrograms further
contain ions characteristic of the monomer component and enable fingerprint identification and chemical interpretation to be undertaken.
Thus, ions at m/z 41 (methacrylates), 55 (acrylates), 51, 78, 103,
104 (styrene), and 31 and 56 (n-butanol) are characteristic ions
from acrylic paints. Alkyd paints display mass pyrograms with component ions at m/z 41 (methacrolein), 27 and 56 (acrolein), and 38,
50, 74, 76, 104, 148 (phthalic anhydride).

   The reproducibility of mass pyrograms has been studied [84],
and may depend upon sample type. Acrylic paint analyses are readily
repeatable, but under the analytical conditions (glass column interface) alkyds showed product segregation which caused poorer reproducibility. Bituminous paints differed markedly between manufacturers.

Less information is provided by pyrolytic methods when refinishing paints are involved. Many of these are based on nitrocellulose which does not provide sufficient structural variation for adequate differentiation of samples.

Household

The main source of household paint flakes are decorative gloss paints which are usually alkyd based. A large proportion (ca. 60%) of these are white and are difficult to distinguish from one another. Pyrolysis of paint flakes yields products such as acrolein, methacrolein, allyl alcohol, benzene, and a complex mixture of saturated and unsaturated hydrocarbons, depending on the detailed composition of the polymer [72]. It has been shown that the ratio of acrolein (from glycerol) to methacrolein (from penta-erythritol) is reproducible for identical samples (in contrast to more variable products such as acrolein:benzene) and varies between paints from different manufacturers. This ratio is thus of significance in establishing the origin of the paint. Provided that single layer samples were available, Py-GC was highly effective in the discrimination of paint fragments, and was very much more efficient than emission spectrometry in achieving this goal [68]. Pyrograms from acrylate and styrene-acrylate emulsion paints have also been briefly reported [85], and an extensive series of pyrograms from Japan Lacquer has been reproduced [86].

Art

The continuously varying formulations of modern paints is sufficient to allow individualization of samples in many cases. This property may also be used to establish the authenticity of works of art. The composition of the natural oil and resin media used before the advent of synthetic bases differs markedly from modern pigments (e.g., alkyds, acrylics, and phenol-formaldehyde resins). These differences may be readily revealed by analytical pyrolysis, which enables structural elucidation of the polymeric matrix to be achieved using minute

amounts of sample, an important consideration when valuable works of
art are at risk. The differences are particularly clear cut when
modern forgeries of the works of old masters are discovered. The
confirmation that supposed paintings by the Dutch artists Vermeer and
de Hoogh are modern forgeries by Van Meegeren has been achieved by
Py-GC analysis and illustrates the approach [87,88].

The pyrograms derived from a synthetic resin and the resin
binder solution found in Van Meegeren's studio are recorded in Fig.
9.23, together with a pyrogram from the white lead paint sample
taken from one of the disputed paintings. The resemblance between
these pyrograms is dramatic and component identification confirms
the organic component to be a phenol-formaldehyde resin. The phenolic pyrolysis products thus confirm the contemporary origin of the
paintings. The presence of a possible diluent is also revealed in
the pyrogram from the painting "Woman Reading Music." This painting
was taken from the artist's studio and presumably had not undergone
the thermal development which was used to provide the finished
paintings with a hard, craquelure texture. Pyrolysis of typical
bases used by the Dutch masters, e.g., gum mastic, sandarac, copal,
colophony, encrusted linseed oil, was also undertaken. Pyrograms
were characteristic and showed no correspondence to the Van Meegeren
polymer.

9.2.2. Fibers

Fibers are a further source of important forensic evidence, and both
Py-GC [67,72,78,89,90] and Py-MS [73,91] techniques have been used.
Various types of fibers (cellulosic, polyolefins, acrylic, polyester,
polyamide) may readily be distinguished from each other by Py-GC and
in many cases individualization may be demonstrated. Py-MS has also
proven to be an applicable technique which provides characteristic
mass pyrograms from a range of natural and synthetic fibers. Although reproducibility of some polymers (e.g., aromatic polyamides)
was poor, a large number of mass pyrograms were reproduced to
illustrate the validity of the technique. Thus, different nylon

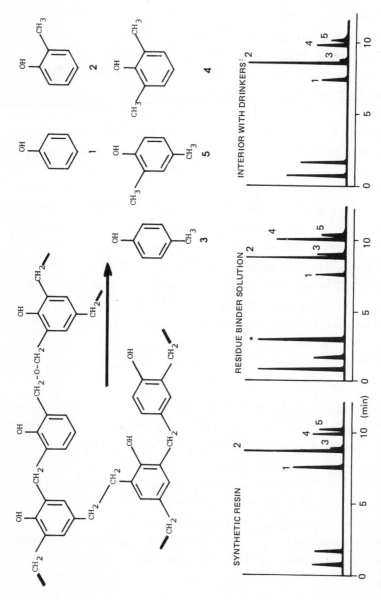

FIGURE 9.23 Pyrolysis of a phenol-formaldehyde resin binder used in the fake Vermeer and De Hoogh paintings (asterisk indicates peak possibly due to a thinner). (Adapted from Ref. 86.)

polymers gave characteristic profiles which enabled all but two
aromatic samples to be clearly differentiated. The remaining pair
could be distinguished by pyrolyzing at a lower temperature, which
yielded N,N-dimethylacetamide (possibly added during processing) from
one of the samples. Fibers based upon cellulose acetate, polyesters,
acrylic and modified acrylic polymers, polyesters, and polyolefins
were also studied. It should be noted that not all fibers yield
distinctive pyrograms. A pure acrylic fiber (polyacrylonitrile)
yielded a typical mass pyrogram with the major ion at m/z 66 ($^+$CH=CH-
CH_2CN) with other components at m/z 105 and 119. An acrylonitrile-
vinylpyrrolidone copolymer gave an identical mass pyrogram. Natural
fibers such as silk, wool, cotton, and rayon also yielded differen-
tiable pyrograms. Py-CIMS of acrylic fibers has also been reported
with significant difference observed between the mass pyrograms from
three types of fiber [74].

9.2.3. Hair

Although of considerable forensic interest, the individualization of
hair has not received extensive study by pyrolysis techniques. This,
in part, no doubt stems from the variability of individual hairs and
the expectation that the pyrograms from hair will be essentially
characterized by amino acid fragments from the major keratin component. Individual variation may well reside in minor or poorly resolved components of the pyrogram [92,93]. Such doubts appear well
founded and a high degree of similarity was found between pyrograms
from hair and from fingernails--both keratinaceous structures. Differentiation between hair and other proteins (casein, gelatin) was
clearer. It was suggested that a major cause of poor characterization was the failure to resolve a sufficient number of components,
particularly those of low intensity. A capillary column was indeed
used to illustrate its potential, but the methodology was not
developed further. A furnace pyrolyzer was also used in this work
[93]. It would be interesting to extend these results further,
through the introduction of more modern pyrolyzers, capillary columns,

component identification, and data handling, to see if a greater measure of individualization is now available. Py-MS studies, too, may have some contribution to make, and a mass pyrogram has been briefly reported [91].

9.2.4. Plastics

The forensic importance of plastics has been illustrated by reference to three criminal cases [67]. In addition, bar diagrams enabling characterization of a wide range of samples have been reported [94]. More recent work has involved the reproduction of pyrogram libraries [72,95] with the identification of components [72]. The effect of morphology on the Py-GC of PVC has also been reported [96].

9.2.5. Adhesives

Adhesive materials cover a wide range of structural types. They include animal and fish glues, celluloses, rubbers, poly(vinyl acetate) and copolymers, epoxy resins, and phenol-formaldehyde resins. Their forensic importance results from the tenacity with which they bind to samples, and identification may be of importance in the analysis of homemade bombs, or to establish ownership of disputed articles. Py-GC [97] and Py-MS [73] studies of a wide range of commercial adhesives have been reported. Of 179 adhesives monitored, all but 14 could be classified by Py-GC or IR. Py-MS also showed a high degree of discrimination and readily distinguished between urea-formaldehyde resins and animal or fish glues, which were difficult to classify by IR. Typical products are recorded in Table 9.1. An antique fish-skin glue has also been identified by Py-GC [98].

9.2.6. Chewing Gums

The pyrolysis of spent chewing gums displays peaks characteristic of synthetic elastomers. Typical products are butylene, derived from poly(isobutylene) and acetic acid from poly(vinyl acetate). Additives such as paraffin wax or candellila wax provide further

TABLE 9.1 Typical Pyrolysis Products from Adhesives

Peak	Component	Peak	Component
1	Butene	11	Dimethylsulfide
2	Butadiene	12	Butanol
3	Isoprene	13	Butyl acrylate
4	Acetaldehyde	14	Dipentene
5	Acetone	15	Styrene
6	Chloroprene	16	Acetic acid
7	Methyl acrylate	17	Pyrrole
8	Benzene	18	Indene
9	Vinylcyclohexene	19	Chloroprene dimer
10	Toluene		

Source: Ref. 97.

characterization [99]. Pyrograms showing the correspondence of chewing gums from the throat of a murdered child, the scene of the murder and the suspect have been reproduced, and show a high degree of correspondence. Unused gums yielded pyrograms dominated by carbohydrate fragments [99] and flavoring agents such as limonene and cinnamaldehyde [78].

9.2.7. Other Studies

Pyrolysis of putties yields mass pyrograms characterized by the low masses of a typical hydrocarbon ion series [73]. Little discrimination between samples was evident. Natural and artificial leathers may be readily distinguished [100], and pyrolysis methylation has been used to identify fats [101]. Here, the profile of fatty acid esters obtained from fats in the stomach of a murder victim was obtained. This was compared with profiles from the fats obtained from all fish and chip shops in the neighborhood, and enabled the shop supplying the meal to be identified. The use of acetone extracts from soil and wood samples have also been described briefly.

REFERENCES

1. Irwin, W.J. and Slack, J.A., The Identification of Formulated
    Drugs by Py-GC-MS, in C.E.R. Jones and C.A. Cramers (Eds.),
    Analytical Pyrolysis, Elsevier, Amsterdam, 1977, pp. 107-116.

2. Irwin, W.J. and Slack, J.A., Py-GC-MS Study of Medicinal Sul-
    phonamides, *J. Chromatogr., 139*(1977)364-369.

3. Slack, J.A. and Irwin, W.J., Analysis of Sulphonamides and their
    Metabolites by Py-GC-MS, *Proc. Analyt. Div. Chem. Soc., 14*(1977)
    249-251.

4. Irwin, W.J. and Slack, J.A., Detection of Sulphonamides in
    Urine by Py-GC-MS, *J. Chromatogr., 153*(1978)526-529.

5. Irwin, W.J., The Identification of White Uncoded Tablets by Py-
    GC-MS, *Fourth Int. Symp. Analyt. Appl. Pyrol.* (Abstracts),
    Budapest, 1979, p. 54.

6. Hofmann, V., Przybylski, M., Ringsdorf, H. and Ritter, H.,
    Pharmacologically Active Polymers. II. Polymeric Sulphonamides
    as Potential Antibacterials and Carriers for Antitumor Agents,
    Makromol. Chem., 177(1976)1791-1813.

7. Auterhoff, H. and Schmidt, U., Pyrolyse und Hydrolyse von Pyrim-
    idinsulfonamid-Derivaten, *Dt. Apoth. Ztg., 144*(1974)1581-1583.

8. Szinai, S.S. and Roy, T.A., Py-GC Identification of Food and
    Drug Ingredients. I. Saccharin, *J. Chromatogr. Sci., 14*(1976)
    327-330.

9. Roy, T.A. and Szinai, S.S., Py-GC Identification of Food and
    Drug Ingredients. II. Qualitative and Quantitative Analysis of
    Penicillins and Cephalosporins, *J. Chromatogr. Sci., 14*(1976)
    580-584.

10. Levy, E.J. and Paul, D.G., The Application of Controlled Partial
    Gas Phase Thermolytic Dissociation to the Identification of GC
    Effluents, *J. Gas Chromatogr., 5*(1967)136-145.

11. Muller, M.D., Seibl, J. and Simon, W., Characterisation of Some
    Penicillins and Cephalosporins by Py-MS, *Analyt. Chim. Acta,
    100*(1978)263-269.

12. Schmid, P.P. and Simon, W., A Technique for Curie-point Py-MS
    with a Knudsen Reactor, *Analyt. Chim. Acta, 89*(1977)1-8.

13. Ohya, K. and Sano, M., Analysis of Drugs by Pyrolysis. I. Se-
    lected Ion Monitoring Combined with a Pyrolysis Method for the
    Determination of Carpronium Chloride in Biological Samples,
    Biomed. Mass Spectrom., 4(1977)241-247.

14. Sano, M., Ohya, K. and Shintani, S., Analysis of Drugs by Py-
    rolysis. II. An Improved Method for the Determination of Car-
    pronium Chloride in Plasma by Selected Ion Monitoring, *Biomed.
    Mass Spectrom., 7*(1980)1-6.

15. Ohya, K., Yotsui, Y. and Sano, M., Intermolecular Methyl Trans-
    fer on Pyrolysis of Carpronium Chloride, *Org. Mass Spectrom.,*
    14(1979)61-65.

16. Bianchi, W., Boniforti, L. and di Domenico, A., Analysis of
    Some Quaternary Ammonium Compounds by GC-MS, in A. Frigerio and
    N. Castagnoli (Eds.), Mass Spectrometry in Biochemistry and
    Medicine, Raven Press, New York, 1974, pp. 183-196.

17. Choi, P., Criddle, W.J. and Thomas, J., Determination of Cetri-
    mide in Typical Pharmaceutical Preparations Using Py-GC, *Analyst,*
    104(1979)451-455.

18. Martuns, M.A. and Heyndrickx, A., Determination of Paraquat in
    Urine by Py-GC, *J. Pharm. Belg., 29*(1974)449-454.

19. Cannard, A.J. and Criddle, W.J., A Rapid Method for the Simul-
    taneous Determination of Paraquat and Diquat in Pond Waters by
    Py-GC, *Analyst, 100*(1975)848-853.

20. Beutler, J.A., Varano, A. and DerMarderosian, A., Pyrolysis
    Analysis of the Herbicide Paraquat on Cannabis by Coupled GC-IR,
    J. Forensic Sci., 24(1979)808-813.

21. Daradics, L., Studiul Structurii Substantelor Tensioactive Prin
    Py-GC. I. Tensidele Cationactive, *Rev. Chim., 29*(1978)268-270.

22. Janak, J., Identification of the Structure of Non-volatile Or-
    ganic Substances by GC of Pyrolytic Products, *Nature, 185*(1960)
    684-686.

23. Janak, J., Identification of Organic Substances by the GC Analy-
    sis of their Pyrolysis Products, in R.P.W. Scott (Ed.), *Gas
    Chromatography,* Butterworths, London, 1960, pp. 387-400.

24. Nelson, D.F. and Kirk, P.L., Identification of Substituted Bar-
    bituric Acids by GC of their Pyrolysis Products, *Analyt. Chem.,*
    34(1962)899-903.

25. Nelson, D.F. and Kirk, P.L., Identification of the Pyrolysates
    of Substituted Barbituric Acids by GC, *Analyt. Chem., 36*(1964)
    875-878.

26. Ericcson, I., Py-GC--A Potential Technique for Characterisation
    of Complex Organic Material, *Acta Path. Microbiol. Scand. (Sect.
    B. Suppl.), 259*(1977)37-42.

27. Barbour, W.M., Refinement of Pyrolysis Techniques for GC, *J.
    Gas Chromatogr., 3*(1965)228-231.

28. Kingston, C.R. and Kirk, P.L., Some Statistical Aspects of Py-
    GC in the Identification of Alkaloids, *Bull. Narcot., 17*(1965)
    19-25.

29. van Binst, G., Dewaersegger, L. and Martin, R.H., Application
    de la GC a l'Etude des Produits de Pyrolyse D'alkaloides In-
    dolique. II. Interpretation des Pyrogrammes et Discussion,
    J. Chromatogr., 25(1966)15-28.

30. Stanford, F.G., An Improved Method of Pyrolysis in GC, *Analyst,* *90*(1965)266-269.

31. Dimitrov, C.R., Petsev, N.D. and Daskalov, R.M., Identification of Narcotics and Psychotropics through Py-GC, *Compt. Rend. Acad. Bulg. Sci., 31*(1978)1027-1030.

32. Li Wan Po, A. and Irwin, W.J., High-performance Liquid Chromatography. Techniques and Applications, *J. Clin. Hosp. Pharm., 5* (1980)107-144.

33. Heller, S.R. and Milne, G.W.A., EPA-NIM MS Data Base, Nat. Stand. Ref. Data Ser., Nat. Bur. Stand. (1978).

34. Schmeltz, I. (Ed.), *The Chemistry of Tobacco and Tobacco Smoke*, Plenum Press, New York, 1972.

35. Schmeltz, I. and Hoffman, D., Nitrogen-containing Compounds in Tobacco and Tobacco Smoke, *Chem. Rev., 77*(1977)295-311.

36. Brown, E.V. and Ahmad, I., Alkaloids of Cigarette Smoke Condensate, *Phytochem., 11*(1972)3485-3490.

37. Agurell, S. and Leander, K., Stability, Transfer and Absorption of Cannabinoid Constituents of Cannabis (Hashish) During Smoking, *Acta Pharm. Suecica, 8*(1971)391-402.

38. Severson, R.F., Schlotzhauer, W.S., Arrendale, R.F., Snook, M. E. and Higman, H.C., Correlation of Polynuclear Aromatic Hydrocarbon Formation Between Pyrolysis and Smoking, *Beitr. Tabakforsch., 9*(1977)23-37.

39. Edmunds, F.S. and Johnstone, R.A., Constituents of Cigarette Smoke. Park IX. The Pyrolysis of Polyenes and the Formation of Aromatic Hydrocarbons, *J. Chem. Soc., 1965,* 2892-2897.

40. Chopra, N.M. and Osborne, N.B., Systematic Studies on the Breakdown of p,p'-DDT in Tobacco Smokes. II. Isolation and Identification of Degradation Products from the Pyrolysis of p,p'-DDT in a Nitrogen Atmosphere, *Analyt. Chem., 43*(1971)849-853.

41. Chopra, N.M. and Domanski, J.J., Systematic Studies on the Breakdown of p,p'-DDT in Tobacco Smokes. III. Isolation and Identification of the Non-volatile Degradation Products of p,p'-DDT in p,p'-DDT-treated Tobacco Smokes, *Beitr. Tabakforsch,* *6*(1972)139-143.

42. Chopra, N.M., Campbell, B.S. and Hurlay, J.C., Systematic Studies in the Breakdown of Endosulfan in Tobacco Smokes: Isolation and Identification of the Degradative Products from the Pyrolysis of Endosulfan I in a Nitrogen Atmosphere, *J. Agric. Food Chem., 26*(1978)255-258.

43. Balasubrahmanyam, S.N. and Quin, L.D., Pyrolytic Degradation of Nornicotine and Myosmine, *Tob. Sci., 6*(1962)133-136.

44. Rosa, N., Py-GC Estimation of Tobacco Alkaloids and Neophytadiene, *J. Chromatogr., 171*(1979)419-423.

45. Posthumus, M.A., Nibbering, N.M.M. and Boerboom, A.J.H., A Comparative Study of the Pyrolytic and EI-induced Fragmentations of 4-Phenylbutanoic Acid and Some Analogues, *Org. Mass Spectrom., 11*(1976)907-919.

46. Slack, J.A. and Irwin, W.J., Role of Py-GC-MS in the Analysis of Drugs, *Proc. Analyt. Div. Chem. Soc., 14*(1977)215-217.

47. Irwin, W.J. and Slack, J.A., The Use of Py-GC-MS as a Screening Technique for Drugs and their Metabolites, *Proc. Symp. Mass Spectrom. (Tubingen), 1*(1977)369-378.

48. Irwin, W.J. and Slack, J.A., The Identification of Ibuprofen and Analogues in Urine by Py-GC-MS, *Biomed. Mass Spectrom., 5* (1978)654-657.

49. Brodasky, T.F., Py-GC in the Differentiation and Characterisation of Antibiotics, *J. Gas Chromatogr., 5*(1967)311-318.

50. Burrows, H.J. and Calam, D.H., Py-GC of Polyene Antifungal Antibiotics: The Nature of Candicidin, Levorin and Trichomycin, *J. Chromatogr., 53*(1970)566-571.

51. Westley, J.W., Evans, Jr. R.H. and Stempel, A., GC Determination of Antibiotics X-537A, Lasalocid, *Analyt. Biochem., 59*(1974)574-582.

52. Fontan, C.R., Jain, N.C. and Kirk, P.L., Identification of the Phenothiazines by GC of their Pyrolysis Products, *Mikrochim. Acta, 1964*, 326-332.

53. Lanzarini, G., Morselli, L., Pifferi, P.G. and Giumanini, A.G., Py-GC-MS for the Identification of Anthocyanins, *J. Chromatogr., 130*(1977)261-266.

54. Schmid, J.P., Schmid, P.P. and Simon, W., Instrumentation for Curie-point Py-GC Using High-resolution Glass Open-Tubular Columns, *Chromatographia, 9*(1976)597-600.

55. Jankowski, K., Macquet, J.P. and Butout, J.L., MS Study of DNA-cis-Platin Complexes, *Biochem. Biophys. Res. Communs., 92* (1980)68-74.

56. Reiner, E., The Role of Py-GC in Biomedical Studies, in C.E.R. Jones and C.A. Cramers (Eds.), *Analytical Pyrolysis*, Elsevier, Amsterdam, 1977, pp. 49-56.

57. Nicholson, J.D., Derivative Formation in the Quantitative GC Analysis of Pharmaceuticals, *Analyst, 103*(1978)1-28, 193-222.

58. van den Heuvel, W.J.A. and Zaccei, A.G., GC in Drug Analysis, *Adv. Chromatogr., 14*(1976)199-263.

59. Kralovsky, J. and Matousek, P., GC of Aromatic Acids after Pyrolysis of their Trimethylphenylammonium Salts, *J. Chromatogr., 147*(1978)404-407.

60. Kossa, W.C., MacGee, J., Ramachandran, S. and Webber, A.J., Pyrolytic Methylation-GC: A Short Review, *J. Chromatogr. Sci., 17* (1979)177-187.

61. Wickramsinghe, J.A.F. and Shaw, S.R., GC Behaviour of Buformin Hydrochloride, Phenformin Hydrochloride and Phenylbiguanide. The Pyrolytic Formation of Substituted 2,4,6-Triamino-1,3,5-Triazenines from Biguanides, *J. Chromatogr., 71*(1972)265-273.

62. Wickramsinghe, J.A.F. and Shaw, S.R., GC Determination of Tolazamide in Plasma, *J. Pharm. Sci., 60*(1971)1669-1672.

63. Winkler, V.W., Regnier, F.E., Yoder, J.M. and Macy, L.R., GLC Analysis of Menadione Bisulfite Addition Compounds by On-Column Pyrolysis to Menadione, *J. Pharm. Sci., 61*(1972)1462-1465.

64. Levitt, M.J., Josimovich, J.B. and Broskin, K.D., Analysis of Prostaglandins by ECD GC. I. Thermal Decomposition of Heptofluorobutyrate Methyl Esters of F1_α and F2_α, *Prostaglandins, 1* (1972)121-131.

65. Dell, D., Boreham, D.R. and Martin, B.K., Estimation of 4-Butoxyphenylacetohydroxamic Acid Utilising the Lossen Rearrangement, *J. Pharm. Sci., 60*(1971)1368-1370.

66. Radecka, C. and Nigam, I.C., Reaction GC. III. Recognition of Tropane Structure in Alkaloids, *J. Pharm. Sci., 56*(1967)1608-1611.

67. Wheals, B.B. and Noble, W., Forensic Applications of Py-GC, *Chromatographia, 5*(1972)553-557.

68. Wheals, B.B., Forensic Applications of Analytical Pyrolysis Techniques, in C.E.R. Jones and C.A. Cramers (Eds.), *Analytical Pyrolysis*, Elsevier, Amsterdam, 1977, pp. 89-97.

69. Kirk, P.L., Identification by Means of Pyrolysis Products, *J. Gas Chromatogr., 5*(1967)11-14.

70. May, R.W., Pearson, E.F., Porter, J. and Scothern, M.D., A Reproducible Py-GC System for the Analysis of Paints and Plastics, *Analyst, 98*(1973)364-371.

71. Levy, E.J., The Analysis of Automotive Paints by Py-GC, in C.E.R. Jones and C.A. Cramers (Eds.), *Analytical Pyrolysis,* Elsevier, Amsterdam, 1977, pp. 319-335.

72. May, R.W., Pearson, E.F. and Scothern, D., *Pyrolysis-Gas Chromatography*, Chemical Society, London, 1977.

73. Hughes, J.C., Wheals, B.B. and Whitehouse M.J., Py-MS. A Technique of Forensic Potential, *Forensic Sci., 10*(1977)217-228.

74. Saferstein, R. and Manura, J.J., Py-MS--A New Forensic Science Technique, *J. Forensic Sci., 22*(1977)748-756.

75. Extranuclear Laboratories, Inc., Pittsburgh, Pennsylvania.

76. Hughes, J.C., Wheals, B.B. and Whitehouse, M.J., Simple Technique for the Py-MS of Polymeric Materials, *Analyst, 102*(1977) 143-144.

77. VG Micromass (Organic) Ltd., Tudor Road, Altrincham, Cheshire, United Kingdom.

78. Wuepper, J.L., Py-GC-MS Identification of Intractable Materials, *Analyt. Chem.*, *51*(1979)997-1000.

79. Wheals, B.B. and Noble, W., The Py-GC Examination of Car Paint Flakes as an Aid to Vehicle Characterisation, *J. Forensic Sci. Soc.*, *14*(1974)23-32.

80. Stewart, Jr., W.D., Py-GC Analysis of Automobile Paints, *J. Forensic Sci.*, *19*(1974)121-129.

81. Jain, N.C., Fontan, C.R. and Kirk, P.L., Identification of Paints by Py-GC, *J. Forensic Sci. Soc.*, *5*(1965)102-109.

82. Stewart, W.D., Jr., Py-GC Techniques for the Analysis of Automobile Finishes: Collaborative Study, *J. Assoc. Offic. Analyt. Chem.*, *59*(1976)35-41.

83. MacLeod, N., Quantitative Analysis by Py-GC of Thermosetting Acrylic Resins Used in Automotive Enamels, *Chromatographia, 5* (1972)516-520.

84. Hickman, D.A. and Jane, I., Reproducibility of Py-MS Using Three Different Pyrolysis Systems, *Analyst, 104*(1979)334-347.

85. Cook, C.D., Py-GC in Paint Analysis, *Paint Manuf., 44*(1975)19-21.

86. Kadosaka, T. and Morino, H., Identification of Japan Lacquer Films by Py-GC, *Rep. Cent. Customs Lab. (Kanzei Chuo Buneki-shoho), 19*(1978)103-117.

87. Breek, R. and Froentjes, W., Application of Py-GC on Some of Van Meegeren's Faked Vermeers and Pieter de Hooghs, *Studies Conserv., 20*(1975)183-189.

88. Froentjes, W. and Breek, R., Een nieuw onderzoek naar de Identiteit van het bindmiddel van Van Meegeren, *Chem. Weekblad., 1977*, 583-589.

89. Gokcen, U. and Cates, D.M., Thermal Analysis of Fibres and Fibre Blends. Pt. II. Py-GC, *Appl. Polym. Symp., 2*(1966)15-24.

90. Crighton, J.S., Characterisation of Textile Materials by Thermal Degradation: A Critique of Py-GC and Thermogravimetry, in C.E.R. Jones and C.A. Cramers (Eds.), *Analytical Pyrolysis*, Elsevier, Amsterdam, 1977, pp. 337-349.

91. Hughes, J.C., Wheals, B.B. and Whitehouse, M.J., Py-MS of Textile Fibres, *Analyst, 103*(1978)482-491.

92. DeForest, P.R., The Potential of Py-GC for the Pattern Individualisation of Macromolecular Materials, *J. Forensic Sci., 19*(1974)113-120.

93. DeForest, P.R. and Kirk, P.L., Forensic Individualisation of Hair, *Criminologist, 8*(1973)35-45.

94. Nelson, D.F., Yee, J.L. and Kirk, P.L., The Identification of Plastics by Py-GC, *Microchem. J., 6*(1962)225-231.

95. Coakley, J.E. and Berry, H.H., *Identification of Elastomers by Py-GC*, US Nat. Tech. Info. Service, No. 730470 (15:8:1971).

96. Leisztner, L., Gal, S., Szanto, J. and Kovacs, L., Influence
     of the Morphology of Samples in the Pyrolysis of PVC, J. Ther-
     mal Anal., 13(1978)141-147.

97. Noble, W., Wheals, B.B. and Whitehouse, M.J., The Characterisa-
     tion of Adhesives by Py-GC and IR Spectroscopy, Forensic Sci.,
     3(1974)163-174.

98. CDS Application Note 100873, Verification of Antique Glues by
     Solids Pyrolysis, Chemical Data Systems, Oxford, Pennsylvania.

99. Lloyd, J.B.F., Hadley, K. and Roberts, B.R.G., Py-GC over
     Hydrogenated Graphitised Carbon Black. Differentiation of
     Chewing Gum Bases for Forensic Purposes, J. Chromatogr., 101
     (1974)417-423.

100. Sekikawa, Y., Kuwata, S., Kadosaka, T. and Morino, H., Iden-
     tification of Natural and Artificial Leather by Py-GC, Rep.
     Cent. Customs Lab. (Kanzei Chuo Bunsekishoho), 19(1978)89-94.

101. Lloyd, J.B.F. and Roberts, B.R.G., GC Characterisation of Cook-
     ing Fats with Reference to a Case of Murder, J. Chromatogr., 77
     (1973)228-232.

Chapter 10

Organic Geopolymers

10.1. INTRODUCTION

The complex polymeric nature of many of the carbon sources in the geosphere has posed considerable problems in characterization. Comprehensive profiles are particularly difficult and time-consuming to obtain by traditional analytical techniques, due to the intractability of the samples. Pyrolysis methods are advantageous in that a rapid analysis of a series of samples is possible, and the pyrogram may be interpreted chemically to reveal structural differences between samples. Moreover, as the amount of sample required for an analysis is small, the heterogeneity within collected samples may be studied. Important applications of analytical pyrolysis now involve soil chemistry, with considerable emphasis on fertility and humification, and the characterization of fossil-fuel source rock. The impetus for much of this work has been provided by the US space exploration program. Here, early and continuing efforts in taxonomy, biological molecules and organic geochemistry have resulted in Py-GC-MS studies being undertaken on the surface of the planet Mars.

In addition to these three main areas of interest, the study of environmental pollutants is now underway. Pyrolysis of trapped air particules has enabled fingerprint characterization to be achieved and the sources of atmospheric pollutants (e.g., industrial processes, power stations, fires) may be established. Py-GC [1] and Py-MS [2] techniques have been reported.

TABLE 10.1   Pyrolysis Products from Californian Desert Soil

PEAK	COMPONENT	PEAK	COMPONENT	PEAK	COMPONENT
1	Ethylene	18	Pentanone	43	Ethylbenzene
2	Propene	18	Propionitrile	44	Dimethylbenzene
3	Butene	19	Me-butadiene	45	2-Methylpyrazine
4	Methylpropane	20	Methacrylonitrile	46	Styrene
4	Methanethiol	21	Pentanal	47	Furfural
4	Dimethyl sulphide	22	Butyronitrile	49	C_8-Diene
5	Methylbutene	23	Benzene	50	Propylbenzene
6	Methanol	24	2-Butenal	52	Methylstyrene
6	Propionaldehyde	25	Dimethylfuran	54	Methylfurfural
7	Acrolein	26	Methylpentene	56	C_9-Alkene
8	Acetone	28	Hexene	57	Furfuryl alcohol
8	Furan	30	Dimethyl sulphide	58	Benzonitrile
9	2-Me-propionaldehyde	31	C_6-Alkene	59	Indene
10	Butanone	32	Pyrrole	60	Phenylacetonitrile
11	Dimethylbutene	34	Toluene	61	Phenol
12	2-Me-propenal	35	Methylpyrrole	62	Cresol
13	Acetonitrile	37	C_7-Alkene	63	Cresol
14	Acrylonitrile	40	Dimethylpyrrole	64	Xylenol
15	Pentene	41	Pyridine	66	Alkylbenzene
16	2-Me-furan	42	Picoline	68	Indole
17	Butyraldehyde				

Source: Ref. 3, from the *Journal of Chromatographic Science*, by permission of Preston Publications, Inc.

10.2.   SOIL AND SOIL HUMIC COMPOUNDS

The structural significance of the pyrolysis products from soil fractions has meant that component identification has been an important
aid in many applications. The product distribution from Californian
desert soil (0.34% organic carbon) is listed in Table 10.1. The
products may be clearly recognized as originating from biological
macromolecules (Chap. 7) with large contributions from furan (carbohydrate), benzene, and toluene (proteins). The strongly retained
indole (protein) is also a significant marker compound [3].

In fertile soils a much higher level of organic material is
present. Chemical and biochemical transformations of decaying plant
tissue result in characteristic profiles which may range from little
altered vegetation through mature earths, which have undergone extensive decomposition with mineral incorporation, to the leached,
nutrient-deficient and acidic podzols which characterize heath and
bog. Variations within a soil profile are determined by the genetic
horizon and by the humus type. The pyrolysis of different soils

yields characteristic product distributions which reflect their macromolecular structure. It is important that the morphology and genetic horizon of samples is adequately recorded to maximize this
information.

The major components of cell walls are carbohydrates (cellulose
and hemicellulose) and lignins (a three-dimensional polymer of phe-
nylpropanes cross-linked to benzylethers and covalently bound to
hemicellulose) [4]. Lignins are most resistant to decay and may be
expected to contribute strongly to the pyrograms of partially decayed humus. The Py-GC of lignins has been reported, and serves as
a useful comparison [5-8]. The pyrograms are characterized by methoxyphenol fragments and the product distribution depends upon the
lignin source [9]. Typical products are recorded in Table 10.2.
Ion series from Py-MS studies reflect this distribution (Table 10.3)
[10,11]. Mor humus, typically found in the L, F, H, and A horizons
of podzols, consists of partially broken down plant material containing low-molecular-weight phenolic compounds. These are not involved
in the formation of stable complexes with other soil components, so
that mineral leaching is evident. Under anoxic conditions lignins
maintain their structure over vast periods of time and a high degree
of correspondence between the pyrograms of modern lignin and those
from demineralized fossil wood 2×10^8 years old [12,13] was found.
Lignin pyrolysis may also be used as a source of phenols [14]. Normal degradation of plant remains involves the formation of humic
substances. These are relatively stable, dark-colored polymers with
variable constitution. The lignin residues are generally not detectable, but are now incorporated into a heavily transformed macromolecular phenolic-proteinaceous complex, possibly with associated
polysaccharides. Mull humus typifies this stage of development.
This is characteristic of the A and B horizons of brown earth soils
and the stable clay-humus complex prevents leaching. The pyrograms
from such samples reflect their maturity and may be used to reveal
the extent of the humification processes. The approach has been ably
demonstrated by the Py-GC-MS and Py-MS studies undertaken by Bracewell
and coworkers on peaty podzols.

TABLE 10.2 Components[a] of Lignin Pyrolysates

PEAK	$M^{+\cdot}$	COMPONENT	PEAK	$M^{+\cdot}$	COMPONENT
1	60	Acetic acid	23	164	cis-isoEugenol
2	96	Furfural	24	154	2,6-Dimethoxyphenol
3	116	Indene	25	168	Methyl 2,6-dimethoxyphenol
4	130	Methylindene	26	164	trans-isoEugenol
5	118	Methylbenzofuran	27	120	Vinylphenol
6	88	Butyric acid	28	134	Methyl vinylphenol
7	62	Ethylene glycol	29	148	C_2-Alkyl vinylphenol
8	132	C_2-Alkylbenzofuran	30	182	Ethyl 2,6-dimethoxyphenol
9	124	2-Methoxyphenol	31	196	Propyl 2,6-dimethoxyphenol
10	138	Methyl 2-methoxyphenol	32	134	Propenylphenol
11	152	C_2-Alkyl 2-methoxyphenol	33	194	Allyl 2,6-dimethoxyphenol
12	94	Phenol	34	152	Vanillin
13	108	o-Cresol	35	180	Vinyl 2,6-dimethoxyphenol
14	152	C_2-Alkyl 2-methoxyphenol	36	194	Propenyl 2,6-dimethoxyphenol
15	166	C_3-Alkyl 2-methoxyphenol	37	166	$C_9H_{10}O_3$
16	122	Xylenol	38	180	$C_{10}H_{12}O_3$
17	108	p-Cresol	39	194	$C_{11}H_{14}O_3$
18	166	C_3-Alkyl 2-methoxyphenol	40	182	Syringaldehyde
19	136	C_3-Alkylphenol	41	182	$C_9H_{10}O_4$
20	164	Eugenol	42	210	Sinapyl alcohol
21	122	Xylenol	43	178	Coniferylaldehyde
22	150	Vinyl 2-methoxyphenol			

[a]Additional components include hydrocarbon gases, aromatic hydrocarbons, and condensed aromatics such as naphthalene and methylnaphthalenes [12,13].

Source: Ref. 9.

TABLE 10.3 Ion Series from Lignin Py-MS

Series	m/z
Phenols	94, 108, 120, 122
2-Methoxyphenols	124, 138, 150, 152, 164
2,6-Dimethoxyphenols	154, 168, 180, 182, 194

Source: Refs. 10 and 11.

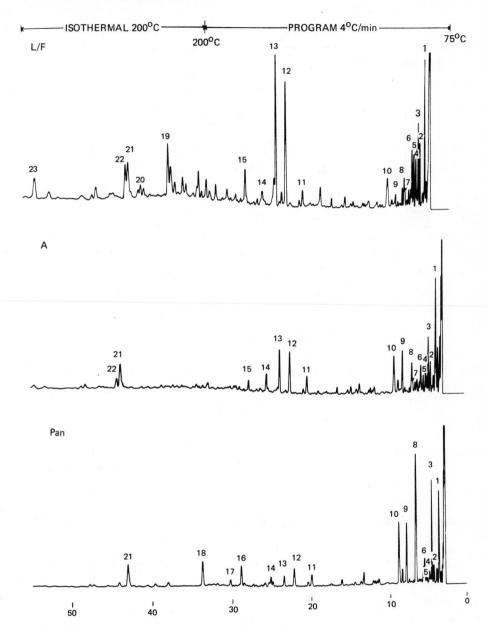

FIGURE 10.1   Pyrograms from surface (L/H):   A horizon (A) and B horizon (Pan) of a peaty podzol.   (Components are identified in Table
10.4.)   (From Ref. 15.)

TABLE 10.4   Pyrolysis Products from a Peaty Podzol

PEAK	M$^{+\cdot}$	COMPONENT	PEAK	M$^{+\cdot}$	COMPONENT
1	44	Acetaldehyde	13	96	Furfural
2	68	Furan	14	67	Pyrrole
3	58	Acetone	15	110	5-Methylfurfural
4	56	Acrolein	16	103	Benzonitrile
5	82	2-Methylfuran	17	120	Acetophenone
6	32	Methanol	18	128	Naphthalene
7	96	2,5-Dimethylfuran	19	124	2-Methoxyphenol
8	78	Benzene	20	138	4-Methyl-2-methoxyphenol
9	41	Acetonitrile	21	94	Phenol
10	92	Toluene	22	98	Dihydropyrone
11	82	Cyclopent-2-enone	23	150	4-Vinyl-2-methoxyphenol
12	60	Acetic acid			

Source: Ref. 15.

Podzol profiles usually show several distinct genetic soil horizons. Analysis of individual horizons leads to characteristic pyrograms. This is illustrated in Fig. 10.1 for the peaty organic surface (L, F, and H horizons), the mineral A horizon and the B horizon pan. The component identification is recorded in Table 10.4 [15]. The surface layers provide a large yield of furfural (from polysaccharides) and a series of methoxyphenols which are characteristic of the lignin-derived residues. The methoxyphenols are absent from A horizon pyrograms, which also display reduced carbohydrate content. Dihydropyrone, however, is a component of both pyrograms. The B horizon pyrogram contains benzonitrile, acetophenone, and naphthalene as characteristic components. These products are found only in the pyrogram of monomorphic, translocated B horizon organic matter and may be used to differentiate this from the polymorphic form. The variation of selected products with depth is illustrated in Fig. 10.2. The intensity of the carbohydrate derivatives decreases rapidly with depth (furfural and 5-methylfurfural), acetaldehyde and acetonitrile increase with depth, and aromatic hydrocarbons, marking translocated material, are abundant only within narrow limits. Pyrrole (from polypeptides) and cyclopentenone (from polycarboxylic acid chains) also vary markedly with maximal contribution from the A and upper B horizons.

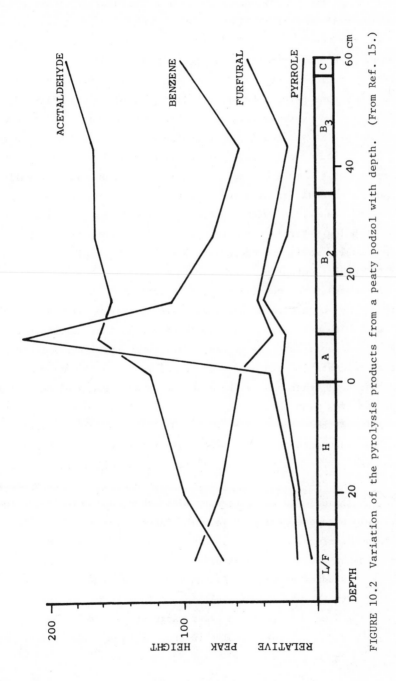

FIGURE 10.2 Variation of the pyrolysis products from a peaty podzol with depth. (From Ref. 15.)

It is probable that compounds such as furfural and 5-methylfur-
fural are major components in the pyrolysis of the relatively undecomposed mor humus type, while pyrrole and cyclopentenone are more
abundant from the heavily transformed mull humus. A pyrolysis ratio
(intensity ratio of furfural + 5-methylfurfural: pyrrole + cyclopentenone) may thus be used to study soil type [16]. In particular,
mull and mor humus types (e.g., podzols, brown podzolic soils, and
brown earths) may be readily distinguished and an index which reveals
mull humus type is obtained. This correlates well with base status,
mineral-organic complexes and, in the case of brown podzolic soils,
with environmental features too. Little correlation with other soil
parameters was evident with podzols (mor humus)--the pyrolysis ratio
probably being determined by the structure of plant tissue--but for
brown podzolic soils and brown earths analytical pyrolysis was superior to carbon content or C/N ratio in classifying soil classes.

In contrast to peaty podzols, a variation of humification with
depth may not be apparent in peats. Here, particle size is of significance with maturation being reflected in an increasing content of
dark, extractable material of small diameter. Pyrolysis methods are
also appropriate in this case and the degree of humification of size
separated soil fractions may be assessed by analogous methods [11,17].
Pyrograms from peat show marked similarities to those from the surface (L/H) of the peaty podzol (Fig. 10.1), with styrene, eluting
prior to cyclopentenone, also present in the large-size (>5 mm) fractions. As particle size decreases (to 5 to 50 μm), increased degradation is indicated by a decrease in the abundance of methoxyphenols
and by an increase in the intensity of benzene, acetonitrile, toluene, cyclopentenone, and pyrrole. The intensities of the carbohydrate
fragments (furfural and 5-methylfurfural) and of phenol are also enhanced in the smaller sized material [11].

Py-MS studies of whole soil samples have also been reported [11].
These confirm the utility of analytical pyrolysis in the study of soil
types and indeed they reveal a wider range of pyrolysis products than
the Py-GC approach. This enables further differences in humification

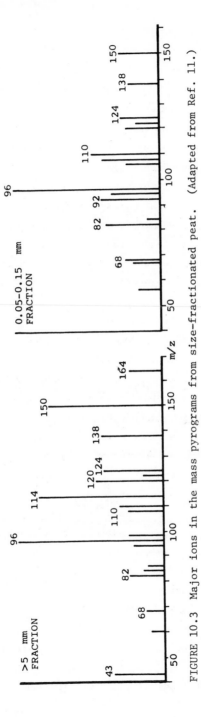

FIGURE 10.3 Major ions in the mass pyrograms from size-fractionated peat. (Adapted from Ref. 11.)

to be exposed. This is illustrated for the size fractionated peat
samples in Fig. 10.3. Significant differences are observed and summation of the intensities of the various ion-series enabled quantitative comparisons to be undertaken. Thus, the lignin-derived methoxyphenols (Table 10.3) decrease in intensity as humification proceeds,
whereas the phenol series (m/z 94, 108, 120) shows no such trend.
The ratio of methoxyphenols:simple phenols was therefore used as an
index of humification which ranged from 2.09 for the large particles
to 0.66 for those comprising the 0.15 to 0.25-mm cut. Larger values
were recorded for a peat with less humification (5.55 to 1.52). Mass
pyrograms were also obtained from various horizons within a peaty
podzol (Fig. 10.4) and these, too, showed characteristic ion-series,
depending upon the degree of humification. The ion-series and the
resulting ratios which were used to compare the mass pyrograms are
displayed in Table 10.5.

The abundance of aromatic hydrocarbons increases as humification
progresses, whereas no clear trend is observed with the aliphatic
series. Condensed aromatics (naphthalenes) are also detected from B
horizon samples. Thus, an increase in the aromatic:aliphatic hydrocarbon ratio monitors the maturity of the sample. The extent of pyrrole formation also increased with the degree of humification and
this, too, followed the expected pattern (Table 10.5). Also evident
in the mass pyrogram were series of peaks derived from carbohydrates.
These may be classified as furans, which arise from many carbohydrate-
based structures, and as dianhydro ions from pentoses (m/z 114), deoxyhexoses (m/z 128), and hexoses (m/z 144). The dianhydro series
represents partially degraded material with considerable amounts of
intact polysaccharide residues. Humification possibly converts
these into more stable structures which pyrolyze to form furans. The
decrease in the intensity of the dianhydro ion series reflects the
degree of humification. It should be noted that this information is
not available with the Py-GC-MS technique, due to the low volatility
of the sugar residues. The intensity of the ion at m/z 114 in the
surface peat samples (Fig. 10.3) reveals this information loss. The
wider product range observed with Py-MS provides an increased

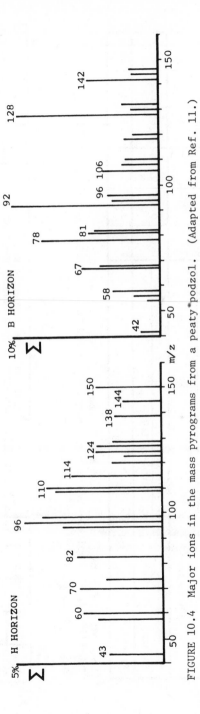

FIGURE 10.4 Major ions in the mass pyrograms from a peaty podzol. (Adapted from Ref. 11.)

TABLE 10.5 Ion Series Used to Assess Degree of Humification of Peat and Peaty Podzol Samples[a]

ION SERIES	M/Z	SERIES RATIO					
		PEAT			PODZOL		
		1	2	3	H	A	B
Methoxyphenols	124, 138, 150, 152, 164 154, 168, 180, 182, 194 :	2.09	1.12	0.78	1.37	0.53	0.10
Phenols	94, 108, 120						
Aromatic Hydrocarbons	78, 92, 106[b] :						
Aliphatic Hydrocarbons	42, 56, 58, 70, 72, 84, 86, 100, 112	0.30	0.76	0.98	0.27	0.45	2.04
Polysaccharides	114, 128, 144 :						
Furans	68, 82, 96, 110	0.65	0.25	0.16	0.68	0.29	0.14
Pyrroles	67, 81, 95	1.30	3.00	5.40	4.70	8.40	10.2

[a]1: 5 mm; 2: 0.5 to 1 mm; 3: 0.05 to 0.15 mm.
[b]B horizon from podzol also shows naphthalene (128) and methylnaphthalene (142).
Source: Ref. 11.

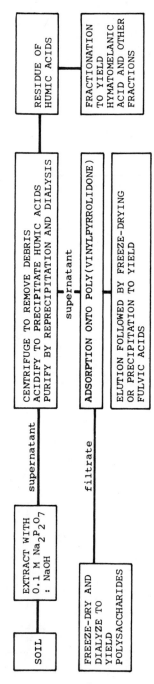

FIGURE 10.5 Extraction of soil organic fractions. (From Ref. 34.)

sensitivity for the elucidation of soil types. Significantly shorter
analysis times add to the potential advantages. Nevertheless, some
loss of information occurs in a Py-MS analysis. In particular, the
cyclopentenone and 5-methylfuran have the same mass (m/z 82) and
cannot be resolved by low resolution work. It would be of interest
to supplement the data analysis described here with the multivariate
approach, including nonlinear mapping of the distance function and
factor analysis, which may well provide significantly improve discrimination between samples.

Earlier work in this area was concerned with the development of
Py-MS [18-20] and Py-GC-MS [21] methods for whole soil analysis and
included a study of a climosequence of seven soils [22]. The ratio
of furfural and 5-methylfurfural: pyrrole and cyclopentenone was
again used to classify soil types, and significant correlations with
altitude, precipitation, and mean annual temperature were observed.
The pyrolysis ratio of A horizons varied from 1.2 for a brown-gray
earth to 3.82 for a high-country podzolized yellow-brown earth.

Analytical pyrolysis may also be applied to the study of isolated
soil fractions (Fig. 10.5). Initial studies were concerned with humic
acids but, although similarities and differences between pyrograms
were evident, no structural conclusions were drawn [23]. Extensions
of this work used the yields of benzene and toluene to monitor the
effect of extractant and soil history on the pyrograms of 24 humic
acids [24,25]. Here, it was shown that less mature humic acids
(sodium hydroxide extracted) yielded more benzene and toluene than
the older, darker, and condensed polycarboxylic acid analogs (pyro-
phosphate extracted). Full identification of the components in the
pyrograms from humic and fulvic acids have now appeared [26-30].
Products are similar to those obtained from whole soils with fulvic
acids yielding simpler pyrograms containing mainly polysaccharide
derivatives, together with some aromatics and few nitrogen derivatives. For example, an orthic humoferric podzol fulvic acid pyrogram
contained acetic acid, furfural, phenol, and cresols as major peaks,
together with smaller amounts of indene, methylfurfural, naphthalene,
methylnaphthalenes, 2-methoxyphenol (guaiacol), and furfuryl alcohol

[27]. The variation of fulvic acids dependent upon extraction method and polysaccharide content has been studied, and quantitative assessments are possible [31,32]. In contrast, the humic acids revealed a range of products derived from proteins and lignins, with pyrrole, 2-methoxyphenol, phenol, cresol, isoeugenol and vinylphenol being significant peaks. Polysaccharide residues were also indicated by the presence of acetic acid, furfural and methylfurfural. Pyrograms of the low-boiling fractions (i.e., up to phenol) which illustrate the variation in humic acids from different sources have also been reported [33].

Comparative Py-MS studies on whole soils and isolated fractions have been reported. The major ion series from the A_1 horizon of a brown soil on granite are displayed in Fig. 10.6, and are listed in Table 10.6 [34]. These characteristic series enable the structural features of the various fractions to be exposed. These include the reduced proportion of furans and other polysaccharide residues in humic acid mass pyrograms, compared to whole soil, whereas the protein-derived peaks (ammonia, sulfides, pyrroles, and indoles) and phenols show increased intensities. Mass pyrograms from fulvic acids varied with the separation procedure, but showed a high proportion of polysaccharide-derived ions. The large proportion of polysaccharide residues found in soil, despite the rapid degradation of carbohydrates under normal environmental conditions, supports the proposals concerning the microbiological biosynthesis of soil polysaccharides [35]. In contrast, lignin (believed to be more resistant to degradation) appears to be rapidly decomposed for only low intensities of methoxyphenol ions were detected.

Further Py-MS studies on whole soils have described results from ancient river clay soils and from plaggen soils [36]. The origin of characteristic ion series (Table 10.6) were discussed, and differences between soil types were exposed. Mass pyrograms were satisfactorily obtained from clays, with as little as 0.4% organic matter. For a recent and a Pleistocene sediment a rank order of discriminating peaks (based on the F-test--the ratio of outer to inner variance) was

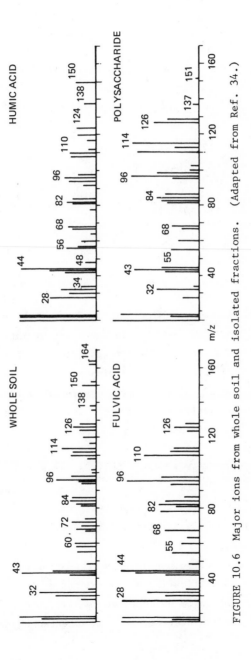

FIGURE 10.6 Major ions from whole soil and isolated fractions. (Adapted from Ref. 34.)

TABLE 10.6   Py-MS Ion Series from Soils and Soil Fractions

Series	m/z	Series	m/z
Aliphatic	28,42,56,58,70, 72,84,86,100,112	Phenols	94,108,120, 122,136,150
Benzenes	78,92,106	Methoxyphenols	124,138,150, 152,164
Naphthalenes	128,142		
Pyridines	79,93,107	Dimethoxyphenols	154,168,180, 182,194
Pyrroles	67,81,95		
Indoles	117,131,145	Sulphides	34,48,62
Furans	68,82,84,96,98, 110,112	Pyrrolidones[a]	85,111
		Alcohols[a]	32,46
Polysaccharides	43,60,72,102, 114,126,128,144	Amino carbohydrates	95,109, 125,137,151

[a]Impurities.

Source: Ref. 34.

provided.   This showed that carbohydrate residues were richer in the more recent (Holocene) sediments, with m/z 96 (furfural) being highly discriminant (F = 4.21, N = 4).   In contrast, aromatic hydrocarbons predominate more in the older sediments (Pleistocene) and m/z 78 (benzene) was also of useful discriminant value (F = 2.11).   A scatter diagram, e.g., plotting m/z 96 versus m/z 78 for each sample, was sufficient to partition the two classes of sample.   Nonlinear mapping of the multidimensional distances allowed clear classification to be achieved.   Such tests were more effective than pore volume and shearing stress measurements, while no useful conclusions at all were possible with determinations of total organic matter and clay content.   Similar results were reported from the Plaggen soil samples. These, too, were readily partitioned and Py-MS revealed an incorrectly identified sample.

Studies are now also beginning to appear on soil maturation processes.   The identification of components or ion-series enables structural information to be obtained readily, and variations with age may be followed.   Interest has been centered on fungal melanins.

These are dark-colored polymers synthesized by soil microorganisms
which resemble soil humic acid in many ways. They are random phenolic polymers into which have also been incorporated protein and
polysaccharide residues. A typical study has compared the Py-MS of
fungal melanins, model polymers (prepared from oxidative coupling
(phenolase) of various phenols either alone or together with amino
acids or peptose), and humic acids [37,38]. Fungal melanins were
characterized by the major ion series representing homologous sulfides, pyrroles, benzenes, phenols, and indoles (Table 10.6). Variations which were dependent upon culture conditions and organism were
evident. Mass pyrograms differed substantially from those obtained
from proteins or polysaccharides, but were strikingly similar to mass
pyrograms derived from a model polymer which was prepared from a complex phenolic mixture and peptone. Both sets of mass pyrograms
showed considerable correspondence to the ion series found from humic
acid pyrolysis, although different relative abundances reflected the
individuality of samples. In particular, soil humic acids showed
somewhat more pronounced benzene and methoxyphenol ions, whereas
phenols and indoles were stronger in melanin mass pyrograms. The
fact that soil humic acid provides methoxyphenol ions which are,
nevertheless, quite weak, confirms the view that few compounds are
present in this fraction which may directly originate from lignins.
The processes have been modeled to some degree by monitoring straw
compost. When composted for various periods of time, straw mass pyrograms still contained the lignin-type ion series. When nitrate was
added, further degradation and N incorporation was possible. This
material then revealed the more typical sulfide, pyrrole, benzene,
and indole ion series, although methoyxphenols were still more abundant than in humic acids.

Other reports on Py-MS [39,40] and thermofractography (TAS) [41]
of isolated soil fractions have appeared and Py-GC-MS studies of fungal melanins have been concerned with the identification of the pyrolysis products [42-45]. These confirm the Py-MS work and establish
analytical pyrolysis as a valuable technique in soil analysis.

10.3. FOSSIL FUELS AND THEIR SOURCES

The continuing search for liquid and solid fossil fuels has resulted
in considerable interest in the origin, environment of deposition,
maturation and potential petroleum-generating capacity of the organic components of sedimentary rocks [45]. Oil shales are of particular importance due to the large available reserves. The organic
residues dispersed throughout sedimentary deposits are known as
kerogens. These are probably derived from biopolymers by various
polymerization and condensation reactions, and are believed to be
the source material for petroleum and natural gas. Type I kerogens
are rich in hydrogen (7 to 12%), (they are probably derived principally from the lipids of water-borne organisms and they are thought
to be primarily petroleum-producing residues). Normal oil-generating
kerogen is type 2. Type 3 and type 4 kerogens are hydrogen depleted,
with a more condensed organic matrix. These include materials such
as coals, which probably are responsible for gas generation with pro-
longed burial times. The elucidation of the structural components of
kerogens plays an important part in assessing the fuel potential of
strata but the analysis is restricted by the intractable nature of
the polymer, perhaps less than 10% is extractable under normal conditions. Pyrolysis methods have significant advantages over many
alternative techniques in this situation, and their application in
geochemical areas is becoming extensive. Such studies range from
large-scale pyrolysis of kerogen or source rock [46-49] through
chromatographic [50-52] or spectroscopic [53] analysis of pyrolysates,
to the integrated analytical pyrolysis approach [54-56].

    The products identified from a Py-GC-MS analysis of Fig Tree
Shale are recorded in Table 10.7 [3]. Major components are toluene
(48), di- (67) and tri- (86) methylbenzenes, and naphthalene (123),
with significant contributions also from benzene (31), dimethyl- (70),
ethylmethyl- (78), and C_6 alkyl- (128) benzenes and ethylnaphthalene
(129). The products are principally aromatic and aliphatic hydrocarbons, with only small yields of O-, N-, or S-containing compounds and
contrast markedly with those from soil pyrolysis (Table 10.1). Under

TABLE 10.7 Pyrolysis Products from Fig Tree Shale

PEAK	COMPONENT	PEAK	COMPONENT	PEAK	COMPONENT
1	Carbon dioxide	30	Dimethylpentane	73	isoPropylbenzene
1	Methane	31	Benzene	74	C_3-Alkylbenzene
2	Ethane	35	Thiophene	75	C_3-Alkylbenzene
3	COS	37	C_7H_{14}	76	Propylbenzene
4	Propene	38	C_7H_{14}	77	$C_{10}H_{22}$
4	isoButene	39	C_7H_{14}	78	Methylethylbenzene
5	Hydrogen sulphide	40	Methylhexene	79	Methylstyrene
5	Butene	41	Heptene	81	$C_{10}H_{20}$
6	Butane	42	C_7H_{14}	83	$C_{10}H_{20}$
7	Methanethiol	44	C_7H_{12} Alkyne	85	$C_{11}H_{24}$
8	Methylbutene	45	Butanol	86	Trimethylbenzene
8	Pent-1-ene	45	C_8H_{18}	87	$C_{11}H_{24}$
9	isoPentane	47	C_8H_{18}	88	Methylisopropylbenzene
9	Methanol	48	Toluene	89	C_4-Alkylbenzene
10	Pent-2-ene	49	C_8H_{18}	91	C_4-Alkylbenzene
11	Acetone	50	Methylthiophene	92	C_4-Alkylbenzene
11	Carbon disulphide	51	C_8H_{16}	94	C_4-Alkylbenzene
14	Methylbutadiene	53	C_8H_{16}	96	C_4-Alkylbenzene
15	Pentadiene	54	C_8H_{16}	97	C_4-Alkylbenzene
16	Ethanol	55	C_8H_{16}	99	C_4-Alkylbenzene
17	Dimethylbutene	57	C_8H_{16}	102	C_4-Alkylbenzene
18	C_6H_{12}	58	C_8H_{16}	104	C_4-Alkylbenzene
19	Hexane	59	C_8H_{16}	116	C_5-Alkylbenzene
20	Acetonitrile	64	Ethylbenzene	117	C_5-Alkylbenzene
21	Methylpentene	66	C_9H_{20}	118	C_5-Alkylbenzene
22	Methylpentane	66	C_9H_{18}	123	Naphthalene
23	C_6H_{12}	67	Dimethylbenzene	126	Methylnaphthalene
24	C_6H_{12}	70	Dimethylbenzene	128	C_6-Alkylbenzene
26	C_6H_{12}	72	Styrene	129	Ethylnaphthalene

Source: Ref. 3.

different chromatographic conditions significantly larger products have been identified [57]. These include intense C_8 alkylbenzene peaks and homologous series of ions up to about C_{35}. Acyclic aliphatic residues, probably derived from isoprenoids such as the phytyl side chain of cholesterol, are also found [58-60]. The most important of these is prist-1-ene which was produced from about one-third of kerogen samples examined [61]. Alicyclic residues such as terpanes and steranes up to C_{33} have also been described [62,63]. The pyrolysis profile of different kerogens is characterized by these various products and gives a rapid assessment of their potential.

One approach is to monitor the volatile pyrolysis residues without a full separation of the species. Thermal extraction (250°C for 30 s) of whole rock, kerogen or maceral concentrates (1 to 30 mg) followed by pyrolysis (programmed 50 to 1200°C at 140°C min^{-1} or isothermally at 500° or 700°C) was undertaken [64]. The pyrolysis vapors were collected and fractionated by heating the trap successively to 110°C (equivalent to C_1-C_{15} alkanes) and 300°C (C_{15}-C_{35}). The volatile fractions were passed through an FID and the total response to each was determined. The data were normalized for total volatility per unit weight and plots of K/(K + R) versus (K/R)^2, where K is the normalized fraction of pyrolysate passing through the trap at 110°C and R is the normalized fraction passing through the trap at 300°C, were displayed (cf. Production Index [48]). These data enabled the classification of samples to be undertaken. Thus, in the case of macerals, found as microscopic components of coal, it was shown that the exinite macerals such as sporinites (derived from pollen and spores) and alginites (derived from algae) yielded larger proportions of the higher molecular weight fractions (~80%) than did vitrinites (derived from cellulose and lignin by anaerobic degradation) which were usually in the range 50 to 65%. Additionally, vitrinites gave 2 to 3 times less volatile material than did the exinite samples. Samples from boreholes were similarly classified and the majority of calcareous mudstones and mudstones from a petroleum-bearing basin showed similarities to the exinite macerals. A three-point classification system, using, additionally, volatiles released at 25°C (C_1-C_8), was also investigated. Triangular coordinates, based upon the percentage yield of the three fractions, were used for the data plot and clusters of samples belonging to the three maceral groups were identified--although some overlap between sporinite and vitrinite zones was evident. This partition was due to the substantially greater yield of smaller fragments from vitrinites.

More information is available if a chromatographic profile of the kerogen or source rock is available. Step-wise pyrolysis enables both soluble and insoluble components to be monitored in the same run [54]. In broad terms, it has been proposed that

Larger molecules ($>C_8$) indicate molecules originally incorporated into kerogen.

Smaller molecules are characteristic of a transformed kerogen core.

The relative amounts of alkenes reveal the bonding of residues to the core.

The degree of aromatization may be indicated by the ratios of components such as methane, ethane, ethylene, and benzene.

The C_{11}-C_{28} residues are comparable to petroleum source-rock extractables and give a basis for the characterization of kerogens.

Kerogen from Green River shale was found to yield mainly hydrocarbons equivalent to C_8 or larger, with a peak at C_{17}. In contrast, a hydrogen-deficient kerogen yielded mainly methane on pyrolysis and no residues above C_8. Such profiles enable reliable characterization of kerogens to be achieved [65,66]. Py-hydrogenation-GC techniques have also been used to advantage [67].

The pyrograms from kerogens are characterized by an abundance of fragments. This complicates the interpretation of data. In particular, to determine the extent of maturation, aliphatic and aromatic residues must be separated and identified. Moreover, each component may be present in small amounts only. These may be difficult to quantify precisely and may only be partially resolved from adjacent peaks. Although packed columns, and even no column at all, yield data which are useful for the characterization of kerogens, a further technique, which allows a detailed probe of molecular structure, is available. This involves Py-GC-MS in which the mass spectrometer is used as a specific ion detector (mass fragmentography). The appearance of the chromatogram is substantially simplified, peaks mainly represent structurally related components with analogous mass spectral fragmentation processes, and a considerable improvement in sensitivity is available. Fragments which have been monitored in this way include m/z 91 (tropylium ion, alkyl benzenes), 105

(polysubstituted alkyl benzenes), 141 (benzotropylium ion, alkylnaphthalenes), 142 ($M^{+\cdot}$, methylnaphthalenes), 156 ($M^{+\cdot}$, C_2-alkyl-
naphthalenes), and 170 ($M^{+\cdot}$, C_3-alkylnaphthalenes) [54,56]. The
approach is illustrated in Fig. 10.7, which displays bar-graph pyrograms from an alginite and a vitrinite kerogen, together with the
corresponding mass fragmentograms. The analysis of these data may
be complicated by the variation in pyrolysis yield. The alginite
samples gave about 80% of the organic matter as volatile derivatives.
With vitrinite this dropped to less than 40%. Qualitative differences are, however, readily apparent. The high proportion of aliphatic residues in the alginite, for example, contrasts markedly
with the higher aromatic content of the vitrinite. These are also
detected on the m/z 141 mass fragmentogram as a homologous alkene-
alkane series due to a $C_{10}H_{21}^+$ fragment which appears in the MS
fragmentation of these aliphatic residues. This gives further information on the aromatic-aliphatic ratio of products from the various kerogens. These results parallel those from the fractionation
technique [64], but give a much more detailed picture of the kerogen
characteristics. The technique was also used to detect naphthalene
fragments in crude oils. Here, the method was rapid and reliable
and has possible application for the characterization of oil-spills
and other sources of pollution by petroleum fractions. Thermal distillation to liberate volatile residues and pyrolysis to characterize nonvolatile components has already been used in this way. The
increasing ratio of volatiles to total of products is a measure of
marine pollution and the pyrogram serves as a useful fingerprint
tool [68].

Steranes and terpanes have been used as marker compounds in
natural product organic geochemistry [69-72]. Such products are
also released through pyrolysis of kerogen and may be used to correlate bitumen, oil, and kerogen samples and to compare natural and
artificial maturation processes [62,63]. Extensions of the mass
fragmentography method [73] have been found useful in focusing on
the essential data [62,63] and the use of on-line data acquisition
enabled recall of mass chromatograms or specific mass fragmentograms

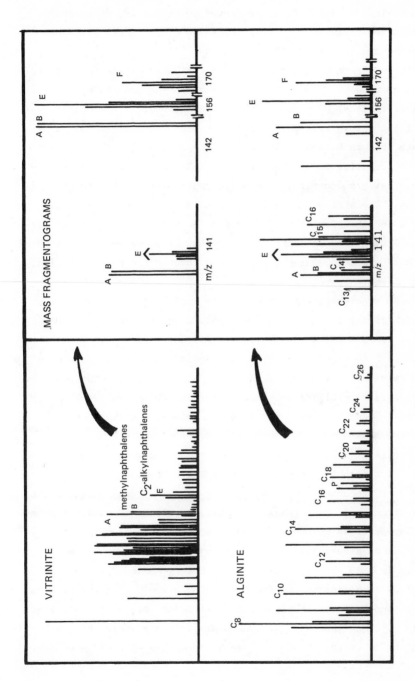

FIGURE 10.7 Bar-graph pyrograms and mass fragmentograms from a vitrinite and an alginite kerogen (A, 2-methylnaphthalene; B, 1-methylnaphthalene; E, C_2-alkylnaphthalenes; F, C_3-alkylnaphthalenes; C_{8-26}, aliphatic residues; P, Prist-1-ene). (Adapted from Ref. 55.)

on demand [68]. Useful ions were found to be m/z 191 (terpanes),
205 (methylhopanes), 217 (steranes), 239 and 253 (monoaromatized
steranes), and 259 (rearranged steranes). Figure 10.8 displays
typical mass fragmentograms showing the presence of the terpane and
sterane residues. These are also present in the associated bitumen
and oil fractions, although the relative intensities are modified,
e.g., (1) the terpane:sterane ration ($C_{27}-C_{30}$) is larger in the pyrolysate than in bitumen, and (2) C_{30} terpanes decrease in the pyrolysate, relative to bitumen, while $C_{27}-C_{29}$ and lower terpanes increase. Pyrogram comparisons confirm the indigenous nature of
bitumens. The stereochemistry at the C_5 and C_{17} centers is essentially maintained during pyrolysis and the ratio of the $17\alpha:17\beta$ terpanes allows different source rocks to be characterized (Fig. 10.9).
Rearranged steranes, which are produced through natural maturation,
are not found in pyrolysates.

Py-MS studies of kerogen have been limited to date, but the
potential of the technique has been ably demonstrated with a comparative Py-GC:Py-MS study [56]. Significant differences between
samples were obtained, including substantial carbohydrate residues
in the fulvic acid fraction of Messel shale (of age 5×10^7 years!).
Homologous alkene-alkane peaks characterized the Py-GC pyrogram,
with prist-1-ene appearing as a single, intense peak between the
hepta- and octa-decane:decene pairs. Terpanes and steranes were
also detected. Lignite mass pyrograms resembled closely those from
lignin, with methoxy phenols predominating. The facile EI fragmentation of hydrocarbons may mean that Py-MS techniques underestimate
the importance of these residues. The low voltage EI mass pyrograms
were characterized by even mass ions (molecular ions) which, in the
case of Green River shale revealed a major olefin:diene series, up
to C_{11}. In this case it would appear that Py-MS and Py-GC are complementary techniques, with Py-MS offering a rapid fingerprint
analysis and Py-GC-MS-DS providing an efficient probe for molecular
structure. The importance of a mechanistic appreciation of pyrolysis events to capitalize on such an approach has been stressed [49,
50].

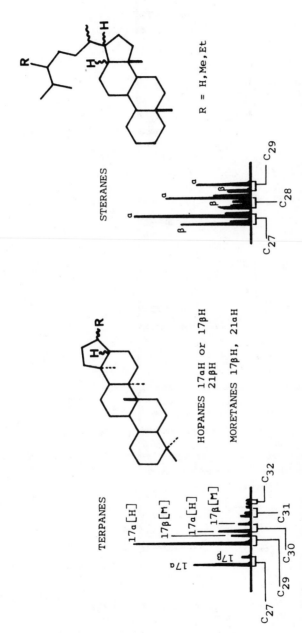

FIGURE 10.8 Mass fragmentograms showing steranes and terpanes released on the pyrolysis of Kreyenhagen shale. (Sterane stereochemistry refers to position 5.) (Adapted from Ref. 63.)

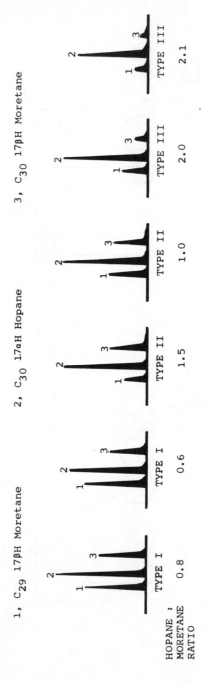

FIGURE 10.9 Differentiation of source rock groups by terpane ratio. (Adapted from Ref. 63.)

Laser pyrolysis of oil-shales is a further method of assessing their oil-bearing potential. In contrast to other pyrolysis methods, the intense laser irradiation yields a range of small, molecular weight compounds which contain typical plasma-tagged products [73-78]. Py-GC quantification of these products enables a rapid assessment of various properties of the shale. Thus, acetylene [77,78] and the acetylene:ethylene ratio [73,74] correlates closely with the oil yield from the shale. Other parameters, such as organic carbon, carbonate, organic hydrogen and water content were estimated from the yields of other volatile products such as carbon monoxide, hydrogen and methane. The noncombustible gases were monitored here with a beta-induced luminescence detector to provide a full profile of the pyrolysis products. The production of these small molecules, together with other residues such as propene and cyclopropane, enables analysis times to be quite short, but error limits were rather wide. These procedures lead to a rapid method for the screening of samples, although the laser pyrolysis unit is not widely available. Laser Py-MS techniques are also being exploited and complement the Py-GC approach. A mass pyrogram from a Devonian shale illustrates the product distribution (Fig. 10.10).

The application of laser techniques to the analysis of coal samples has also been demonstrated [79-82]. Coal probably consists of a range of substituted cyclic and polycyclic aromatic compounds, linked by short aliphatic bridges and bearing terminal COOH groups. This highly condensed structure requires a large and rapid influx of energy for efficient pyrolysis. Laser irradiation provides such heating easily, and may also be focused on small areas of a sample to study distributional effects. The volatile products from coal pyrolysis are, again, mainly low molecular weight species. A typical distribution from Gilsonite is displayed in Table 10.8. This material is believed to be composed of complex porphyrin condensates, and although ammonia was the sole nitrogen-containing product detected, the presence of acidic side chains is clearly indicated [75]. A bituminous coal yielded benzene and toluene, together with various plasma-tagged products, the most abundant of these being diacetylene.

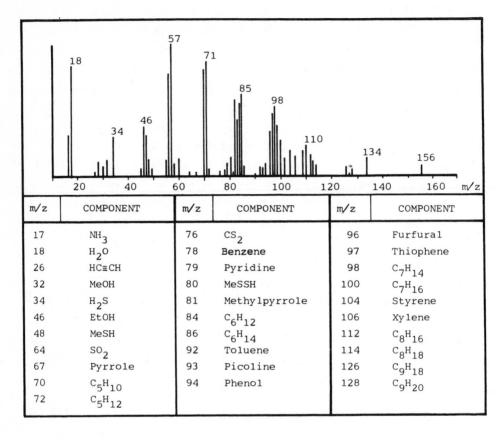

m/z	COMPONENT	m/z	COMPONENT	m/z	COMPONENT
17	NH_3	76	CS_2	96	Furfural
18	H_2O	78	**Benzene**	97	Thiophene
26	HC≡CH	79	Pyridine	98	C_7H_{14}
32	MeOH	80	MeSSH	100	C_7H_{16}
34	H_2S	81	Methylpyrrole	104	Styrene
46	EtOH	84	C_6H_{12}	106	Xylene
48	MeSH	86	C_6H_{14}	112	C_8H_{16}
64	SO_2	92	Toluene	114	C_8H_{18}
67	Pyrrole	93	Picoline	126	C_9H_{18}
70	C_5H_{10}	94	Phenol	128	C_9H_{20}
72	C_5H_{12}				

FIGURE 10.10   Laser Py-MS of a Devonian shale.   (From Ref. 76.)

TABLE 10.8   Laser Pyrolysis Products from Gilsonite

Peak	Component	Peak	Component	Peak	Component
1	Methane	6	But-1-ene	14	Hex-1-yne
1	Carbon monoxide	7	But-1-yne	15	HC≡C-CH_2-CH=CH-CH_3
1	Ammonia	8	HC≡C-CH_2OH	16	Hept-1-yne
2	Acetylene	9	Diacetylene	18	C_8H_{10}
2	Carbon dioxide	10	HC≡C-CH(CH_3)$_2$	19	R-COOH
3	Propylene	11	Pent-1-ene	20	R-COOH
4	Mixture	12	Pent-1-yne	21	$C_7H_{12}O_2$
5	Propyne	13	HC≡C-CH_2-COOH		

Source: Ref. 75.

TABLE 10.9 Mass Fragmentography of Coal Pyrolysate

Formula	Series	Components
C_nH_{2n+2}	86,100,114,...,422	Paraffins
C_nH_{2n}	84,98,112,...	Alkenes
C_nH_{2n-2}	82,96,110,...	Dialkenes, cycloalkenes
C_nH_{2n-6}	78,92,106,...,386	Benzenes
$C_nH_{2n-6}O$	94,108,122,...	Phenols
C_nH_{2n-8}	132,146,160,...,202	Benzoalicyclics
C_nH_{2n-10}	106,130,144,...	Indenes
C_nH_{2n-12}	128,142,156,...	Naphthalenes
C_nH_{2n-16}	166,180,194,...	Fluorenes
C_nH_{2n-18}	178,192,206,...	Anthracenes, phenanthrenes
C_nH_{2n-20}	218,232,246,...	Monoalicyclic anthracenes and phenanthrenes
C_nH_{2n-22}	202,216,230,...	Pyrenes, fluoranthenes
C_nH_{2n-24}	228,242,256,...	Tetraaromatics

Py-GC-MS-DS studies of coal reveal the presence of the polycyclic residues [83]. Mass fragmentograms of various characteristic
ions enabled the product distribution to be revealed. The ion
series are displayed in Table 10.9. The presence of terpanes and
steranes (m/z 191, 206) was also demonstrated. The relative intensities of the various series of ions differed with the maturity of
the coal, with the aromatic residues predominating in a Carboniferous
coal (2.9×10^8 years), while aliphatics were more important in
younger deposits (e.g., tertiary 6×10^6 years). These differences
are illustrated in Fig. 10.11 for the alkane:naphthalene fragments.
Steranes and terpanes, too, are useful marker compounds and decrease
in abundance as the coal matures. In view of these products, mechanistic studies of the pyrolysis of aromatic molecules are welcome [84].
The detailed structural information which can be extracted from studies of this type is immense and the potentialities of this approach
will surely be realized rapidly as MS-DS systems become more widely
available.

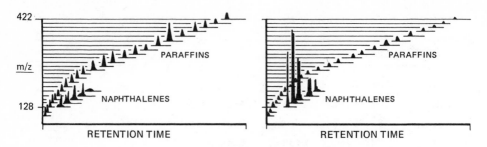

FIGURE 10.11 Mass fragmentograms showing variation of naphthalenes and alkanes in two coal pyrolysates. (Naphthalene intensities reduced for clarity.) (Adapted from Ref. 83.)

Other applications of analytical pyrolysis to solid fuel sources are many and include bitumens [85-90], lignites [91,92], further work on coals [93-95], and sulfur compounds in petroleum [96,97,98]. Py-hydrogenation-GC [99] and direct mass spectrometry of coal samples [100-102] have also been described.

10.4. EXTRATERRESTRIAL SOURCES

Modern cosmological theories, the vast accretions of organic matter in interstellar space (e.g., formaldehyde and cyanoacetylene) and the abiotic synthesis of biological molecules suggest that, on a universal scale, biological systems may be common. Experimental justification for such proposals is limited to spectroscopic and radio investigations because of the vast distances involved, but studies on the analysis of extraterrestrial organic matter have been eagerly undertaken to provide evidence on molecular evolution within the solar system. Samples have included meteorites collected on the surface of the earth, lunar rocks transported to earth for analysis, and martian soils analyzed on the surface of that planet.

The pyrolysis products from the Murray meteorite are recorded in Table 10.10 [3]. Naphthalene is the major large pyrolysis product, and other aromatic (peaks 37,65,98) and aliphatic (peaks 9,10, 19,20,33,45,91) hydrocarbons give significant peaks. Such residues are derived by evaporation from the complex aromatic-aliphatic

TABLE 10.10 Pyrolysis Products from the Murray Meteorite

PEAK	COMPONENT	PEAK	COMPONENT	PEAK	COMPONENT
1	Carbon dioxide	27	C_7H_{14}	67	Octan-2-one
1	Methane	29	Dithiabutane	68	C_3-Alkylbenzene
2	Ethane	31	C_7H_{16}	69	C_3-Alkylbenzene
2	COS	33	C_7H_{14}	70	C_3-Alkylthiophene
2	Sulphur dioxide	35	Pyrrole	71	C_3-Alkylbenzene
2	Propene	36	C_7H_{12} Alkyne	72	C_4-Alkylbenzene
3	Methanethiol	37	Toluene	73	C_4-Alkylbenzene
4	Hydrogen sulphide	38	Methylthiophene	74	C_4-Alkylbenzene
4	Methylpropene	39	Methylcyclohexene	75	C_4-Alkylbenzene
5	Methylbutene	40	C_8H_{16}	76	C_4-Alkylthiophene
6	Acetone	41	C_8H_{16}	77	C_4-Alkylbenzene
7	Ethanol	42	C_7H_{12} Cycloalkene	78	Phenylbutene
8	Ethylthiol	44	C_8H_{16}	79	C_4-Alkylthiophene
8	Furan	45	C_8H_{18}	80	Methylindane
9	Pentene	45	C_8H_{16}	81	C_5-Alkylbenzene
10	Isoprene	46	C_8H_{16}	82	$C_{11}H_{24}$
10	Carbon disulphide	46	C_8H_{16} Cycloalkane	83	C_5-Alkylbenzene
12	Pentadiene	47	Octene	84	C_5-Alkylthiophene
12	Methylpropanal	48	Dimethylhexadiene	85	C_5-Alkylbenzene
13	Dimethylbutene	49	C_8H_{16}	86	C_5-Alkylbenzene
13	Methylfuran	50	Methylnorbornene	87	C_5-Alkylbenzene
14	Butanone	51	Ethylbenzene	88	C_5-Alkylbenzene
15	Methylpentene	53	Dimethylthiophene	89	Phenylhexane
16	Dimethylbutyne	54	Xylene	90	Naphthalene
17	Hexene	56	Ethylthiophene	91	$C_{12}H_{26}$
18	Benzene	57	Dimethylthiophene	91	C_6-Alkylbenzene
18	Thiophene	58	Styrene	92	Triethylbenzene
19	C_6H_{10} Alkyne	59	Vinylthiophene	93	Methylnaphthalene
20	C_7H_{14}	60	isoPropylbenzene	94	Biphenyl
21	C_7H_{14}	61	Propylthiophene	95	C_6-Alkylbenzene
22	Butanol	62	Propylbenzene	97	Acenaphthalene
23	Methylhexyne	63	Trimethylthiophene	98	Ethylnaphthalene
23	Dimethylfuran	64	Benzaldehyde	99	C_6-Alkylbenzene
24	C_7H_{14}	65	Methylethylbenzene	100	Dodecane
25	C_7H_{14}	66	Methylstyrene		

Source: Ref. 3, from the Journal of Chromaographic Science, by permission of Preston Publications, Inc.

condensate which forms the bulk of the organic matter in meteoritic
rock. The high sulfur content is revealed by a range of products
including thiophene and alkyl derivatives. Surprisingly, trace
amounts of furans were also detected, although contamination may be
a possible origin for these. This product distribution effectively
parallels that obtained from high-resolution Py-MS and Py-GC-MS
studies [103]. The Py-GC-MS-DS system which was used allowed a wide
range of products to be detected, and aliphatic residues up to hep-
tadecane were identified. Stepped pyrolysis was used in this work,
which enabled a full profile of the organic constituents to be obtained. Most products were released at under 300°C. The Py-MS
technique was also applied to the Holbrook meteorite [103]. In contrast to the Murray meteorite, little volatile material was present
(benzene, acetone and sulphur dioxide) and most products were produced by pyrolysis. At temperatures of 300 to 480°C aromatic residues (alkyl-benzenes, naphthalenes, acenaphthalene, styrene), sulphur compounds (alkyl-thiophenes, benzothiophenes) and other hetero-
atomic species (benzofuran, benzonitrile) were the major products.
Processes such as sulfur-catalyzed dehydrogenation or thermal aromatization were excluded, which suggests that a highly aromatic polymer is present in this meteorite.

The stepped pyrolysis technique [104] has found many applications
in the determination of organic residues within an inorganic matrix.
Typically, the GC column is held at subambient temperatures (-60°C)
while the sample is subjected to heat (ranging from ambient, 150°,
300°, 460°, to 480°C), maintained for 5-10 mins. Carrier gas carries
the products through to be trapped by the column. Chromatographic
separation is achieved by a programmed temperature rise, after which
the cycle is repeated at a higher pyrolysis temperature.

This technique ensures that volatiles are properly detected
(all compounds up to $C_{15}H_{32}$ and significant proportions of those up
to $C_{17}H_{36}$ may be lost when solvent extraction is used), while nonex-
tractable residues (perhaps up to 70% of the indigenous organic
matter) are characterized by their pyrolysis products. The Allende

[105,106] and Murchison [106] meteorites have been characterized in this way. Comparisons reveal that the Allende meteorite contains less organics, but is richer in alkanes than the Murchison sample. The alkanes are liberated at low temperatures and 83% of the organics are released at 150°C. Typical profiles are recorded in Table 10.11.

Studies of meteorites have included the Orgueil [107], the Allen Hills meteorite [108] and other work on the Murchison [109-111] and Allende [112] meteorites. The study of amino acids as stable markers in terrestrial and extraterrestrial sources is also under active investigation [113].

The success of the Apollo lunar missions [114] and the Viking Landings on Mars [115] have now provided further information on the distribution of organic compounds within the solar system. A variety of mass spectrometric techniques, including Py-MS and Py-GC-MS were used to analyze both lunar fines and rock [116]. The major carbon-containing component was found to be carbon monoxide, together with smaller amounts of carbon dioxide and traces of methane. Various organic residues were also detected, but it would appear that none are indigenous to the lunar surface [117,118].

The most advanced studies to date have involved the fully automated Py-GC-MS-DS analysis of martian soils. The approach has been described in Chap. 3 [119-124] and involved unmanned landings on the surface of Mars on the twentieth of July 1976 (Viking I) and on the third of September 1976 (Viking II). The analytical systems were able to provide much information concerning the physical and chemical properties of Mars. The GC-MS modules, for example, measured the composition of the martian atmosphere and quantitatively detected rare gases [125-127]. It was with some disappointment therefore, that the search for organic residues in the martian surface proved fruitless. Stepped pyrolysis (200°, 350°, 500°C) was used, but despite success with test samples [121] only water and carbon dioxide were detected in surface and subsurface samples from the Chryse Planitia [122] and Utopia Planitia [123] regions. More complex

TABLE 10.11 Major Products from the Stepped Pyrolysis of Allende and Murchison Meteorites[a]

COMPONENT	ALLENDE			MURCHISON		
	50°	150°	300°	150°	300°	430°
Methane					1.4	60.5
Ethylene					5.3	37.2
Butenes					1.6	5.2
Benzene					0.9	4.1
Thiophene					1.0	6.8
Toluene	0.1		0.1	0.6	0.8	4.4
1-Methylthiophene					0.6	2.2
2-Methylthiophene					0.9	3.3
Octane	0.1					
Xylene	0.3	0.1	0.1			
Nonane	0.4	0.6	0.1			
Decanes		1.6				
C_3-Alkylbenzenes	0.3	2.9	0.1		3.1	4.9
Decane	0.8	6.3	0.3			
Dimethylthiophene					1.0	3.4
Decene		1.2				
Styrene			0.1		0.3	1.3
Undecanes		3.6				
Methylstyrene			0.8			
C_4-Alkylbenzene		8.0				
Undecane		5.4	4.1	0.2	1.7	0.2
Dimethylstyrene			0.8			
C_5-Alkylbenzenes		1.0				
Naphthalene		0.4	1.6		5.5	5.5
TOTAL ORGANIC PRODUCTS	5	50	14	0.8	24.1	138.8

[a]The table gives the amount of volatile product released in ppm at each pyrolysis temperature.

Source: Refs. 105 and 106.

TABLE 10.12 Viking Py-GC-MS System: Detection Limits

COMPONENT	PARTS PER 10^9		COMPONENT	PARTS PER 10^9	
	LANDER 1	LANDER 2		LANDER 1	LANDER 2
Methanol	10-90 ppm	<300-3000	Toluene	<0.5-5	<3-30
Ethanol	10-90 ppm	<9-90	Naphthalene	<0.05-0.5	<0.0015-0.015
Formaldehyde	10-90 ppm	<1200-12000	Acetone	<10-50	<250-2500
Ethane	10-90 ppm	<1200-12000	Furan	<0.1-1	<0.05-0.5
Propane	10-90 ppm	<3-30	Methylfuran	<0.2-2	<0.15-1.5
Butane	<1-10	<3-30	Acetonitrile	<1-10	<0.5-5
Hexane	<1-10	<0.5-5	Benzonitrile	<0.2-2	<0.015-0.15
Octane	<1-10	<0.15-1.5	Thiophene	<0.1-0.5	<0.015-0.15
Benzene	<0.5-5	<8-80	Methylthiophene	<0.1-0.5	<0.015-0.15

Source: Ref. 123.

molecules (methyl chloride, fluoroethers) were detected and identified, but were of earthly origin. The detection limits for typical
molecules which might have been expected to be produced from pyrolysis of an organic matrix are recorded in Table 10.12. That such
levels were not achieved on the surface of Mars was a considerable
surprise and contrasts markedly with the organic constituents of
meteorites. These results have evoked much discussion on the fate
of organic compounds on Mars [128]. The conflicting theories (e.g.,
photochemical dissociation, oxidation, dilution of meteoritic
sources) cannot be resolved at present, and it may well be that
future analysis of deeper or colder soils, which may protect organic
residues from the destructive forces on the surface, will initiate a
study of martian organic chemistry.

REFERENCES

1. Kunen, S.M., Voorhees, K.J., Hill, A.C., Hileman, F.D. and Os-
    borne, D.N., Chemical Analysis of the Insoluble Carbonaceous
    Components of Atmospheric Particles with Py-GC-MS, *Proc. Ann.
    Meet. Air Pollut. Control Assoc., 70*(1977)77.

2. Voorhees, K.J., Hileman, F.D. and Kunen, S.M., *Proc. 26th Ann.
    Conf. Mass Spectrom. Allied Topics*, St. Louis, 1978, p. 437.

3. Simmonds, P.G., Shulman, G.P. and Stembridge, C.H., Organic
    Analysis by Py-GC-MS. A Candidate Experiment for the Biologi-
    cal Exploration of Mars, *J. Chromatogr. Sci., 7*(1969)36-41.

4. Bolker, H.I. and Brenner, H.S., Polymeric Structure of Spruce
    Lignin, *Science, 170*(1970)173-176.

5. Watanabe, Y. and Kitao, K., Py-GC of Lignin, *Wood Res., 38*
    (1966)40-46.

6. Stahl, E., Karig, F., Brogman, U., Nimz, H. and Becker, H.,
    Thermofractography of Lignins and its Use for Rapid Analysis on
    the Ultra-microscale, *Holzforsch., 27*(1973)89-92.

7. Stieglitz, L. and Leger, W., Die Anwendung der Py-GC zur Analy-
    tik von schwerfluchtigen organischen wasser inhaltsstoffen, *Vom
    Wasser, 45*(1975)233-251.

8. Faix, O. and Schweers, W., Vergleichende Untersuchungen an
    Polymermodellen des Lignins verschiedener Zusammensetzungen.
    7. Pyrolyse, *Holzforsch., 29*(1975)224-229.

9. Martin, F., Saiz-Jimenez, C. and Gonzalez-Vila, F.J., Py-GC-MS
    of Lignins, *Holzforsch., 33*(1979)210-212.

10. Meuzelaar, H.L.C., Haider, K., Nagar, B.R. and Martin, J.P., Comparative Studies of Py-MS of Melanins, Model Phenolic Polymers and Humic Acids, *Geoderma, 17*(1977)239-252.

11. Bracewell, J.M., Robertson, G.W. and Williams, B.L., Py-MS Studies of Humification in a Peat and a Peaty Podzol, *J. Anal. Appl. Pyrol., 2*(1980)53-62.

12. Sigleo, A.C., Degraded Lignin Compounds Identified in Silicified Wood 200 Millions Years Old, *Science, 200*(1978)1054-1056.

13. Sigleo, A.C., Organic Geochemistry of Silicified Wood, Petrified Forest National Park, Arizona, *Geochim. Cosmochim. Acta, 42* (1978)1397-1405.

14. Iatridis, B. and Gavalas, G.R., Pyrolysis of Precipitated Kraft Lignin, *Ind. Eng. Chem. Prod. Res. Dev., 18*(1979)127-130.

15. Bracewell, J.M. and Robertson, G.W., Pyrolysis Studies on Humus in Freely Drained Scottish Soils, in C.E.R. Jones and C.A. Cramers (Eds.), *Analytical Pyrolysis*, Elsevier, Amsterdam, 1977, pp. 167-178.

16. Ragg, J.M., Bracewell, J.M., Logan, J. and Robertson, L., Some Characteristics of the Brown Forest Soils of Scotland, *J. Soil Sci., 29*(1978)228-242.

17. Tate, K.R. and Churchman, G.J., Organo-mineral Fractions of a Climosequence of Soils in New Zealand Tussock Grasslands, *J. Soil Sci., 29*(1978)331-339.

18. Bracewell, J.M., Characterisation of Soils by Pyrolysis Combined with MS, *Geoderma, 6*(1971)163-168.

19. Bracewell, J.M. and Robertson, G.W., Humus Type Discrimination Using Pattern Recognition of the MS of Volatile Pyrolysis Products, *J. Soil Sci., 24*(1973)421-428.

20. Bracewell, J.M., Robertson, G.W. and Stephen, G.J.M., Humus Type Discrimination from MS by a Simplified Statistical Treatment, *J. Soil Sci., 26*(1975)62-65.

21. Bracewell, J.M. and Robertson, G.W., A Py-GC Method for Discrimination of Soil Humus Types, *J. Soil Sci., 27*(1976)196-205.

22. Bracewell, J.M., Robertson, G.W. and Tate, K.R., Py-GC Studies on a Climosequence of Soils in Tussock Grasslands, New Zealand, *Geoderma, 15*(1976)209-215.

23. Nagar, B.R., Examination of the Structure of Soil Humic Acids by Py-GC, *Nature, 199*(1963)1213-1214.

24. Kimber, R.W.L. and Searle, P.L., Py-GC of Soil Organic Matter. I. Introduction and Methodology, *Geoderma, 4*(1970)47-55.

25. Kimber, R.W.L. and Searle, P.L., Py-GC of Soil Organic Matter. 2. The Effect of Extractant and Soil History on the Yields of Products from Pyrolysis of Humic Acids, *Geoderma, 4*(1970)57-71.

26. Martin, F., Saiz-Jimenez, C. and Cert, A., Py-GC-MS of Soil
     Humic Fractions: I. The Low Boiling Point Compounds, *Soil Sci.
     Soc. Amer. J.*, *41*(1977)1114-1118.

27. Martin, F., Saiz-Jimenez, C. and Cert, A., Py-GC-MS of Soil
     Humic Fractions: II. The High Boiling Point Compounds, *Soil
     Sci. Soc. Amer. J.*, *43*(1979)309-312.

28. Martin, F., Saiz-Jimenez, C. and Gonzalez-Vila, F.J., Degrada-
     tion Thermique de Lignines et Acides Humiques, *Compt. Rend. J.
     Internat. d'Etude l'Assemblee Generale (Nancy)*, *17-19*(May 1978)
     241-247.

29. Martin, F., Characteristicas Analiticas y productos de Pirolisis
     de Acidos Humicos Pardos y Grises extraidos de un Vertisol,
     Agrochimica, *21*(1977)190-199.

30. Wershaw, R.L. and Bohner, G.E., Pyrolysis of Humic and Fulvic
     Acids, *Geochim. Cosmochim. Acta*, *33*(1969)757-762.

31. Martin F., Effects of Extractants on Analytical Characteristics
     and Py-GC of Podzol Fulvic Acids, *Geoderma*, *15*(1976)253-265.

32. Martin, F., The Determination of Polysaccharides in Fulvic
     Acids by Py-GC, in C.E.R. Jones and C.A. Cramers (Eds.),
     Analytical Pyrolysis, Elsevier, Amsterdam, 1977, pp. 179-187.

33. Martin, F., Py-GC of Humic Substances from Different Origin,
     Z. Pflazenernahr. Bodenk., *4*(1975)407-416.

34. Saiz-Jimenez, C., Haider, K. and Meuzelaar, H.L.C., Comparisons
     of Soil Organic Matter and its Fractions by Py-MS, *Geoderma*, *22*
     (1979)25-37.

35. Cheshire, M.V., Greaves, M.P. and Mundie, C.M., Decomposition of
     Soil Polysaccharides, *J. Soil Sci.*, *25*(1974)483-498.

36. Halma, G., Miedema, R., Posthumus, M.A., van de Westeringh, W.
     and Meuzelaar, H.L.C., Characterisation of Soil Types by Py-MS,
     Agrochima, *22*(1978)372-382.

37. Meuzelaar, H.L.C., Haider, K., Nagar, B.R. and Martin, J.P.,
     Comparative Studies of Py-MS of Melanins, Model Phenolic Poly-
     mers and Humic Acids, *Geoderma*, *17*(1977)239-252.

38. Haider, K., Nagar, B.R., Saiz, C., Meuzelaar, H.L.C. and Martin,
     J.P., Studies on Soil Humic Compounds, Fungal Melanins and Model
     Polymers by Py-MS, *Soil Organic Matter Studies (IAEA)*, *2*(1977)
     213-220.

39. Nagar, B.R., Waight, E.S., Meuzelaar, H.L.C. and Kistemaker, P.
     G., Studies on the Structure and Origin of Soil Humic Acids by
     Curie-point Pyrolysis in Direct Combination with Low-voltage MS,
     Plant Soil, *43*(1975)681-685.

40. Saiz-Jimenez, C., Martin, F., Haider, K. and Meuzelaar, H.L.C.,
     Comparison of Humic and Fulvic Acids from Different Soils by Py-
     MS, *Agrochima*, *22*(1978)353-359.

41. Gonzalez-Vila, F.J., Martin, F., Saiz-Jimenez, C. and Nimz, A. H., Thermofractography of Humic Substances, *J. Thermal Anal.*, *15*(1979)279-284.

42. Saiz-Jimenez, C., Martin, F. and Cert, A., Low Boiling-point Compounds Produced by Pyrolysis of Fungal Melanins and Model Phenolic Polymers, *Soil Biol. Biochem.*, *11*(1979)305-309.

43. Saiz-Jimenez, C. and Martin, F., Py-GC of the Pigment of *Eurotium echinulatum* Delacr., *Humus et Plants, 6*(1975)11-16.

44. Tsuchida, H., Komoto, M., Kato, H., Kurata, T. and Fujimaki, M., Identification of Heterocyclic N-Compounds Produced by Pyrolysis of the Non-dializable Melanoidin, *Agr. Biol. Chem.*, *40*(1976) 2051-2056.

45. Abelson, P.H., Organic Matter in the Earth's Crust, *Ann. Rev. Earth Planet Sci.*, *6*(1978)325-351.

46. Harwood, R.J., Oil and Gas Generation by Laboratory Pyrolysis of Kerogen, *Amer. Assoc. Petrol. Geol. Bull.*, *61*(1977)2082-2102.

47. Tissot, B., Durand, B., Espitalie, J. and Combaz, A., Influence of Nature and Diagenesis of Organic Matter in Formation of Petroleum, *Bull. Amer. Assoc. Petrol. Geol.*, *58*(1974)499-506.

48. Claypool, G.E. and Reed, P.R., Thermal Analysis Technique for Source Rock Evaluation: Quantitative Estimate of Richness and Effects of Lithographic Variation, *Bull. Amer. Assoc. Petrol. Geol.*, *60*(1976)608-626.

49. Barker, C., Pyrolysis Techniques for Source Rock Evaluation, *Bull. Amer. Assoc. Petrol. Geol.*, *58*(1974)2349-2361.

50. Urov, K., Thermal Decomposition of Kerogens. Mechanism and Analytical Application, *J. Anal. Appl. Pyrol.*, *1*(1980)323-338.

51. Klesment, I., Investigation of Aliphatic Structures of Oil Shales by Pyrolysis and Chromatographic Methods, *J. Anal. Appl. Pyrol.*, *2*(1980)63-77.

52. Robillard, M.V., Siggia, S. and Uden, P.C., Effect of Oxygen on Composition of Light Hydrocarbons Evolved in Oil Shale Pyrolysis, *Analyt. Chem.*, *51*(1979)435-439.

53. Robin, P.L. and Rouxhet, P.G., Characterisation of Kerogens and Study of their Evolution by IR Spectroscopy: Carbonyl and Carboxyl Groups, *Geochim. Cosmochim. Acta, 42*(1978)1341-1349.

54. Leventhal, J.S., Stepwise Py-GC of Kerogen in Sedimentary Rocks, *Chem. Geol.*, *18*(1976)5-20.

55. Solli, H., Larter, S.R. and Douglas, A.G., The Analysis of Kerogens by Py-GC-MS Using Selective Ion Monitoring. 2. Alkylnaphthalenes, *J. Anal. Appl. Pyrol.*, *1*(1980)231-241.

56. Maters, W.L., Meent, V.d.V., Schuyl, P.J.W., de Leeuw, J.W., Schenck, P.A. and Meuzelaar, H.L.C., Curie-point Pyrolysis in Organic Geochemistry, in C.E.R. Jones and C.A. Cramers (Eds.), *Analytical Pyrolysis*, Elsevier, Amsterdam, 1977, pp. 203-216.

57. Larter, S.R., Solli, H. and Douglas, A.G., Analysis of Kerogens
     by Py-GC-MS Using Selective Ion Detection, *J. Chromatogr., 167*
     (1978)421-431.

58. Giraud, A., Application of Py-GC to Geochemical Characterisa-
     tion of Kerogen in Sedimentary Rock, *Bull. Amer. Assoc. Petrol.
     Geol., 54*(1970)439-451.

59. Ishiwatari, R., Ishiwatari, M., Rohrback, B.G. and Kaplan, I.R.,
     Thermal Alteration Experiments on Organic Matter from Recent
     Marine Sediments in Relation to Petroleum Genesis, *Geochim.
     Cosmochim. Acta, 41*(1977)815-828.

60. Philip, P.R., Calvin, M., Brown, S. and Young, E., Organic Geo-
     chemical Studies on Kerogen Precursors in Recently Deposited
     Algal Mats and Oozes, *Chem. Geol., 22*(1978)207-231.

61. Larter, S.R., Solli, H., Douglas, A.G., de Large, F. and de
     Leeuw, J.W., Occurrence and Significance of Prist-1-ene in
     Kerogen Pyrolysates, *Nature, 279*(1979)405-408.

62. Gallegos, E.J., Terpane-Sterane Release from Kerogen by Py-GC-
     MS, *Analyt. Chem., 47*(1975)1524-1528.

63. Seifert, W.K., Steranes and Terpanes in Kerogen Pyrolysis for
     Correlation of Oils and Source Rocks, *Geochim. Cosmochim. Acta,
     42*(1978)473-484.

64. Larter, S.R., Horsfield, B. and Douglas, A.G., Pyrolysis as a
     Possible Means of Determining the Petroleum Generating of Sedi-
     mentary Organic Matter, in C.E.R. Jones and C.A. Cramers (Eds.),
     Analytical Pyrolysis, Elsevier, Amsterdam, 1977, pp. 189-202.

65. Larter, S.R. and Douglas, A.G., Low Molecular Weight Aromatic
     Hydrocarbons in Coal Maceral Pyrolysates as Indicators of Dia-
     genesis and Organic Matter Type, in W.E. Krumbein (Ed.), *En-
     vironmental Biogeochemistry and Geomicrobiology*, Ann Arbor Sci.,
     Ann Arbor, Michigan, 1978, pp. 373-386.

66. Espitalie, J., Laporte, J.L., Madec, M., Marquis, P., Leplat,
     P., Paulet, J. and Boutefeu, A., Rapid Method for the Charac-
     terisation of Source Rocks--Their Petroleum Potential and De-
     gree of Evolution, *Rev. Inst. Fr. Petrol., 32*(1977)23-42.

67. McHugh, D.J., Saxby, J.D. and Tardiff, J.W., Py-Hy-GC of Car-
     bonaceous Material from Australian Sediment, Part II. Kerogens
     from Some Australian Coals, *Chem. Geol., 21*(1978)1-14.

68. Whelan, J.K., Hunt, J.M. and Huc, A.Y., Application of Thermal-
     Distillation-Pyrolysis to Petroleum Source Rock Studies and
     Marine Pollution, *J. Anal. Appl. Pyrol., 2*(1980)79-96.

69. Seifert, W.K. and Moldowan, J.M., Applications of Steranes,
     Terpanes and Monoaromatics to the Maturation, Migration and
     Source of Crude Oils, *Geochim. Cosmochim. Acta, 42*(1978)77-95.

70. Pym, J.G., Ray, J.E., Smith, G.W. and Whitehead, E.V., Petroleum
     Triterpane Fingerprinting of Crude Oils, *Analyt. Chem., 47*(1975)
     1617-1622.

71. Seifert, W.K., Moldowan, J.M., Smith, G.W. and Whitehead, E.V., First Proof of Structure of a C_{28}-Pentacyclic Triterpane in Petroleum, *Nature, 271*(1978)436-437.

72. Wardroper, A.M.K., Brooks, P.W., Humberson, M.J. and Maxwell, J.R., Analysis of Steranes and Triterpanes in Geolipid Extracts by Automatic Classification of MS, *Geochim. Cosmochim. Acta, 41* (1977)499-510.

73. Hanson, R.L., Brookins, D. and Vanderborgh, N.E., Stoichiometric Analysis of Oil Shales by Laser Py-GC, *Analyt. Chem., 48*(1976) 2210-2214.

74. Vanderborgh, N.E., Laser Induced Pyrolysis Techniques, in C.E.R. Jones and C.A. Cramers (Eds.), *Analytical Pyrolysis*, Elsevier, Amsterdam, 1977, pp. 235-248.

75. Vanderborgh, N.E., Fletcher, M.A. and Jones, C.E.R., Laser Pyrolysis of Carbonaceous Rocks, *J. Anal. Appl. Pyrol., 1*(1979) 177-186.

76. Jones, C.E.R. and Vanderborgh, N.E., Elucidation of Geomatrices by Laser Py-GC and Py-MS, *J. Chromatogr., 186*(1979)831-841.

77. Biscar, J.P., On-line Laser Pyrolysis Cell for GC, *Analyt. Chem., 43*(1971)982-983.

78. Biscar, J.P., Laser Pyrolysis of Oil Shales, *J. Chromatogr., 56* (1971)348-352.

79. Vastole, F.J., Pirone, A.J., Given, P.H. and Dutcher, R.R., Analysis of Coal with the Laser Py-MS, *Amer. Chem. Soc. Div. Fuel Chem. Preprints, 11*(1967)229-231.

80. Joy, W.K., Ladner, W.R. and Pritchard, E., Laser Heating of Pulverised Coal in the Source of a TOF MS, *Fuel, 49*(1970)26-38.

81. Hanson, R.L., Plasma Quenching Reactions with Laser Pyrolysis of Graphite and Coal in Helium or Hydrogen, *Carbon, 16*(1978) 159-162.

82. Hanson, R.L. and Vanderborgh, N.E., Characterisation of Coals Using Laser Py-GC, in C. Karr (Jr.), (Ed.), *Analytical Methods for Coal and Coal Products, Vol. 3*, Academic Press, New York, 1979, pp. 73-103.

83. Gallegos, E.J., *in* Analytical Chemistry of Liquid Fuel Sources: Tar Sands, Oil Shale, Coal and Petroleum, P.C. Uden, S. Siggia and H.B. Jensen, Eds., *Advances in Chemistry Series* No. 170; American Chemical Society, Washington, D.C., 1978, pp. 13-36.

84. Benjamin, B.M., Raaen, V.F., Maupin, P.H., Brown, L.L. and Collins, C.J., Thermal Cleavage of Chemical Bonds in Selected Coal-related Structures, *Fuel, 57*(1978)269-272.

85. Ramljack, Z., Deur-Siftar, D. and Solc, A., Characterisation of Bitumens by Py-GC, *J. Chromatogr., 119*(1976)445-450.

86. Leplat, P., Application of Py-GC to the Study of the Non-
     volatile Petroleum Fractions, *J. Gas Chromatogr.*, 5(1967)128-
     135.

87. Posadov, I.A., Pokonova, Y.V., Popov, O.G. and Proskuryakov,
     V.A., Study of the Chemical Structure of Petroleum Asphalt-
     enes by Pyrolytic Methods, *J. Appl. Chem. USSR*, 50(1977)1516-
     1518.

88. Knotnerus, J., Constitution of Asphaltic Bitumen. Character-
     isation of Bitumens by a Combination of Py-Hy-GC, *Ind. Eng.
     Chem. (Prod. Res. Develop.)*, 6(1967)43-52.

89. Poxon, D.W. and Wright, R.G., The Characterisation of Bitumens
     using Py-GC, *J. Chromatogr.*, 61(1971)142-144.

90. Glajch, J.L., Lubkowitz, J.A. and Rogers, L.B., Py-GC Applied
     to Coal Tar and Petroleum Pitches, *J. Chromatogr.*, 168(1979)
     355-364.

91. Giam, C.S., Goodwin, T.E., Giam, P.Y., Rion, K.F. and Smith,
     S.G., Characterisation of Lignites by Py-GC, *Analyt. Chem.*, 49
     (1977)1540-1543.

92. Suuberg, E.M., Peters, W.A. and Howard, J.B., Product Composi-
     tion and Kinetics of Lignite Pyrolysis, *I&EC Process Design
     Develop.*, 17(1978)37-46.

93. Gray, D., Cogoli, J.G. and Essenhigh, R., Problems in Pulver-
     ised Coal and Char Combustion in C.G. Massey (Ed.), Coal
     Gasification, *Advances in Chemistry Series 13,* American
     Chemical Society, Washington, D.C., 1974, pp. 72-91.

94. Romovacek, J. and Kubat, J., Characterisation of Coal Substance
     by Py-GC, *Analyt. Chem.*, 40(1968)1119-1126.

95. Bannerjee, N.N., Ghosh, B. and Nair, C.S.B., Effect of Argon
     and Nitrogen Atmospheres and Cracking Catalyst on Flash Pyroly-
     sis of Makum Coal, *Fuel*, 56(1977)192-194.

96. Granger, A.F. and Ladner, W.R., The Flash Heating of Pulver-
     ised Coal, *Fuel*, 49(1970)17-25.

97. Giraud, A. and Bestougeff, M.A., Characterisation of High
     Molecular Weight Sulphur Compounds in Petroleum by Py-GC, *J.
     Gas Chromatogr.*, 5(1967)464-470.

98. Drushel, H.V., Sulphur Compound Distributions in Petroleum Us-
     ing an In-Line Reactor or Pyrolyser Combined with GC and a
     Microcoulometric Detector, *Analyt. Chem.*, 41(1969)569-576.

99. McHugh, D.J., Saxby, J.D. and Tardiff, J.W., Py-Hy-GC of Car-
     bonaceous Material from Australian Sediments, Pt. 1. Some
     Australian Coals, *Chem. Geol.*, 17(1976)243-259.

100. Holden, H.W. and Robb, J.C., A Study of Coal by MS. II. Ex-
     tracts and Extractable Pyrolysis Products, *Fuel*, 39(1960)485-
     494.

101. Karn, F.S., Friedel, R.A. and Sharkey (Jr.), A.G., Studies of the Solid and Gaseous Products from Laser Pyrolysis of Coal, *Fuel, 51*(1972)113-115.

102. Vahrman, M. and Watts, R.M., The Smaller Molecules Obtainable from Coal and Their Significance. Pt. 5. Composition of Vacuum Tars, *Fuel, 51*(1972)130-134.

103. Hayes, J.M. and Biemann, K., High-Resolution MS Investigations of the Organic Constituents of the Murray and Holbrook Chondrites, *Geochim. Cosmochim. Acta, 32*(1968)239-267.

104. Levy, R.L., Wolf, C.J. and Oro, J., A GC Method for the Characterisation of the Organic Content Present in an Inorganic Matrix, *J. Chromatogr. Sci., 8*(1970)524-526.

105. Levy, R.L., Wolf, C.J., Grayson, M.A., Gilbert, J., Gelpi, E., Updegrove, W.S., Zlatkis, A. and Oro, J., Organic Analysis of the Pueblito de Allende Meteorite, *Nature, 227*(1970)148-150.

106. Levy, R.L., Grayson, M.A. and Wolf, C.J., The Organic Analysis of the Murchison Meteorite, *Geochim. Cosmochim. Acta, 37*(1973) 467-483.

107. Bandurski, E.L. and Nagy, B., The Polymer-like Organic Material in the Orgeuil Meteorite, *Geochim. Cosmochim. Acta, 40*(1970) 1397-1406.

108. Holzer, G. and Oro, J., The Organic Composition of the Allan Hills Carbonaceous Chondrite (77306) as Determined by Py-GC-MS and other Methods, *J. Mol. Evol., 13*(1979)265-270.

109. Hayatsu, R., Matsuoka, S., Scott, R.G., Studier, M.H. and Anders, E., Origin of Organic Matter in the Early Solar System. VII. The Organic Polymer in Carbonaceous Chondrites, *Geochim. Cosmochim. Acta, 41*(1977)1325-1339.

110. Oro, J., Gilbert, J., Lichtenstein, H., Wikstrom, S. and Flory, D.A., Amino-acids, Aliphatic and Aromatic Hydrocarbons in the Murchison Meteorite, *Nature, 230*(1971)105-106.

111. Holzer, G. and Oro, J., Pyrolysis of Organic Compounds in the Presence of Ammonia. The Viking Mars Lander Site Alteration Experiment, *Org. Geochem., 1*(1977)37-52.

112. Studier, M.H., Hayatsu, R. and Anders, E., Origin of Organic Matters in Early Solar System. V. Further Studies of Meteoritic Hydrocarbons and a Discussion of Their Origin, *Geochim. Cosmochim. Acta, 36*(1972)189-215.

113. Kvenvolden, K.A., Advances in the Geochemistry of Amino Acids, *Ann. Rev. Earth Planet Sci., 3*(1975)183-212.

114. P.H. Abelson (Ed.), Moon Issue, *Science, 167*(1970)418-784.

115. P.H. Abelson (Ed.), The Viking Mission, *Science, 194*(1976) 1274-1353.

116. Oro, J., Gilbert, J., Updegrove, W., McReynolds, J., Ibanez, J., Gil-Av, E., Flory, D. and Zlatkis, A., GC and MS Methods Applied to the Analysis of Lunar Samples from the Sea of Tranquility, *J. Chromatogr. Sci., 8*(1970)297-308.

117. Burlingame, A.L., Calvin, M., Han, J., Henderson, W., Reed, W. and Simoneit, B.R., Lunar Organic Compounds: Search and Characterisation, *Science, 167*(1970)751-752.

118. Murphy, R.C., Preti, G., Nafissi, V.M.M. and Biemann, K., Search for Organic Material in Lunar Fines by MS, *Science, 167*(1970)755-757.

119. Johnson, R.D. and Davis, C.C., Py-Hydrogen FID of Organic Carbon in a Lunar Sample, *Science, 167*(1970)759-760.

120. Anderson, D.M., Biemann, K., Orgel, L.E., Oro, J., Owen, T., Shulman, G.P., Toulmin III, P. and Urey, H.C., MS Analysis of Organic Compounds, Water, and Volatile Constituents in the Atmosphere and Surface of Mars: The Viking Mars Lander, *Icarus, 16*(1972)111-138.

121. Biemann, K., Test Results on the Viking GC-MS Experiment, *Origins of Life, 5*(1974)417-430.

122. Biemann, K., Oro, J., Toulmin III, P., Orgel, L.E., Nier, A.O., Anderson, D.M., Simmonds, P.G., Flory, D., Diaz, A.V., Rushneck, D.R. and Biller, J.A., Search for Organic and Volatile Inorganic Compounds in Two Surface Samples from the Chryse Planitia Region of Mars, *Science, 194*(1976)72-76.

123. Biemann, K., Oro, J., Toulmin III, P., Orgel, L.E., Nier, A.O., Anderson, D.M., Simmonds, P.G., Flory, D., Diaz, A.V., Rushneck, D.R., Biller, J.E. and LaFleur, A.L., The Search for Organic Substances and Inorganic Volatile Compounds in the Surface of Mars, *J. Geophys. Res., 82*(1977)4641-4658.

124. Rushneck, D.R., Diaz, A.V., Howarth, D.W., Rampacek, J., Olsen, K.W., Dencker, W.D., Smith, P., McDavid, L., Tomassian, A., Harris, M., Bulota, K., Biemann, K., LaFleur, A.L., Biller, J.E. and Owen, T., Viking GC-MS, *Rev. Sci. Instrum., 49*(1978)817-834.

125. Owen, T. and Biemann, K., Composition of the Atmosphere at the Surface of Mars: Detection of Argon-36 and Preliminary Analysis, *Science, 193*(1976)801-803.

126. Biemann, K., Owen, T., Rushneck, D.R., LaFleur, A.L. and Howarth, D.W., The Atmosphere of Mars Near the Surface: Isotope Ratios and Upper Limits on Noble Gases, *Science, 194*(1976)76-78.

127. Owen, T., Biemann, K., Rushneck, D.R., Biller, J.E., Howarth, D.W. and LaFleur, A.L., The Atmosphere of Mars: Detection of Krypton and Xenon, *Science, 194*(1976)1293-1295.

128. Anders, E. and Biemann, K., Mars and Earth: Origin and Abundance of Volatiles, *Science, 198*(1977)453-465.

Appendix 1

Abbreviations

A	Eddy diffusion term, van Deemter equation
A'	Peak area
AW	Acid-washed chromatographic support
α	Relative retention factor
B	Solute diffusion term, van Deemter equation
BSA	Bovine serum albumin
C	Mass transfer term, van Deemter equation
C_g	Resistance to mass transfer in gaseous phase
C_l	Resistance to mass transfer in liquid phase
CAMS	Collisional activation mass spectrometry
CCGC	Capillary-column gas chromatography
CIMS	Chemical-ionization mass spectrometry
D_g	Solute diffusivity in carrier gas
D_l	Solute diffusivity in stationary phase
d_f	Stationary phase thickness
d_p	Support particle diameter
DADI	Direct analysis of daughter ions
DMCS	Dimethylchlorosilazane
DS	Data system
E	Electric field strength, column coating efficiency
E_u	Minimum detectable quantity
ECD	Electron-capture detector
EIMS	Electron-impact mass spectrometry
F_c	Carrier gas flow rate

FDMS	Field-desorption mass spectrometry
FID	Flame ionization detector
FIMS	Field-ionization mass spectrometry
GC	Gas chromatography
H	Height equivalent of a theoretical plate
H	Magnetic field strength
h	Peak height
HMDS	Hexadimethyldisilazane
I	Peak intensity, Kovats retention index
k	Capacity factor
L	Column length
LAMMA	Laser microprobe mass analysis
LPTDMS	Linear-programmed thermal-degradation mass spectrometry
MIKE	Mass-selection ion kinetic energy analysis
MS	Mass spectrometry
m/z	Mass-to-charge ratio
Mass pyrogram	Pyrolysate mass spectrum
N	number of theoretical plates
N_B	Instrumental noise
P	Peak ratio
P_i	Column inlet pressure
P_o	Column outlet pressure
Py-CIMS	Pyrolysis chemical-ionization mass spectrometry
Py-FDMS	Pyrolysis field-desorption mass spectrometry
Py-FIMS	Pyrolysis field-ionization mass spectrometry
Py-GC	Pyrolysis gas chromatography
Py-GC-MS	Pyrolysis gas-chromatography mass spectrometry
Py-GC-MS-DS	Pyrolysis gas-chromatography mass-spectrometry data system
Py-Hy-GC	Pyrolysis-hydrogenation gas chromatography
Py-IR	Pyrolysis infrared spectroscopy
Py-MS	Pyrolysis mass spectrometry
Pyrogram	Pyrolysate chromatogram
R	Retention factor
R_p	Resolving power

R_s	Resolution
r	Radius of column bore, flight tube radius
SCOT	Support-coated open-tubular column
SSA	Specific surface area
σ_d	Detector sensitivity
σ_r	Recorder sensitivity
T_{eq}	Equilibrium pyrolysis temperature
TAS	Thermal application thin-layer chromatography
THT	Total heating time
T_R	Programmed retention temperature
TRT	Temperature rise time
t	Retention time
t'	Net retention time
U	Carrier gas velocity
V	Accelerating voltage, chart recorder speed
v	Ion velocity
W	Peak width at base
$W_{\frac{1}{2}}$	Peak width at half-height
W_1	Weight of liquid phase in column
w_A	Weight of solute
WCOT	Wall-coated open-tubular column

Appendix 2

Checklists

A2.1 Pyrolysis

	HEATED FILAMENT	CURIE POINT	FURNACE	LASER
PYROLYSER	Boosted or fixed voltage	Wire, tube or foil	Solids or volatiles	High or low power
	Ribbon, wire or coil	Composition of conductor	Dynamic or static	Pulsed or continuous
	Composition of conductor	Dimensions of conductor	Pulsed or continuous	Duration
	Sample holder	Position of conductor	Open or packed tube	Area of exposure
	Sample holder material	R$_f$ power	Chamber material	Additives
	Preparation and cleaning of conductor and chamber		Preparation and cleaning of sample holder and chamber	
TEMPERATURE	TRT	Teq	Pyrolysis temperature	Heating duration
	Cooling profile		Variation of pyrogram with temperature	
SAMPLE	Preparation	State	Weight	Loading method
	Solvent	Internal standard	Position within pyrolyser	
PRODUCTS	Identification	Primary or secondary	Reproducibility	Specificity
	Contamination from pyrolyser		Dependence upon flow rate and carrier gas	
ANALYTICAL DEVICE	On-line or off-line	GC, MS, or GCMS	Position of pyrolyser	Injector temperature
	Connection of pyrolyser - dead volume		Special techniques (FIMS,FDMS, CAMS)	

A2.2 Gas Chromatography

	Packed – coated or porous polymer		Capillary – WCOT or SCOT	
COLUMN	Glass or SS	ID (2r)	Length (L)	Inlet-outlet splitting
	Nature	Deactivation		
SUPPORT	Nature		Particle size (d_p)	Particle shape
	Free-fall density		Packed density	
STATIONARY PHASE	Nature	Polarity	Loading	Thickness (d_f)
	Temperature range	Weight (W_L)	Volume (V_L)	Density
CARRIER GAS	Nature	Flow-rate (F_C)	Linear velocity (U)	Mass flow control
TEMPERATURE	Isothermal temp.	Programmed: initial T	Heating rate	Final period
DETECTOR	Flame-ionisation	Alkali-bead	Electron-capture	Thermal-conductivity
	Temperature	MS range	Separator	Gas flows
TUNING	Standards	Retention times	Internal standards	Resolution
	Loading	Column degeneration	Peak ratio	Pyrolyser tuned
IDENTIFICATION	Retention times	MS	Pyrolysis	Isolation
QUANTIFICATION	Base-line separation	Base-line drift	Internal standards	Normalisation
	Attenuation	Manual or integrator	Computer	Mass fragmentography

A2.3 GC Parameters

P A R A M E T E R		VALUE
Column length	L	
Column radius	r	
Flow-rate	F_C	
Retention time (air)	t_O	
Gas hold-up volume	$V_M = t_O F_C$	
Inlet pressure	P_i	
Outlet pressure	P_O	
j	$j = \dfrac{3[(P_i/P_O)^2 - 1]}{2[(P_i/P_O)^3 - 1]}$	
Weight of liquid on column	W_L	
Volume of liquid on column	V_L	

If an air peak (t_O) is not apparent, this parameter may be calculated from a knowledge that the plot of log t' versus carbon number for homologous n-alkanes is linear.

Three paraffins are chosen with chain lengths (n) so that

$$n_2 = \frac{n_1 + n_3}{2}$$

The retention time of an unretained solute may then be calculated from

$$t_O = \frac{t_2^2 - t_1 t_3}{t_1 - 2t_2 + t_3}$$

where t_1, t_2, and t_3 are the retention times of the three n-alkanes with n_1, n_2, and n_3 carbon atoms.

A2.4 GC Parameters: Calculation

P A R A M E T E R		COMPONENT A	COMPONENT B
Retention time	t		
Corrected retention time	$t' = t - t_0$		
Relative retention time	$t_r = t'_A / t'_B$		
Peak width	W		
Half-height peak width	$W_{\frac{1}{2}}$		
Peak height	h		
Peak area	$A' = h.W_{\frac{1}{2}}$		
Peak ratio	$P = A'_A / A'_B$		
Retention volume	$V_R = t.F_c$		
Adjusted retention volume	$V'_R = V_R - V_M$		
Net retention volume	$V_N = j.V'_R$		
Specific retention volume	$V_g = 273 V_N / W_L T$		
Partition coefficient	$K = V_N / V_L$		
Retention factor	$R = t_0 / t$		
Capacity factor	$k = t' / t_0$		
Relative retention factor	$a = k_B / k_A$		
Plate number	$N = 5.54(t/W_{\frac{1}{2}})^2$		
Effective plate number	$N_{eff} = 5.54(t'/W_{\frac{1}{2}})^2$		
HETP	$H = L / N_{eff}$		
Resolution	$R_s = 2(t_B-t_A) / (W_A+W_B)$		
Kovats Index	$I = 100 \left[\dfrac{\log(t'/t'_n)}{\log(t'_{n+1}/t'_n)} \right] + 100n$		
Minimum HETP	$H_{min} = r\left[\dfrac{1 + 6k + 11k^2}{3(1+k)^2} \right]^{\frac{1}{2}}$		
Coating efficiency	$E = 100H / H_{min}$		
Retention temperature	T_R		
Isothermal retention time at T_R	t_i		
Resolution	$R_s = \dfrac{T_{R_B} - T_{R_A}}{t_{i_A} + t_{i_B}} \cdot \dfrac{N^{\frac{1}{2}}}{2W_L \Delta}$		
Heating rate	Δ		

A2.5 Pyrolysis Mass Spectrometry

PYROLYSIS	Curie-point	Heated filament	Laser	Probe
INTERFACE	Expansion chamber	Heated line	Insertion	Temperature
IONISATION	EI	CI	FI	FD
	Ionisation voltage	Source pressure	Reactant gas	
SEPARATION	Magnetic sector	Double focusing	Quadrupole	ToF
	Mass range		Resolution	
DETECTION	Analogue	Ion-counting	Photoplate	Data system
	Scan speed	Scans	Background subtraction	
TUNING	MS standardisation		Pyrogram standardisation	
SPECIAL TECHNIQUES	High-resolution		CAMS	

Appendix 3

Computer Programs

Programs DISTANCE, LEARN, and SIMCO are written in Microsoft BASIC
to run on a TRS-80 Level II system. In the listings the printer
has transcribed exponentiation (↑) as [.

Program STAT is written in FORTRAN IV to run on an ICL 1904S
system.

A3.1 DISTANCE

```
100 CLS:DIM D(20,20),IN(20,10),W(10)
110 PRINT"CALCULATION OF MULTIDIMENSIONAL EUCLIDEAN DISTANCES"
120 INPUT"HOW MANY SAMPLES";M
130 INPUT"HOW MANY DIMENSIONS";N
140 FOR I=1TON:PRINT "WEIGHT PEAK";I;:INPUT "  ";W(I):NEXTI
150 SW=0:FOR I=1TON:SW=SW+W(I):NEXTI
160 FOR I=1TOM:FOR J=1TON
170 PRINT"SAMPLE";I;"   PEAK";J;:INPUT "   ";IN(I,J)
180 NEXTJ:NEXTI
190 FOR I=1TOM-1:FOR J=I+1TOM:D(I,J)=0
200 FOR K=1TON:D(I,J)=D(I,J)+((IN(I,K)-IN(J,K))[2)*W(K):NEXTK
210 D(I,J)=SQR((D(I,J))/SW)
220 PRINT"D(";I;",";J;")=";D(I,J)
230 NEXTJ:INPUT"PRESS ENTER TO CONTINUE";Z:NEXTI
240 FOR I=MTO2STEP-1:PRINT USING"#######";I;:NEXTI:PRINT
250 FOR I=1TOM-1:PRINTI;
260 FOR J=MTOI+1STEP-1:PRINT USING"####.##";D(I,J);
270 NEXTJ:PRINT:NEXTI:END
```

This program calculates multidimensional euclidean distances.

    Input data: Intensity data from Table 5.12.

    Output data: Table 5.13.

A3.2 LEARN*

```
100 DIM DA(50,5),LI(50),W(6),IT(50),IP(50),NS(50),A$(50)
110 CLS:PRINT"LINEAR LEARNING MACHINE PROGRAM"
120 INPUT "NUMBER IN TRAINING SET ";NT
130 INPUT "NUMBER IN PREDICTION SET";NP
140 INPUT "NUMBER OF FACTORS";NU
150 N1=NT+NP:WI=0.1:TS=0.75:QU=500:MM=NU+1
160 PRINT"TRAINING SET":FORI=1TONT:PRINTI;"CLASS ";;
170 INPUT LI(I):FORJ=1TONU:PRINT"FACTOR";J;:INPUTDA(I,J)
180 NEXTJ:NEXTI
190 PRINT"PREDICTION SET":FORI=1+NTTON1:PRINTI-NT;
200 INPUT "NAME";A$(I-NT):INPUT "CLASS";LI(I)
210 FOR J=1TONU:PRINT"FACTOR";J;:INPUT DA(I,J):NEXTJ:NEXTI
220 FORI=1TONT:IT(I)=I:NEXTI:FORI=1TONP:IP(I)=I+NT:NEXTI
230 FORI=1TONU:W(I)=WI:NEXTI:W(MM)=WI
240 GOSUB 250::GOSUB 470:TS=0.0:GOSUB 470:GOTO 110
250 CLS:PRINT"TRAINING SET";NT;"  DEAD ZONE";TS
260 NF=0:KV=0
270 KK=0:IF KV<=0 GOTO 290
280 N2=KV:GOTO 300
290 N2=NT:FORI=1TONT:NS(I)=IT(I):NEXTI
300 FORIR=1TON2:I=NS(IR):S=W(MM)
310 FORJ=1TONU:S=S+DA(I,J)*W(J):NEXTJ
320 IF LI(I)>0 GOTO 340
330 IF (S+TS)>0 GOTO 360ELSE GOTO 400
340 IF (S-TS)>0 GOTO 400
350 C=2*(TS-S):GOTO 370
360 C=2*(-TS-S)
370 XX=0:FORJ=1TONU:XX=XX+DA(I,J)[2:NEXTJ
380 C=C/XX:FORJ=1TONU:W(J)=W(J)+C*DA(I,J):NEXTJ
390 W(MM)=W(MM)+C:KK=KK+1:NS(KK)=I:NF=NF+1
400 PRINTUSING"####";KK;
410 NEXTIR:PRINT:KV=KK:IF (NF-QU)>=0 GOTO 440
420 IF (N2-NT)<>0 GOTO 270
430 IF KV<>0 GOTO 270
440 FOR J=1TOMM:PRINT" W(";J;") =";W(J):NEXTJ:PRINT
450 PRINT"FEEDBACK=";NF:INPUT"PRESS ENTER TO CONTINUE";Z
460 CLS:RETURN
470 L0=0:L1=0:KW=0:N0=0:N1=0:Z0=0
480 FORII=1TONP:I=IP(II):S=W(MM)
490 FORJ=1TONU:S=S+DA(I,J)*W(J):NEXTJ
500 PRINTI-NT;" ";A$(I-NT);" ";LI(I);"  =  ";S
510 Z0=Z0+1:IF Z0<11 GOTO 530
520 INPUT "PRESS ENTER TO CONTINUE";Z:Z0=0
530 IFABS(S)-TS>0 GOTO 550
540 KW=KW+1:GOTO 600
550 IF LI(I)>0 GOTO 580
560 N0=N0+1:IF(-S-TS)>0  GOTO 600
570 L0=L0+1:GOTO 600
580 N1=N1+1:IF(S-TS)>0GOTO 600
590 L1=L1+1
600 NEXTII:LT=L0+L1+KW:JW=N0+N1+KW
610 P=100-LT/JW*100:P0=100-L0/N0*100:P1=100-L1/N1*100
620 PRINT"DEAD ZONE";TS," CORRECT";JW-LT," INCORRECT";LT,
```

A3.2 LEARN* (continued)

```
630 PRINT" TRAPPED";KW
640 PRINT"    ERRORS:-","TOTAL   ";LT;"/";JW;"=";P;"% CORRECT"
650 PRINT,"  +     "L1;"/";N1;"=";P1;"% CORRECT"
660 PRINT,"  -     ";L0;"/";N0;"=";P0;"% CORRECT"
670 INPUT"PRESS ENTER TO CONTINUE";Z:RETURN
```

*Based on Ref. 1, Chap. 5.

This program undertakes linear learning machine calculations to partition samples into one of two classes.

    Input data: Intensity data in Table 5.12.

    Output data: Table 5.16.

A3.3 SIMCO

```
100 DIM X1(50),Y1(50),X2(50),Y2(50),M$(10),S1(10)
110 CLS:PRINT"FIT FACTOR AND SIMILARITY COEFFICIENT CALCULATION"
120 INPUT"CHOOSE OPTION, 1-SIMILARITY, 2-FIT, 3-STOP";O1
130 IF O1>3 GOTO 3000
140 IF O1>3 OR O1<1 GOTO 120
150 ZZ=0:I3=0:N3=10:FORI=1TON3:S1(I)=0:M$(I)="":NEXTI
160 INPUT"HOW MANY PEAKS IN THIS DATA SET";N1
170 FORI=1TON1:PRINTI;"GIVE M/Z OR RT. THEN INTENSITY";
180 INPUTX1(I),Y1(I):NEXTI:INPUT"GIVE % ERROR ON RT.";E1
190 GOSUB 400:GOSUB 500
200 READ N$:IF N$="****" GOTO 280
210 READ N2:FORI=1TON2:READ X2(I),Y2(I):NEXTI
220 I3=I3+1:IF O1=1 GOSUB 600
230 IF O1=2 GOSUB 700
240 PRINTI3;N$;TAB(20)"MATCH FACTOR";F1
250 ZZ=ZZ+1:IF ZZ<14 GOTO 270
260 INPUT"PRESS ENTER TO CONTINUE";Z:ZZ=0
270 GOSUB 900:GOTO 200
280 RESTORE:INPUT"PRESS ENTER TO CONTINUE";Z
290 CLS:PRINT"BEST MATCHES FOR THESE DATA ARE:"
300 FORI=1TON3:PRINTI;M$(I);TAB(20)"MATCH FACTOR";S1(I):NEXTI
310 INPUT"PRESS ENTER TO CONTINUE";Z:CLS:GOTO 120
400 'SORT INPUT PEAKS INTO ASCENDING M/Z OR RT. SERIES
410 FORJ=1TON1:A=X1(J):B=Y1(J):K=J
420 FORI=JTON1:IF A-X1(I)<=0 GOTO 440
430 A=X1(I):B=Y1(I):K=I
440 NEXTI:X1(K)=X1(J):Y1(K)=Y1(J)
450 X1(J)=A:Y1(J)=B:NEXTJ:RETURN
500 'NORMALISE INTENSITY TO TOTAL SUM OF 1000
510 T1=0:FORI=1TON1:T1=T1+Y1(I):NEXTI
520 FORI=1TON1:Y1(I)=Y1(I)/T1*1000:NEXTI:RETURN
600 'CALCULATE SIMILARITY COEFFICIENT (%)
610 S=0:N=0:I1=1:I2=1
620 IF I1>N1 OR I2>N2 GOTO 690
630 IF X1(I1) > X2(I2)+X2(I2)*E1/100 GOTO 670
640 IF X1(I1) < X2(I2)-X2(I2)*E1/100 GOTO 680
650 Q=Y1(I1)/Y2(I2):IF Q>1 THEN Q=1/Q
660 S=S+Q:N=N+1:I1=I1+1:I2=I2+1:GOTO 620
670 I2=I2+1:GOTO 620
680 I1=I1+1:GOTO 620
690 F1=S/(N1+N2-N)*100:RETURN
700 ' CALCULATE FIT FACTOR
710 SN=0:SD=0:I1=1:I2=1:NA=N1:NB=N2
720 IF N1>N2 GOTO 750
730 IF N2>N1 GOTO 760
740 GOTO 780
750 NB=N2+1:X2(NB)=X1(N1):Y2(NB)=0:GOTO 780
760 NA=N1+1:X1(NA)=X2(N2):Y1(NA)=0:GOTO 780
770 IF I1>NA OR I2>NB GOTO 850
780 IF X1(I1) > X2(I2)+X2(I2)*E1/100 GOTO 810
790 IF X1(I1) < X2(I2)-X2(I2)*E1/100 GOTO 820
800 GOTO 830
810 YA=0:YB=Y2(I2):I2=I2+1:GOTO 840
820 YB=0:YA=Y1(I1):I1=I1+1:GOTO 840
830 YA=Y1(I1):YB=Y2(I2):I1=I1+1:I2=I2+1
840 SN=SN+(YA-YB)[2:SD=SD+(YA[2+YB[2):GOTO 770
850 F1=1000*(1-SN/SD):RETURN
```

A3.3 SIMCO (continued)

```
900 'SAVE BEST MATCHES
910 FOR I=1TON3:IF F1>S1(I) GOTO 930
920 NEXTI:RETURN
930 FOR J=N3-1TOISTEP-1:S1(J+1)=S1(J):M$(J+1)=M$(J):NEXTJ
940 S1(I)=F1:M$(I)=N$:RETURN
1000 ' DATA FILE HERE. IDENTIFIER, NO. OF PEAKS THEN X,Y PAIRS
1010 ' INTENSITIES ARE NORMALISED TO A SUM OF 1000
1020 DATA ETMA,6,29,111.9,39,128.8,41,325.4,69,339,99,71.2,114,23.7
1030 DATA MA,6,26,98.2,27,160.3,42,145.5,55,363.3,85,105.4,86,127.3
1040 DATA TOLUENE,6,39,45,51,30,63,30,65,60,91,497,92,338
1050 DATA BUTADIENE,6,27,129,28,91,39,253,50,65,53,185,54,267
1060 DATA MEMA,6,15,45,39,123,41,341,69,283,99,34,100,174
1070 DATA ETA,6,26,83,27,271,29,131,55,452,56,54,100,9
1080 DATAVNLCYCHXEN,6,39,123,54,350,66,137,79,214,80,116,108,60
2000 DATA ****
3000 CLS:END
```

This program calculates similarity index (Table 5.5) or FIT factor
(Table 5.6) for pyrogram comparisons.

Input data:

1 (or 2)

6

39, 45

51, 30

63, 30

65, 60

91, 497

92, 338

0

Output data:

Similarity		Fit	
TOLUENE	100	TOLUENE	1000
MEMA	3.32594	BUTADIENE	112.95
VNLCYCHXEN	3.32594	MEMA	19.0141
ETMA	3.17617	VNLCYCHXEN	18.9449
BUTADIENE	1.61696	ETMA	18.7392

A3.4   STAT

```
      MASTER MAIN
      COMMON/A/X(100),Y(100,10),Z(100)
      COMMON/B/NX,NY,NZ
      COMMON/C/YAV(100),NREP(100)
      COMMON/D/AV,SD,CV,ERR,TX,PX,SE,V
      COMMON/E/TITLE(10)
      DIMENSION Y1(100)
      CALL INPUT
   10 READ (5,20) IOPT
   20 FORMAT (I1)
      IF (IOPT.EQ.0) STOP
      READ (5,25) TITLE
   25 FORMAT (10A8)
      DO 30 I=1,100
      X(I)=0.0
      Z(I)=0.0
      DO 30 J=1,10
   30 Y(I,J)=0.0
      GOTO (40,110,140), IOPT
C     ANALYSIS OF VARIANCE
   40 READ (5,50)NX,NY,NZ
   50 FORMAT (3I10)
      READ (5,60) (X(I),I=1,NX)
   60 FORMAT (8F10.0)
      DO 70 I=1,NX
   70 READ (5,60) (Y(I,J),J=1,NY)
      IF (NZ.EQ.0) GOTO 80
      READ (5,60) (Z(I),I=1,NZ)
   80 CONTINUE
      WRITE (6,997)
  997 FORMAT (1H1,30X,'LINEAR REGRESSION. ANALYSIS OF VARIANCE')
      IF (NY-4) 81,82,83
   81 WRITE (6,998) TITLE
      GOTO 89
   82 WRITE (6,999) TITLE
      GOTO 89
   83 WRITE (6,1000) TITLE
  998 FORMAT (1H0,/1X,10A8/10X,'THE INPUT DATA ARE:'//
     18X,'X',5X,   7(1H-),'REPEAT Y-VALUES',   7(1H-),3X,'MEAN Y',3X,
     2'STD.DEVN',3X,'CV(%)',4X,'95%ERROR', 2X,'T-TEST',4X,'P(%)')
  999 FORMAT (1H0,/1X,10A8/10X,'THE INPUT DATA ARE:'//
     18X,'X',5X, 12(1H-),'REPEAT Y-VALUES', 12(1H-),3X,'MEAN Y',3X,
     2'STD.DEVN',3X,'CV(%)',4X,'95%ERROR', 2X,'T-TEST',4X,'P(%)')
 1000 FORMAT (1H0,/1X,10A8/10X,'THE INPUT DATA ARE:'//
     18X,'X',5X,17 (1H-),'REPEAT Y-VALUES',17 (1H-),3X,'MEAN Y',3X,
     2'STD.DEVN',3X,'CV(%)',4X,'95%ERROR', 2X,'T-TEST',4X,'P(%)')
   89 DO 100 I=1,NX
      DO  90 J=1,NY
   90 Y1(J)=Y(I,J)
      CALL MEAN (Y1 ,NY)
      YAV(I)=AV
  100 WRITE (6,1010) X(I),(Y(I,J),J=1,NY),AV,SD,CV,ERR,TX,PX
 1010 FORMAT (1X,11F10.4,F8.2)
      WRITE (6,1070)
      CALL ANALVAR
      GOTO 10
C     LINEAR LEAST SQUARES - NO REPLICATES NECESSARY
  110 READ (5,120) NX,NZ
  120 FORMAT (2I10)
      READ (5,60)(X(I),Y(I,1),I=1,NX)
      IF (NZ.EQ.0)GOTO 130
      READ (5,66) (Z(I), NREP(I),I=1,NZ)
```

A3.4   STAT (continued)

```
   66 FORMAT ( 4(F10.4,I10))
  130 CALL LEAS@
      GOTO 10
C     MEAN, STANDARD DEVIATION
  140 READ (5,150) NX
  150 FORMAT (I10)
      READ (5,60)(X(I),I=1,NX)
      CALL MEAN (X ,NX)
      WRITE (6,1020) TITLE
 1020 FORMAT (1H1,20X,'CALCULATION OF REPRODUCIBILITY PARAMETERS'/
     11X,10A8//10X,'THE INPUT DATA ARE::')
      WRITE (6,1030) (X(I),I=1,NX)
 1030 FORMAT (1X,12F10.4)
      WRITE (6,1070)
      WRITE (6,1040) AV,SE  ,ERR,SD,V,CV,TX,PX
 1040 FORMAT (5X,'MEAN',6X,'STANDARD',5X,' 95% ',5X,'STANDARD',5X,'VARIA
     1NCE',3X,'COEFFICIENT' ,3X,'T-TEST',7X,'PROBABILITY'/
     216X,'ERROR',7X,'ERROR',5X,'DEVIATION',15X,'OF VARIATION'/
     37(F10.4 ,2X),F10.2)
      WRITE (6,1070)
 1070 FORMAT (/30X,50(1H*)/)
      GOTO 10@
      END
      SUBROUTINE ANALVAR
      COMMON/A/X(100),Y(100,10),Z(100)
      COMMON/B/NX,NY,NZ
      COMMON/E/TITLE(10)
      COMMON/C/YAV(100)
      COMMON/F/A,B
      COMMON/G/T(100)
      DIMENSION YBAR(100)
      SY=0.0
      SYY=0.0
      SX=0.0
      SXX=0.0
      SXY=0.0
      AX=NX
      AY=NY
      AXY=AX*AY
      NXY=AXY-2.0
      DO 10 I=1,NX
   10 YBAR(I)=0.0
      DO 20 I=1,NX
      SX=SX+X(I)
      SXX=SXX+X(I)**2
      DO 20 J=1,NY
      SY=SY+Y(I,J)
      SYY=SYY+Y(I,J)**2
      SXY=SXY+X(I)*Y(I,J)
   20 YBAR(I)=YBAR(I)+Y(I,J)
      SX=SX* AY
      SXX=SXX*AY
      SSX =SXX-SX*SX/AXY
      SSXY=SXY-SX*SY/AXY
      T2=0.0
      DO 30 I=1,NX
   30 T2=T2+YBAR(I)**2
      T2=T2/AY
      T3=SY**2/AXY
      S1=SYY-T3
      S3=SYY-T2
      IF (S3.E@.0.0) S3=1E-6
      S2=T2-T3
      B=SSXY/SSX
      A=SY/AXY-B*SX/AXY
```

A3.4 STAT (continued)

```
      S5=SSXY**2/SSX
      S4=S2-S5
      NDF1=AXY-1.0
        DF1=NDF1
      NDF3=NDF1-NX+1
        DF3=NDF3
      NDF5=1
        DF5=1.0
      NDF4=NDF1-NDF5-NDF3
        DF4=NDF4
      NDF2=NDF1-NDF3
        DF2=NDF2
      V1=S1/DF1
        V2=S2/DF2
      V3=S3/DF3
      V4=S4/DF4
      V5=S5/DF5
        F5=V5/V3
        F4=V4/V3
        F2=V2/V3
        CALLP1(F5,P5, DF5, DF3)
        CALLP1(F4,P4, DF4, DF3)
        CALLP1(F2,P2, DF2, DF3)
        WRITE (6,1000) TITLE
 1000 FORMAT (//1H0,   'ANALYSIS OF VARIANCE'/20X,10A8/)
        WRITE (6,1010)
 1010 FORMAT (1X,'SOURCE OF',9X,  'SUMS OF',9X,'DEGREES OF',9X ,'MEAN',
     1 9X,'VARIANCE',  9X,'PROBABILITY'/
     2 1X,'VARIATION',9X,'SQUARES',9X,  'FREEDOM',11X,'SQUARE',10X,'RATIO
     1'/)
        WRITE (6,1020) S5,NDF5,V5,F5,P5,S4,NDF4,V4,F4,P4,S2,NDF2,V2,F2,
     1P2,S3,NDF3,V3,S1,NDF1,V1
 1020 FORMAT (1X,'REGRESSION',5X,F10.4, 5X,I10,10X,2(F10.4,5X),F10.2/
     11X,'DEVIATION',6X,F10.4,5X,I10,10X,2(F10.4,5X),F10.2/
     21X,'BETWEEN X-S', 4X,F10.4,5X,I10,10X,2(F10.4,5X),F10.2/
     31X,'RESIDUAL', 7X,F10.4,5X,I10,10X,F10.4/
     41X,'TOTAL',10X,F10.4,5X,I10,10X,F10.4/ )
        WRITE (6,1070)
        VY=((SYY-SY*SY/AXY)-(B*B)*(SXX-SX*SX/AXY))/(AXY-2.0)
        VB=VY/(SXX-(SX*SX)/AXY) *AXY
        VA=VB*SXX/AXY
        SA=SQRT(VA/AXY)
        SB=SQRT(VB/AXY)
        ERRA=SA*T(NXY)
        ERRB=SB*T(NXY)
        IF (SA.EQ.0.0) SA=1E-6
        IF (SB.EQ.0.0) SB=1E-6
        TA=A/SA
        TB=B/SB
        FA=TA*TA
        FB=TB*TB
        XY=NXY
        CALL P1(FA,PA,XY,0.0)
        CALL P1(FB,PB,XY,0.0)
        A1=A-ERRA
        A2=A+ERRA
        B1=B-ERRB
        B2=B+ERRB
        WRITE   (6,1030)A,SA,ERRA,A1,A2,TA,PA,B,SB,ERRB,B1,B2,TB,PB
 1030 FORMAT (1X,'PARAMETER', 8X,'VALUE', 8X,'STANDARD', 6X,'95%',
     117X,'95%',12X,'T-TEST', 7X,'PROBABILITY'/
     232X,'ERROR', 8X,'ERROR',14X,'RANGE'/
     31X,'INTERCEPT',5X,3(F10.4,3X),F10.4, ' -',F10.3,2(F10.3,3X)/
     41X,'SLOPE',9X,  3(F10.4,3X),F10.4, ' -',F10.4,2(F10.3,3X)/ )
        WRITE (6,1070)
        TOP=SXY-SX*SY/AXY
```

A3.4   STAT (continued)

```
      COVAR=TOP/(AXY-2.0)
      R=TOP/SQRT((SXX-(SX*SX)/AXY)*(SYY-(SY*SY)/AXY))
      R2=R**2*100.0
      WRITE (6,1040) COVAR,R,R2,VY
1040  FORMAT (1X,'COVARIANCE',10X,'CORRELATION',10X,'R-SQUARED (%)', 6X,
     1'VARIANCE OF Y'/21X,'COEFFICIENT',29X,'ABOUT REGRESSION LINE'/
     24(F10.4,10X)/)
      WRITE (6,1070)
      CALL RESIDS
      WRITE (6,1070)
      IF (NZ.EQ.0) RETURN
      WRITE (6,1050) TITLE
1050  FORMAT (1H1,1X,'ASSAY RESULTS'//20X,10A8//2X,'MEAN Y-VALUE',5X,
     1'INTERPOLATED',3X,'STANDARD',9X,'95%',19X,'95%',13X,'VARIANCE'/
     219X,'X-VALUE',10X,'ERROR',9X,'ERROR',16X,'RANGE',13X,'ABOUT X')
      DO 40   I=1,NZ
      VX=(VY/(B*B))*(((1.0/AY)+(1.0/AXY))+(((Z(I)-SY/AXY)**2*
     1(1.0/((B*B)*(SXX-SX*SX/AXY)))))
      S=SQRT(VX/AY)
      NXY=AXY-2.0
      ERR=S*T(NXY)
      BX=(Z(I)-A)/B
      X1=BX-ERR
      X2=BX+ERR
      WRITE (6,1060) Z(I),BX,S,ERR,X1,X2,VX
1060  FORMAT (1X,4(F10.4, 5X),F10.4,'   -',2F10.4)
  40  CONTINUE
      WRITE (6,1070)
1070  FORMAT (/30X,50(1H*)/)
      RETURN
      END
      SUBROUTINE RESIDS
      COMMON/A/X(100),Y(100,10),Z(100)
      COMMON/B/NX,NY,NZ
      COMMON/C/YAV(100)
      COMMON/E/TITLE(10)
      COMMON/F/A,B
      WRITE (6,10) TITLE
  10  FORMAT (1X,'RESIDUALS'//20X,10A8//
     1 6X,'X',13X,'MEAN Y', 9X,'Y-CALC', 9X,'RESIDS', 8X,'RATIO (%)')
      DO 20 I=1,NX
      YCALC=B*X(I)+A
      Y2=YCALC-YAV(I)
      Y3=YCALC*100.0/YAV(I)
  20  WRITE (6,1000) X(I),YAV(I),YCALC,Y2,Y3
1000  FORMAT (1X,5(F10.4 , 5X))
      RETURN
      END
      SUBROUTINE P1(F,P,ANX,AN2)
      AN2=ANX
      AN1=1.0
      IF (F.GE.1.0)GOTO 10
      IF (F.EQ.0.0) F=1E-6
      F=1.0/F
      A=AN1
      AN1=AN2
      AN2=A
  10  A1=2.0/AN1/9.0
      A2=2.0/AN2/9.0
      Z=ABS(((1.0-A2)*F**0.3333333-1.0+A1)/SQRT(A2*F**0.6666667+A1))
      IF (AN2 .LE. 3.0)Z=Z*(1.0+0.08*Z**4/AN2**3)
      FZ=EXP(-Z*Z/2.0)*0.3989423
      W=1.0/(1.0+Z*0.2316419)
      P=FZ*W*(((((1.330274*W-1.821256)*W+1.781478)*W-0.3565638)*W+
     10.3193815)
      IF (F.LT.1.0)P=1.0-P
```

A3.4 STAT (continued)

```
      P=(1.0-P)*100.0
      RETURN
      END
      BLOCK DATA
      COMMON/G/T(100)
      DATA T /12.71,4.3,3.18,2.78,2.57,2.45,2.37,2.31,2.26,2.23,
     12.20,2.18,2.17,2.15,6*2.13,5*2.09,5*2.06,10*2.04,10*2.02,
     220*2.03,30*1.98/
      END
      SUBROUTINE MEAN (X,NX)
      DIMENSION X(100)
      COMMON/D/AV,SD,CV,ERR,TX,PX,SE,VX
      COMMON/G/T(100)
      SX=0.0
      SXX=0.0
      AX=NX
      DO 10 I=1,NX
      SX=SX+X(I)
   10 SXX=SXX+X(I)**2
      SSX=SX**2
      AV=SX/AX
      VX=SXX/(AX-1.0)-SSX/(AX*(AX-1.0))
      SD=SQRT(VX)
      CV=SD*100.0/AV
      SE=SD/SQRT(AX)
      ERR=SE*T(NX-1)
      IF (SD.EQ. 0.0) SD=1E-6
      TX=AV*SQRT(AX)/SD
      FX=TX**2
      ANX=AX-1.0
      CALL PI(FX,PX,ANX,0.0)
      RETURN
      END
      SUBROUTINE LEASQ
      COMMON/A/X(100),Y1(100 ,10),Z(100)
      COMMON/B/NX,NY,NZ
      COMMON/E/TITLE(10)
      COMMON/F/A,B
      COMMON/C/YAV(100),NREP(100)
      COMMON/G/T(100)
      DIMENSION Y(100)
      SX=0.0
      SY=0.0
      SXX=0.0
      SYY=0.0
      SXY=0.0
      AX=NX
      ANX=AX-2.0
      DO 10 I=1,NX
      Y(I)=Y1(I,1)
      SX=SX+X(I)
      SY=SY+Y(I)
      SXX=SXX+X(I)*X(I)
      SXY=SXY+X(I)*Y(I)
   10 SYY=SYY+Y(I)*Y(I)
      SSX=SX*SX
      SSY=SY*SY
      XBAR=SX/AX
      YBAR=SY/AX
      ANUM=SXY-SX*SY/AX
      DEN1=SXX-SSX/AX
      DEN2=SYY-SSY/AX
      B=ANUM/DEN1
      A=YBAR-(B*XBAR)
      COVAR=(SXY-(XBAR*YBAR*AX))/(AX-2.0)
```

A3.4 STAT (continued)

```
      R=ANUM/SQRT(DEN1*DEN2)
      R2=R*R*100.0
      VY=(SYY-(AX*YBAR*YBAR)-B*B*(SXX-AX*XBAR*XBAR))/(AX-2.0)
      VB=VY/DEN1 *AX
      VA=VB*SXX/AX
      SA=SQRT(VA/AX)
      SB=SQRT(VB/AX)
      ERRA=SA*T(NX-2)
      ERRB=SB*T(NX-2)
      IF (SA .EQ. 0.0) SA=1E-6
      IF (SB .EQ. 0.0) SB=1E-6
      TA=A/SA
      TB=B/SB
      FA=TA*TA
      FB=TB*TB
      CALLP1(FA,PA,ANX,0.0)
      CALLP1(FB,PB,ANX,0.0)
      A1=A-ERRA
      A2=A+ERRA
      B1=B-ERRB
      B2=B+ERRB
      WRITE (6,1000) TITLE
 1000 FORMAT (1H1,30X,'LINEAR LEAST SQUARES ANALYSIS'//1X,10A8/)
      WRITE (6,1020) A,SA,ERRA,A1,A2,TA,PA,B,SB,ERRB,B1,B2,TB,PB
 1020 FORMAT (1X,'PARAMETER', 8X,'VALUE', 8X,'STANDARD', 6X,'95%',
     117X,'95%',12X,'T-TEST', 7X,'PROBABILITY'/
     232X,'ERROR', 8X,'ERROR',14X,'RANGE'/
     31X,'INTERCEPT',5X,3(F10.4,3X),F10.4, ' -',F10.3,2(F10.3,3X)/
     41X,'SLOPE',9X, 3(F10.4,3X),F10.4, ' -',F10.4,2(F10.3,3X)/ )
      WRITE (6,1070)
      WRITE (6,1030)COVAR,R,R2,VY
 1030 FORMAT (1X,'COVARIANCE',10X,'CORRELATION',10X,'R-SQUARED (%)', 6X,
     1'VARIANCE OF Y'/21X,'COEFFICIENT',29X,'ABOUT REGRESSION LINE'/
     24(F10.4,10X)/)
      DO 20 I=1,NX
   20 YAV(I)=Y(I)
      WRITE (6,1070)
      CALL RESIDS
      WRITE (6,1070)
      IF (NZ.EQ.0) RETURN
      WRITE (6,1040) TITLE
 1040 FORMAT (1H1,1X,'ASSAY RESULTS'//20X,10A8//2X,'MEAN Y-VALUE',5X,
     1'INTERPOLATED',3X,'STANDARD',9X,'95%',19X,'95%',13X,'VARIANCE',5X,
     2'REPEATS'/19X,'X-VALUE',10X,'ERROR', 9X,'ERROR',16X,'RANGE',
     313X,'ABOUT X')
      DO 40 I=1,NZ
      IF (NREP(I).EQ.0)NREP(I)=1
      REP=NREP(I)
      VX=VY/(B*B)*(1.0/REP+1.0/AX+((Z(I)-YBAR)**2/(B*B*DEN1)))
      S=SQRT(VX/REP)
      ERR=S*T(NX-2)
      BX=(Z(I)-A)/B
      X1=BX-ERR
      X2=BX+ERR
      WRITE (6,1050) Z(I),BX,S,ERR,X1,X2,VX,NREP(I)
 1050 FORMAT (1X,4(F10.4, 5X),F10.4,' -',2F10.4,I12)
   40 CONTINUE
      WRITE (6,1070)
 1070 FORMAT (/30X,50(1H*)/)
      RETURN
      END
      SUBROUTINE INPUT
      WRITE (6,10)
   10 FORMAT (1H1, 5X,'PROGRAM TO CALCULATE ASSAY DATA BY ANALYSIS OF VA
     1RIANCE, LEAST SQUARES OR REPRODUCIBILITY PARAMETERS'//)
      WRITE (6,20)
```

A3.4 STAT (continued)

```
 20 FORMAT (10X,'THE INPUT DATA ARE AS FOLLOWS:'/)
    WRITE (6,30)
 30 FORMAT (15X,'CARD 1 - IOPT  (I1) '/
   125X,'IOPT=0  STOP RUN'/
   225X,'      =1  ANALYSIS OF VARIANCE - REPLICATE Y RESULTS NEEDED'/
   325X,'      =2  LEAST SQUARES ANALYSIS - NO REPLICATES REQUIRED'/
   425X,'      =3  CALCULATES MEAN, STD. DEVN. ETC. ON REPEATED RESULTS'
   5/)
    WRITE (6,35)
 35 FORMAT (15X,'CARD 2  TITLE  (10A8)'/
   125X,'     TITLE OF CURRENT SAMPLE'/)
    WRITE (6,40)
 40 FORMAT (10X,'IOPT=1   ANALYSIS OF VARIANCE'/
   115X,'CARD 3 - NX,NY,NZ  (3I10)    '/
   225X,'  NX=NUMBER OF X (CONCENTRATION) POINTS IN ASSAY'/
   325X,'  NY=NUMBER OF Y (INTENSITY) REPLICATES FOR EACH X VALUE'/
   425X,'  NZ=NUMBER OF ASSAY POINTS TO BE INTERPOLATED ONTO CALIBRATI
   5ON LINE'/)
    WRITE (6,50)
 50 FORMAT (15X,'CARD 4 - CALIBRATION DATA X  (8F10.4)'/
   126X,'    X (CONCN) DATA -8 VALUES PER CARD--NX POINTS IN TOTAL'/)
    WRITE (6,60)
 60 FORMAT (15X,'CARD 3 -ETC. Y (INTENSITY) DATA FOR CALIBRATION LINE
   1(8F10.4)'/
   225X,'     NOW FOLLOW NX CARDS HOLDING REPLICATE DETERMINATIONS OF
   3Y'/
   425X,'     THE DATA CARDS ARE IN THE SAME ORDER AS THE X-DATA'/
   525X,'     EACH CARD CONTAINS THE NY REPLICATES OF Y AT EACH X-VALU
   6E'/)
 70 FORMAT (15X,'CARD NX+2 -ETC- Z (ASSAY) DATA FOR INTERPOLATION ONTO
   1 CALIBRATION LINE (8F10.4)'/
   225X,'     NOW FOLLOW NZ POINTS FROM UNKNOWN SAMPLES'/)
    WRITE (6,70)
    WRITE (6,80)
 80 FORMAT (10X,'IOPT=2   LEAST SQUARES ANALYSIS'/
   115X,'CARD 3 - NX,NZ  (2I10)'/
   225X,'  NX=NUMBER OF PAIRED X-Y POINTS - CALIBRATION DATA'/
   225X,'  NZ=NUMBER OF ASSAY POINTS (Z) FOR INTERPOLATION'/)
    WRITE (6,90)
 90 FORMAT (15X,'CARD 4 - CALIBRATION DATA X AND Y  (8F10.4)'/
   125X,'     NOW FOLLOW PAIRS OF DATA POINTS X (CONCN) FOLLOWED BY Y
   2(INTENSITY)'/
   325X,'     FOUR PAIRS OF POINTS PER CARD'/)
    WRITE (6,100)
100 FORMAT (15X,'CARD NX/4 +2  - Z (ASSAY) DATA FOR INTERPOLATION ONTO
   1 CALIBRATION LINE   4(F10.4,I10)'/
   225X,'     DATA IS PAIRED - FIRST IS Z FOLLOWED BY NREP-NUMBER'/
   325X,'     OF REPLICATES GIVING THE MEAN VALUE Z   '/
   425X,'     IF NREP=1 THEN THIS PARAMETER MAY BE LEFT BLANK'/)
    WRITE (6,110)
110 FORMAT (10X,'IOPT=3   MEAN, STANDARD DEVIATION ETC.'/
   115X,'CARD 3 - NX  NUMBER OF OBSERVATIONS IN DATA SET (I10)'/
   225X,'  NX=NUMBER OF OBSERVATIONS IN DATA SET'/)
    WRITE (6,120)
120 FORMAT (15X,'CARD 4  X VALUES F10.4' /
   125X,'     NOW FOLLOW NX REPLICATE OBSERVATIONS - EIGHT PER CARD'/)
    WRITE (6,130)
130 FORMAT (15X,'NEW DATA SETS MAY NOW FOLLOW - FROM CARD 1'/
   125X,'     FINISH DATA WITH A BLANK CARD'//)
    RETURN
    END
    FINISH
****
```

A3.4 STAT (continued)
Input data: The input data is fully described in SUBROUTINE INPUT in program STAT.

```
1
ANALYSIS OF VARIANCE.  TEST DATA.
      5        5       12
1.9031   2.0792   2.2553   2.4314   2.6075
1.0      5.0      1.0      1.0      2.0        12.0   14.0   16.0
4.0      1.0      5.0      10.0     8.0
19.0     15.0     12.0     17.0     19.0
25.0     25.0     24.0     17.0     24.0
27.0     26.0     20.0     37.0     29.0
2.0      4.0      6.0      8.0      10.0
20.0     25.0     30.0     35.0
2

LEAST SQUARES ANALYSIS.  TEST DATA.
     25       10
1.9031   5.0      1.0        1.9031   1.0      1.0
1.9031   4.0      2.0        2.0792   1.0      5.0
2.0792   8.0      10.0       2.2553   19.0     15.0
2.2553   17.0     12.0       2.2553   19.0     25.0
2.4314   24.0     25.0       2.4314   17.0     24.0
2.6075   26.0     27.0       2.6075   20.0     37.0
2.6075            29.0
6.0      5        8.0                  5       12.0
10.0     2        10.0                 3       10.0
10.0     6        10.0
3

MEAN, STANDARD DEVIATION CALCULATION.  TEST DATA.
      5
1.0      5.0      4.0        1.0      2.0
0
****
```

A3.4 STAT (continued)

Output:

LINEAR REGRESSION. ANALYSIS OF VARIANCE

ANALYSIS OF VARIANCE. TEST DATA.
THE INPUT DATA ARE:

X	------REPEAT Y-VALUES------					MEAN Y	STD.DEVN	CV(%)	95%ERROR	T-TEST	P(%)
1.9031	1.0000	5.0000	1.0000	1.0000	2.0000	2.0000	1.7321	86.6025	2.1534	2.5820	93.88
2.0792	4.0000	1.0000	10.0000	8.0000	5.0000	5.6000	3.5071	62.6274	4.3603	3.5704	97.56
2.2553	19.0000	15.0000	12.0000	19.0000	17.0000	16.4000	2.9665	18.0883	3.6881	12.3620	99.93
2.4314	25.0000	25.0000	24.0000	17.0000	24.0000	23.0000	3.3912	14.7442	4.2161	15.1658	99.95
2.6075	27.0000	26.0000	20.0000	37.0000	29.0000	27.8000	6.1400	22.0864	7.6336	10.1242	99.88

****

ANALYSIS OF VARIANCE

ANALYSIS OF VARIANCE. TEST DATA.

SOURCE OF VARIATION	SUMS OF SQUARES	DEGREES OF FREEDOM	MEAN SQUARE	VARIANCE RATIO	PROBABILITY
REGRESSION	2380.5000	1	2380.5000	162.3806	95.08
DEVIATION	55.2600	3	18.4200	1.2565	65.49
BETWEEN X-S	2435.7600	4	608.9400	41.5375	99.59
RESIDUAL	293.2000	20	14.6600		
TOTAL	2728.9600	24	113.7067		

****

PARAMETER	VALUE	STANDARD ERROR	95% ERROR	95% RANGE	T-TEST	PROBABILITY
INTERCEPT	-73.4078	7.0926	14.8235	-88.2313 - -58.584	-10.350	100.000
SLOPE	39.1823	3.1259	6.5330	32.6493 - 45.7153	12.535	100.000

****

COVARIANCE	CORRELATION COEFFICIENT	R-SQUARED (%)	VARIANCE OF Y ABOUT REGRESSION LINE
2.6415	0.9360	87.2310	15.1504

****

RESIDUALS

ANALYSIS OF VARIANCE. TEST DATA.

X	MEAN Y	Y-CALC	RESIDS	RATIO (%)
1.9031	2.0000	1.1600	-0.8400	58.0000
2.0792	5.6000	8.0600	2.4400	143.9286
2.2553	16.4000	14.9600	-1.4400	91.2195
2.4314	23.0000	21.8600	-1.1400	95.0435
2.6075	27.8000	28.7600	0.9600	103.4532

****

A3.4 STAT (continued)
(Output)

LINEAR LEAST SQUARES ANALYSIS

LEAST SQUARES ANALYSIS. TEST DATA.

PARAMETER	VALUE	STANDARD ERROR	95% ERROR	95% RANGE		T-TEST	PROBABILITY
INTERCEPT	-73.4078	7.0926	14.8235	-88.2313 -	-58.584	-10.350	100.000
SLOPE	39.1823	3.1259	6.5330	32.6493 -	45.7153	12.535	100.000

****

COVARIANCE	CORRELATION COEFFICIENT	R-SQUARED (%)	VARIANCE OF Y ABOUT REGRESSION LINE
2.6415	0.9340	87.2310	15.1504

****

RESIDUALS

LEAST SQUARES ANALYSIS. TEST DATA.

X	MEAN Y	Y-CALC	RESIDS	RATIO (%)
1.9031	1.0000	1.1600	0.1600	116.0000
1.9031	5.0000	1.1600	-3.8400	23.2000
1.9031	1.0000	1.1600	0.1600	116.0000
1.9031	1.0000	1.1600	0.1600	116.0000
1.9031	2.0000	1.1600	-0.8400	58.0000
2.0792	4.0000	8.0600	4.0600	201.5000
2.0792	1.0000	8.0600	7.0600	806.0000
2.0792	5.0000	8.0600	3.0600	161.2000
2.0792	10.0000	8.0600	-1.9400	80.6000
2.0792	8.0000	8.0600	0.0600	100.7500
2.2553	19.0000	14.9600	-4.0400	78.7368
2.2553	15.0000	14.9600	-0.0400	99.7333
2.2553	12.0000	14.9600	2.9600	124.6667
2.2553	17.0000	14.9600	-2.0400	88.0000
2.2553	19.0000	14.9600	-4.0400	78.7368
2.4314	25.0000	21.8600	-3.1400	87.4400
2.4314	25.0000	21.8600	-3.1400	87.4400
2.4314	24.0000	21.8600	-2.1400	91.0833
2.4314	17.0000	21.8600	4.8600	128.5882
2.4314	24.0000	21.8600	-2.1400	91.0833
2.6075	27.0000	28.7600	1.7600	106.5185
2.6075	26.0000	28.7600	2.7600	110.6154
2.6075	20.0000	28.7600	8.7600	143.8000
2.6075	37.0000	28.7600	-8.2400	77.7297
2.6075	29.0000	28.7600	-0.2400	99.1724

A3.4 STAT (continued)
(Output)

ASSAY RESULTS

ANALYSIS OF VARIANCE. TEST DATA.

MEAN Y-VALUE	INTERPOLATED X-VALUE	STANDARD ERROR	95% ERROR	95% RANGE		VARIANCE ABOUT X
2.0000	1.9245	0.0248	0.0517	1.8728 -	1.9763	0.0031
4.0000	1.9756	0.0239	0.0500	1.9255 -	2.0256	0.0029
6.0000	2.0266	0.0232	0.0486	1.9780 -	2.0752	0.0027
8.0000	2.0777	0.0227	0.0474	2.0303 -	2.1250	0.0026
10.0000	2.1287	0.0222	0.0465	2.0823 -	2.1752	0.0025
12.0000	2.1798	0.0219	0.0458	2.1339 -	2.2256	0.0024
14.0000	2.2308	0.0218	0.0455	2.1853 -	2.2763	0.0024
16.0000	2.2818	0.0218	0.0455	2.2363 -	2.3274	0.0024
20.0000	2.3839	0.0222	0.0465	2.3374 -	2.4304	0.0025
25.0000	2.5115	0.0236	0.0493	2.4622 -	2.5609	0.0028
30.0000	2.6391	0.0257	0.0537	2.5854 -	2.6929	0.0033
35.0000	2.7668	0.0284	0.0594	2.7074 -	2.8261	0.0040

****

ASSAY RESULTS

LEAST SQUARES ANALYSIS. TEST DATA.

MEAN Y-VALUE	INTERPOLATED X-VALUE	STANDARD ERROR	95% ERROR	95% RANGE	VARIANCE ABOUT X	REPEATS
6.0000	2.0266	0.0232	0.0486	1.9780 - 2.0752	0.0027	5
8.0000	2.0777	0.0227	0.0474	2.0303 - 2.1250	0.0026	5
10.0000	2.1287	0.0222	0.0465	2.0823 - 2.1752	0.0025	5
12.0000	2.1798	0.0219	0.0458	2.1339 - 2.2256	0.0024	5
10.0000	2.1287	0.1018	0.2128	1.9159 - 2.3415	0.0104	1
10.0000	2.1287	0.0521	0.1089	2.0198 - 2.2376	0.0054	2
10.0000	2.1287	0.0355	0.0742	2.0545 - 2.2030	0.0038	3
10.0000	2.1287	0.0272	0.0569	2.0718 - 2.1856	0.0030	4
10.0000	2.1287	0.0189	0.0395	2.0892 - 2.1682	0.0021	6
10.0000	2.1287	0.0122	0.0255	2.1033 - 2.1542	0.0015	10

****

CALCULATION OF REPRODUCIBILITY PARAMETERS
MEAN, STANDARD DEVIATION CALCULATION. TEST DATA.

THE INPUT DATA ARE::
1.0000 5.0000 1.0000 1.0000 2.0000

****

MEAN	STANDARD ERROR	95% ERROR	STANDARD DEVIATION	VARIANCE	COEFFICIENT OF VARIATION	T-TEST	PROBABILITY
2.0000	0.7746	2.1534	1.7321	3.0000	86.6025	2.5820	93.88

****

Appendix 4

MS Data: Some Polymer Pyrolysis Products

COMPONENT	MASS	m/z						RELATIVE INTENSITIES					
		1	2	3	4	5	6	1	2	3	4	5	6
Methane	16	17	16	15	14	13	12	1	100	75	8	3	1
Hydrogen cyanide	27	28	27	26	14	13	12	1	100	16	2	2	4
Carbon monoxide	28	29	28	16	14	12		1	100	2	1	5	
Ethylene	28	29	28	27	26	25	24	2	100	62	53	8	2
Ethane	30	30	29	28	27	26	25	27	23	100	31	19	3
Formaldehyde	30	31	30	29	28	14	13	2	86	100	31	4	4
Methanol	32	32	31	30	29	28	15	79	100	7	39	31	12
Acetonitrile	41	42	41	40	39	38	28	3	100	55	19	10	4
Propene	42	42	41	40	39	38	27	70	100	29	74	20	38
Propane	44	44	43	39	29	28	27	29	23	18	100	60	40
Acetaldehyde	44	44	43	42	41	29	28	88	50	15	6	100	9
Dimethyl ether	46	46	45	31	29	28	15	48	100	3	36	7	24
Ethanol	46	46	45	43	31	29	27	33	67	12	100	13	17
Acrylonitrile	53	53	52	51	28	27	26	96	73	32	22	32	100
Butadiene	54	54	53	50	39	28	27	100	69	24	94	34	48
Propionitrile	55	55	54	52	28	27	26	10	62	11	100	18	21
But-1-ene	56	56	55	41	39	28	27	41	19	100	31	26	23
2-Methylpropene	56	56	55	41	39	28	27	51	20	100	40	16	15
Acrolein	56	56	55	29	28	27	26	74	53	37	66	100	54
Butane	58	58	43	41	29	28	27	14	100	26	32	28	21
Acetone	58	58	43	42	29	27	26	28	100	8	6	8	4
Allyl alcohol	58	58	57	39	31	29	27	26	100	23	32	39	19
Acetic acid	60	60	45	43	42	29	15	64	94	100	20	12	35
Crotononitrile	67	67	41	40	39	38	27	34	100	26	50	24	27
Penta-1,3-diene	68	68	67	53	41	40	39	73	100	54	31	28	45
Isoprene	68	68	67	53	41	40	39	71	100	59	26	28	40
Pent-1-ene	70	70	55	42	41	39	27	31	60	100	44	35	32
Methacrolein	70	70	43	42	41	40	39	69	25	17	100	25	70
Crotonaldehyde	70	70	69	41	39	38	29	63	29	100	97	24	26
Pentane	72	72	43	42	41	28	27	9	100	60	45	21	23
Neopentane	72	57	41	39	29	27	15	100	42	13	39	15	5
2-Methylpropanal	72	72	43	41	39	29	27	35	100	69	26	45	69
Butanal	72	72	44	43	41	29	27	66	100	70	60	72	64
Methyl acetate	74	74	59	43	42	31	29	15	6	100	10	3	11
Butan-1-ol	74	56	43	41	31	29	27	86	60	63	100	34	56

MS Data (continued)

COMPONENT	MASS	m/z						RELATIVE INTENSITIES					
		1	2	3	4	5	6	1	2	3	4	5	6
Benzene	78	78	77	52	51	50	39	100	17	18	16	13	10
Hexa-2,4-diene	82	82	67	54	53	41	39	53	100	31	22	49	42
Hex-1-ene	84	84	56	43	42	41	27	25	86	59	75	100	67
Z-2-Methylbut-2-enal	84	84	55	53	29	28	27	84	100	19	62	14	43
Cyclopentanone	84	84	56	55	41	28	27	35	29	100	40	44	27
Methyl acrylate	86	86	85	55	42	27	26	3	18	100	15	65	18
Hexane	86	86	57	56	43	42	41	13	100	54	59	32	62
Chloroprene	88	90	89	88	75	73	53	25	3	83	1	3	100
Toluene	92	92	91	65	63	51	39	68	100	12	6	6	9
Hept-1-ene	98	98	56	55	41	29	27	12	71	57	100	57	46
Tetrafluoroethylene	100	101	100	82	81	50	31	2	87	2	100	24	50
3-Methylhexane	100	100	71	57	43	41	29	3	47	47	100	46	43
Ethyl acrylate	100	100	56	55	29	27	26	2	12	100	29	60	18
Methyl methacrylate	100	100	99	69	41	39	15	51	10	83	100	36	13
2-Isopropoxyethanol	104	89	73	45	43	41	27	32	37	76	100	25	16
Styrene	104	105	104	103	78	77	51	9	100	45	32	17	21
4-Vinylcyclohex-1-ene	108	108	80	79	66	54	39	17	33	61	39	100	35
Oct-1-ene	112	112	56	55	43	42	41	6	64	83	100	63	78
Octane	114	114	85	57	43	41	29	7	30	34	100	38	35
Ethyl methacrylate	114	114	99	69	41	39	29	7	21	100	96	38	33
Butyl acrylate	128	73	56	55	41	29	27	43	54	100	14	8	17
Hexafluoropropene	150	150	131	100	81	69	31	58	94	53	14	100	39

Source: Adapted from Ref. 59 of Chap. 3.

Author Index

Numbers in parentheses are reference numbers and indicate that an author's work is referred to, although the name may not be cited in the text. Italic numbers give the page on which the complete reference is cited.

A

Aanerud, T.W., 337(7)*369*

Abbey, L.E., 26(167)*44*; 243(20) *287*; 418(106,107)*430*

Abelson, P.H., 490(45)*511*; 505 (114,115)*515*

Abrahamson, S., 123(60)*161*

Adam, S., 392(44)*426*

Adams, G.E., 316(146), 317 (156)*329*

Aguilera, C., 302(44)*322*

Agurell, S., 450(37)*468*

Ahlstrom, D.H., 293,316(4)*319*

Ahmad, I., 450(36)*468*

Alajberg, A., 310(107)*326*

Albright, L.F., 7(12)*34*

Albro, P.W., 26(170)*44*

Aldridge, M.H., 295(28)*321*

Alexseeva, K.V., 22(120)*41*; 82(93)*89*; 146(148)*167*; 303 (71)*323*; 310(100)*325*

Alexander, M., 414(100)*430*

Alford, E.D., 343(74)*374*

Alishoyev, V.R., 91(4)*158*; 294(25)*321*

Anderberg, M.R., 263(42)*288*

Anders, E., 505(109,112)*515*; 505 (128)*516*

Anderson, D.H., 70(40)*86*

Anderson, D.J., 414(97)*430*

Anderson, D.M., 24(133)*41*; 123 (68)*162*; 214(82)*234*; 505(120, 122,123)*516*

Andersson, E.M., 26(162)*43*; 294, 311(14)*320*

Ando, J., 307(75)*324*

Andrade, L., 381(5)*424*

Andreoni, R., 140(115)*165*

Anhalt, J.P., 181(26)*230*; 405 (56)*427*

Armitage, F., 74(64)*87*

Arnesano, A., 179(22)*230*; 317 (161)*330*

Arpino, P., 310(107)*326*

Arrendale, R.F., 7(19)*34*; 74(66) *87*; 333(15,16)*370*; 450(38)*468*

Austin, P.R., 351(102)*375*

Auterhoff, H., 437(7)*466*

Autian, J., 172(3)*229*

Averitt, O.R., 55(16)*84*

Azarraga, C.V., 19(82)*38*; 70(36) *85*

I

Subject Index

A

Acceptance sampling, 283
Acetylcholine, 18, 138
Acrylate polymers, 12, 124, 312
Actinomycins, 336
Adhesives, 303, 464
Albumin, 218
Algae, 414
Alginite, 494
Alkaloids, 140, 449
Amino acids, 26, 30, 71, 148,
    227, 334, 415, 421
Amino sugars, 351
Amniotic fluid, 418
Analytical pyrolysis, 33
Anthocyanins, 454
Antibiotics, 453
Arthroderma, 414
Aspergillus, 409, 412
Atomic weights, 204
Automation, 23, 149
Autoscaling, 253, 259

B

Bacillus, 227, 384, 393
Back-flushing, 112
Bacteria, 151, 227, 384, 393
Barbiturates, 140, 447
Base peak, 189
Bitumen, 496
Blood, 421
BMDP, 241, 265, 279

Boundary effects, 145, 296, 307
Brain, 418
Brake linings, 303

C

Cancer cells, 418
Canonical correlation, 277
Capillary columns, 15, 108, 146
Caprolactone, 299
Carbon dioxide laser, 71
Carrier gas, 106
Cellulose, 14, 343
Cell walls, 241, 351, 394, 399,
    414, 475
Cephalosporins, 440
Cetrimide, 139, 445
Characteristicity, 243, 253, 259
Characteristic temperature, 46
Chewing gum, 464
Chi-squared test, 253
Chitin, 351
Chromatopyrography, 294
Chromosorb, 99
cis-Platin, 454
Clay, 486
Climosequence, 485
Cluster analysis, 263
Coal, 499, 502
Coating efficiency, 109
Cockroaches, 95, 414
Coding, 237, 243
Collisional activation MS, 24,
    208